Student Solutions Manual

John Garlow
Tarrant County College

John Tobey Jeffrey Slater

Beginning Algebra

Sixth Edition

PEARSON
Prentice Hall

Upper Saddle River, NJ 07458

Editor-in-Chief: Chris Hoag
Executive Editor: Paul Murphy
Supplement Editor: Christina Simoneau
Executive Managing Editor: Kathleen Schiaparelli
Assistant Managing Editor: Becca Richter
Production Editor: Donna Crilly
Supplement Cover Manager: Paul Gourhan
Supplement Cover Designer: Joanne Alexandris
Manufacturing Buyer: Ilene Kahn

Pearson Prentice Hall
Pearson Education, Inc.
Upper Saddle River, NJ 07458

Printed in the United States of America

10 9 8 7 6 5 4 3 2 1

ISBN 0-13-153060-7 Standalone
0-13-185801-7 Student Study Pack Component

Pearson Education Ltd., *London*
Pearson Education Australia Pty. Ltd., *Sydney*
Pearson Education Singapore, Pte. Ltd.
Pearson Education North Asia Ltd., *Hong Kong*
Pearson Education Canada, Inc., *Toronto*
Pearson Educación de Mexico, S.A. de C.V.
Pearson Education—Japan, *Tokyo*
Pearson Education Malaysia, Pte. Ltd.

CONTENTS

Chapter

Chapter 0

0.1 Exercises

1. $\dfrac{12}{13}$: numerator = 12

3. When two or more numbers are multiplied, each number that is multiplied is called a factor. In 2×3, 2 and 3 are factors.

5. $2\dfrac{2}{3}$

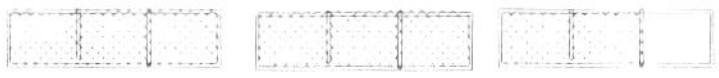

7. $\dfrac{18}{24}=\dfrac{3\times 6}{4\times 6}=\dfrac{3}{4}$

9. $\dfrac{12}{36}=\dfrac{1\times 12}{3\times 12}=\dfrac{1}{3}$

11. $\dfrac{60}{12}=\dfrac{12\times 5}{12\times 1}=5$

13. $\dfrac{17}{6}=2\dfrac{5}{6}$

15. $\dfrac{111}{9}=9\overset{\displaystyle 12}{\overline{\smash{)}111}}=12\dfrac{3}{9}=12\dfrac{1}{3}$

$$\begin{array}{r} 9 \\ \hline 21 \\ 18 \\ \hline 3 \end{array}$$

17. $\dfrac{38}{7}=7\overset{\displaystyle 5}{\overline{\smash{)}38}}=5\dfrac{3}{7}$

19. $3\dfrac{1}{5}=\dfrac{5\times 3+1}{5}=\dfrac{16}{5}$

21. $6\dfrac{3}{5}=\dfrac{5\times 6+3}{5}=\dfrac{33}{5}$

23. $\dfrac{72}{9}=\dfrac{8\times 9}{1\times 9}=\dfrac{8}{1}=8$

25. $\dfrac{3}{11}\times\dfrac{4}{4}=\dfrac{12}{44}$

27. $\dfrac{3}{5}\times\dfrac{7}{7}=\dfrac{21}{35}$

29. $\dfrac{4}{13}=\dfrac{?}{39}\Rightarrow\dfrac{4\times 3}{13\times 3}=\dfrac{12}{39}$

31. $\dfrac{1560}{68}=22\dfrac{64}{68}=22\dfrac{16}{17}$

$$\begin{array}{r} 22 \\ 68\overline{)1560} \\ \underline{136} \\ 200 \\ \underline{136} \\ 64 \end{array}$$

33. $\dfrac{13,200}{64,000}=\dfrac{400\times 33}{400\times 160}=\dfrac{33}{160}$

35. $\dfrac{1}{1+1+2}=\dfrac{1}{4}$

37. $\dfrac{1+1}{1+1+2}=\dfrac{2}{4}=\dfrac{1}{2}$

39. Aaron: $\frac{15}{25}=\frac{3}{5}$

Paul: $\frac{12}{30}=\frac{2}{5}$

Tom: $\frac{28}{35}=\frac{4}{5}$

0.2 Exercises

1. 8 is the LCD of $\frac{3}{4}$ and $\frac{5}{8}$ because 8 is exactly divisible by 4.

3. $\frac{7}{15}$ and $\frac{11}{21}$

$15=3\cdot 5$

$21=3\cdot 7$

$\text{LCD}=3\cdot 5\cdot 7=105$

5. $\frac{7}{10}$ and $\frac{1}{4}$: $10=2\cdot 5$

$4=2\cdot 2$

$\text{LCD}=2\cdot 2\cdot 5=20$

7. $\frac{5}{18}$ and $\frac{7}{54}$: $18=2\cdot 3\cdot 3$

$54=\quad 2\cdot 3\cdot 3\cdot 3$

$\text{LCD}=2\cdot 3\cdot 3\cdot 3=54$

9. $\frac{1}{2}, \frac{1}{18}$ and $\frac{13}{30}$

$2=2$

$18=2\cdot 3\cdot 3$

$30=2\cdot 3\cdot 5$

$\text{LCD}=2\cdot 3\cdot 3\cdot 5=90$

11. $\frac{3}{8}+\frac{2}{8}=\frac{5}{8}$

13. $\frac{5}{14}-\frac{1}{14}=\frac{4}{14}=\frac{2}{7}$

15. $\frac{3}{8}+\frac{5}{6}=\frac{3\cdot 3}{8\cdot 3}+\frac{5\cdot 4}{6\cdot 4}=\frac{9}{24}+\frac{20}{24}=\frac{29}{24}=1\frac{5}{24}$

17. $\frac{5}{7}-\frac{2}{9}=\frac{5\cdot 9}{7\cdot 9}-\frac{2\cdot 7}{9\cdot 7}$

$=\frac{45}{63}-\frac{14}{63}$

$=\frac{31}{63}$

19. $\frac{1}{3}+\frac{2}{5}=\frac{1\cdot 5}{3\cdot 5}+\frac{2\cdot 3}{5\cdot 3}=\frac{5}{15}+\frac{6}{15}=\frac{11}{15}$

21. $\frac{7}{18}+\frac{1}{12}=\frac{7\cdot 2}{18\cdot 2}+\frac{1\cdot 3}{12\cdot 3}=\frac{14}{36}+\frac{3}{36}=\frac{17}{36}$

23. $\frac{2}{3}+\frac{7}{12}+\frac{1}{4}=\frac{8}{12}+\frac{7}{12}+\frac{3}{12}=\frac{18}{12}=\frac{3}{2}=1\frac{1}{2}$

25. $\frac{5}{36}+\frac{7}{9}-\frac{5}{12}=\frac{5}{36}+\frac{7\cdot 4}{9\cdot 4}-\frac{5\cdot 3}{12\cdot 3}$

$=\frac{5}{36}+\frac{28}{36}-\frac{15}{36}$

$=\frac{18}{36}=\frac{1}{2}$

27. $4\frac{1}{3}+3\frac{2}{5}=\frac{13}{3}+\frac{17}{5}=\frac{65}{15}+\frac{51}{15}=\frac{116}{15}=7\frac{11}{15}$

29. $1\frac{5}{24}+\frac{5}{18}=1\frac{15}{72}+\frac{20}{72}=1\frac{35}{72}$

31. $7\frac{1}{6}-2\frac{1}{4}=7\frac{2}{12}-2\frac{3}{12}=6\frac{14}{12}-2\frac{3}{12}=4\frac{11}{12}$

33. $8\frac{5}{7}-2\frac{1}{4}=\frac{61}{7}-\frac{9}{4}=\frac{244}{28}-\frac{63}{28}$
$=\frac{181}{28}=6\frac{13}{28}$

35. $2\frac{1}{8}+3\frac{2}{3}=2\frac{3}{24}+3\frac{16}{24}=5\frac{19}{24}$

37. $11\frac{1}{7}-6\frac{5}{7}=\frac{78}{7}-\frac{47}{7}=\frac{31}{7}=4\frac{3}{7}$

39. $2\frac{1}{8}+6\frac{3}{4}=2\frac{1}{8}+6\frac{6}{8}=8\frac{7}{8}$

41. $\frac{7}{9}+\frac{5}{6}=\frac{14}{18}+\frac{15}{18}$
$=\frac{29}{18}=1\frac{11}{18}$

43. $2\frac{1}{7}+3\frac{11}{14}=\frac{15}{7}+\frac{53}{14}$
$=\frac{30}{14}+\frac{53}{14}$
$=\frac{83}{14}=5\frac{13}{14}$

45. $\frac{16}{21}-\frac{2}{7}=\frac{16}{21}-\frac{6}{21}$
$=\frac{10}{21}$

47. $5\frac{1}{5}-2\frac{1}{2}=\frac{26}{5}-\frac{5}{2}$
$=\frac{52}{10}-\frac{25}{10}$
$=\frac{27}{10}=2\frac{7}{10}$

49. $3\frac{1}{8}+2\frac{2}{3}+4\frac{1}{2}=\frac{25}{8}+\frac{8}{3}+\frac{9}{2}=\frac{75}{24}+\frac{64}{24}+\frac{108}{24}$
$=\frac{247}{24}=10\frac{7}{24}$ miles

51. $8\frac{1}{2}-2\frac{2}{3}-1\frac{3}{4}=\frac{17}{2}-\frac{8}{3}-\frac{7}{4}$
$=\frac{102}{12}-\frac{32}{12}-\frac{21}{12}$
$=\frac{49}{12}=4\frac{1}{12}$ hours

53. height $=2+\frac{1}{2}+3\frac{1}{2}+\frac{1}{2}+3\frac{1}{2}+\frac{1}{2}+1\frac{1}{2}$
$=12$ in.

$A=2+\frac{1}{2}+3\frac{1}{2}+\frac{1}{2}+3\frac{1}{2}+\frac{1}{2}+1\frac{1}{2}=12$ in.

$B=\frac{1}{2}+4\frac{5}{8}+\frac{1}{2}+4\frac{5}{8}+\frac{1}{2}+4\frac{5}{8}+\frac{1}{2}=15\frac{7}{8}$ in.

55. $2\frac{1}{2}-\frac{7}{8}=\frac{5}{2}-\frac{7}{8}=\frac{20}{8}-\frac{7}{8}=\frac{13}{8}=1\frac{5}{8}$

$1\frac{5}{8}$ inches

Cumulative Review

57. $\frac{36}{44}=\frac{9\cdot 4}{11\cdot 4}=\frac{9}{11}$

0.3 Exercises

1. First change each mixed number to an improper fraction. Look for a common factor in the numerator and denominator to divide by ; if one is found, perform the division. Multiply the numerators. Multiply the denominators.

3. $\frac{36}{7}\times\frac{5}{9}=\frac{9\cdot 4}{7}\times\frac{5}{9}=\frac{20}{7}=2\frac{6}{7}$

5. $\frac{17}{18}\times\frac{3}{5}=\frac{17}{6\cdot 3}\times\frac{3}{5}=\frac{17}{30}$

7. $\frac{4}{5}\times\frac{3}{10}=\frac{2\cdot 2}{5}\times\frac{3}{2\cdot 5}=\frac{6}{25}$

9. $\frac{24}{25}\times\frac{5}{2}=\frac{12\cdot 2}{5\cdot 5}\times\frac{5}{2}=\frac{12}{5}=1\frac{7}{5}$

11. $9\times\frac{2}{5}=\frac{9}{1}\times\frac{2}{5}=\frac{18}{5}=3\frac{3}{5}$

13. $\frac{8}{5}\div\frac{8}{3}=\frac{8}{5}\times\frac{3}{8}=\frac{3}{5}$

15. $\frac{3}{7}\div 3=\frac{3}{7}\div\frac{3}{1}=\frac{3}{7}\times\frac{1}{3}=\frac{1}{7}$

17. $10\div\frac{5}{7}=\frac{10}{1}\times\frac{7}{5}=\frac{2\cdot 5}{1}\times\frac{7}{5}=14$

19. $\frac{\frac{7}{8}}{\frac{3}{4}}=\frac{7}{8}\times\frac{4}{3}=\frac{7}{2\cdot 4}\times\frac{2\cdot 2}{3}=\frac{14}{12}=\frac{7}{6}=1\frac{1}{6}$

21. $\frac{\frac{5}{6}}{\frac{7}{9}}=\frac{5}{6}\times\frac{9}{7}=\frac{5}{2\cdot 3}\times\frac{3\cdot 3}{7}=\frac{15}{14}=1\frac{1}{14}$

23. $1\frac{3}{7}\div 6\frac{1}{4}=\frac{10}{7}\div\frac{25}{4}=\frac{10}{7}\times\frac{4}{25}=\frac{2\cdot 5}{7}\times\frac{4}{5\cdot 5}=\frac{8}{35}$

25. $3\frac{1}{3}\div 2\frac{1}{2}=\frac{10}{3}\div\frac{5}{2}$
$=\frac{2\cdot 5}{3}\times\frac{2}{5}$
$=\frac{4}{3}=1\frac{1}{3}$

27. $\frac{\frac{7}{9}}{1\frac{1}{3}}=\frac{\frac{7}{9}}{\frac{4}{3}}=\frac{7}{9}\times\frac{3}{4}=\frac{7}{3\cdot 3}\times\frac{3}{4}=\frac{7}{12}$

29. $\frac{\frac{2}{3}}{1\frac{1}{4}}=\frac{\frac{2}{3}}{\frac{5}{4}}=\frac{2}{3}\times\frac{4}{5}=\frac{8}{15}$

31. $\frac{6}{5}\times\frac{10}{12}=\frac{6}{5}\times\frac{2\cdot 5}{2\cdot 6}=1$

33. $\frac{5}{14}\div\frac{2}{7}=\frac{5}{14}\times\frac{7}{2}=\frac{5}{2\cdot 7}\times\frac{7}{2}=\frac{5}{4}=1\frac{1}{4}$

35. $10\frac{3}{7}\times 5\frac{1}{4}=\frac{73}{7}\times\frac{21}{4}=\frac{73}{7}\times\frac{3\cdot 7}{4}=\frac{219}{4}=54\frac{3}{4}$

37. $2\frac{1}{8}\div\frac{1}{4}=\frac{17}{8}\div\frac{1}{4}$
$=\frac{17}{2\cdot 4}\times\frac{4}{1}$
$=\frac{17}{2}=8\frac{1}{2}$

39. $6\times 4\frac{2}{3}=\frac{6}{1}\times\frac{14}{3}=\frac{2\cdot 3}{1}\times\frac{14}{3}=28$

41. $2\frac{1}{2}\times\frac{1}{10}\times\frac{3}{4}=\frac{5}{2}\times\frac{1}{5\cdot 2}\times\frac{3}{4}=\frac{3}{16}$

43. $2\frac{3}{4}\times 26=\frac{11}{4}\times\frac{26}{1}=\frac{11}{2\cdot 2}\times\frac{2\cdot 13}{1}=\frac{143}{2}$
$=71\frac{1}{2}$ yards

45. $\frac{2}{3}\times 4\frac{1}{5}=\frac{2}{3}\times\frac{21}{5}=\frac{2}{1}\times\frac{7}{5}=\frac{14}{5}=2\frac{4}{5}$ miles

Cumulative Review

47. $\frac{116}{124}=\frac{4\times 29}{4\times 31}=\frac{29}{31}$

0.4 Exercises

1. A decimal is another way of writing a fraction whose denominator is 10;100;1000;10,000;and so on.

3. When dividing 7432.9 by 1000 we move the decimal point <u>3</u> places to the <u>left</u>.

5. $\frac{5}{8}=8\overline{)5.000}=0.625$
$$\begin{array}{r} 0.625 \\ 8\overline{)5.000} \\ \underline{48} \\ 20 \\ \underline{16} \\ 40 \\ \underline{40} \\ 0 \end{array}$$

7. $\frac{3}{15}=15\overline{)3.0}=0.2$
$$\begin{array}{r} 0.2 \\ 15\overline{)3.0} \\ \underline{3.0} \\ 0 \end{array}$$

9. $11\overline{)7.000}=0.\overline{63}$
$$\begin{array}{r} 0.63 \\ 11\overline{)7.000} \\ \underline{66} \\ 40 \\ \underline{33} \\ 7 \end{array}$$

11. $0.8=\frac{8}{10}=\frac{4}{5}$

13. $0.625=\frac{625}{1000}=\frac{5}{8}$

15. $2.6=\frac{26}{10}=\frac{13}{5}=2\frac{3}{5}$

17.
$$\begin{array}{r} 1.71 \\ +0.38 \\ \hline 2.09 \end{array}$$

19.
$$\begin{array}{r} 2.50 \\ 3.42 \\ +\quad 4.90 \\ \hline 10.82 \end{array}$$

21.
$$\begin{array}{r} 46.030 \\ 215.100 \\ +\quad 0.078 \\ \hline 261.208 \end{array}$$

23.
$$\begin{array}{r} 147.18 \\ -15.39 \\ \hline 131.79 \end{array}$$

25.
$$\begin{array}{r} 7.21 \\ \times\ \ 4.2 \\ \hline 1442 \\ \underline{2884\ } \\ 30.282 \end{array}$$

27.
$$\begin{array}{r} 4.23 \\ \times\ 0.025 \\ \hline 2115 \\ 846\ \\ \hline 0.10575 \end{array}$$

29.
$$\begin{array}{r} 169,000 \\ \times\ 0.0013 \\ \hline 507000 \\ 169000\ \\ \hline 219.7000 \end{array}$$
or 219.7

31.
$$\begin{array}{r} 2.64 \\ 3.02\overline{)7.9728} \\ \underline{604}\ \ \\ 1932\ \\ \underline{1812}\ \\ 1208 \\ \underline{1208} \\ 0 \end{array}$$

33.
$$\begin{array}{r} 261.5 \\ 0.002\overline{)0.5230} \\ \underline{4}\ \ \ \\ 12\ \ \\ \underline{12}\ \ \\ 3\ \\ \underline{2}\ \\ 10 \\ \underline{10} \\ 0 \end{array}$$

35.
$$\begin{array}{r} 0.257 \\ 0.08\overline{)0.02056} \\ \underline{16}\ \ \\ 45\ \\ \underline{40}\ \\ 56 \\ \underline{56} \end{array}$$

37. $3.45 \times 1000 = 3450$

39. $0.76 \div 100 = 0.0076$

41. $7.36 \times 10,000 = 73,600. = 73,600$

43. $73,892 \div 100,000 = 0.73892$

45.
$$\begin{array}{r} 23.75 \\ \times\ \ 0.06 \\ \hline 1.4250 \end{array}$$

47.
$$\begin{array}{r} 1.62 \\ 2.005 \\ +\ 8.1007 \\ \hline 11.7257 \end{array}$$

49. $0.027\overline{)0.05724}$ = 2.12

$$\begin{array}{r} 2.12 \\ 0.027\overline{)0.05724} \\ \underline{54} \\ 32 \\ \underline{27} \\ 54 \\ \underline{54} \\ 0 \end{array}$$

51. $0.7683 \times 1000 = 768.3$

53.
$$\begin{array}{r} 2.54 \\ \times\ 9.5 \\ \hline 1270 \\ 2286 \\ \hline 24.130 \end{array}$$

24.13 cm

55. $11.50 \times 18.5 = \$212.75$

57. $1.3 \times 5 = 6.5$ milligrams
no; there was 0.3 milligrams of copper in excess of the standard for 5 liters of water.

Cumulative Review

59. $3\frac{1}{2} \div 5\frac{1}{4} = \frac{7}{2} \div \frac{21}{4}$
$= \frac{7}{2} \cdot \frac{4}{21} = \frac{1 \cdot 2}{1 \cdot 3} = \frac{2}{3}$

61. $\frac{12}{25} + \frac{9}{20} = \frac{48}{100} + \frac{45}{100} = \frac{93}{100}$

How Am I Doing? Sections 0.1-0.4

1. $\frac{15}{55} = \frac{3 \cdot 5}{11 \cdot 5} = \frac{3}{11}$

2. $\frac{46}{115} = \frac{2 \cdot 23}{5 \cdot 23} = \frac{2}{5}$

3. $\frac{15}{4} = 4\overline{)15}$ (quotient 3) $= 3\frac{3}{4}$

4. $4\frac{5}{7} = \frac{4 \cdot 7 + 5}{7} = \frac{28 + 5}{7} = \frac{33}{7}$

5. $\frac{3}{7} = \frac{?}{14} \Rightarrow \frac{3 \cdot 2}{7 \cdot 2} = \frac{6}{14}$

6. $\frac{7}{4} = \frac{?}{20} \Rightarrow \frac{7 \cdot 5}{4 \cdot 5} = \frac{35}{20}$

7. $8 = 2 \cdot 2 \cdot 2$
$6 = 2 \cdot 3$
$15 = 3 \cdot 5$
$\text{LCD} = 2 \cdot 2 \cdot 2 \cdot 3 \cdot 5 = 120$

8. $\frac{3}{7} + \frac{2}{7} = \frac{5}{7}$

9. $\frac{5}{14} + \frac{2}{21} = \frac{5 \cdot 3}{14 \cdot 3} + \frac{2 \cdot 2}{21 \cdot 2} = \frac{15}{42} + \frac{4}{42} = \frac{19}{42}$

10. $2\frac{3}{4}+5\frac{2}{3}=\frac{11}{4}+\frac{17}{3}$
$=\frac{11\cdot 3}{4\cdot 3}+\frac{17\cdot 4}{3\cdot 4}$
$=\frac{33}{12}+\frac{68}{12}$
$=\frac{101}{12}=8\frac{5}{12}$

11. $\frac{17}{18}-\frac{5}{9}=\frac{17}{18}-\frac{5\cdot 2}{9\cdot 2}$
$=\frac{17}{18}-\frac{10}{18}$
$=\frac{7}{18}$

12. $\frac{6}{7}-\frac{2}{3}=\frac{6\cdot 3}{7\cdot 3}-\frac{2\cdot 7}{3\cdot 7}$
$=\frac{18}{21}-\frac{14}{21}$
$=\frac{4}{21}$

13. $3\frac{1}{5}-1\frac{3}{8}=\frac{16}{5}-\frac{11}{8}$
$=\frac{16\cdot 8}{5\cdot 8}-\frac{11\cdot 5}{8\cdot 5}$
$=\frac{128}{40}-\frac{55}{40}$
$=\frac{73}{40}=1\frac{33}{40}$

14. $\frac{25}{7}\times\frac{14}{45}=\frac{5\cdot 5}{7}\times\frac{2\cdot 7}{5\cdot 9}=\frac{10}{9}=1\frac{1}{9}$

15. $2\frac{4}{5}\times 3\frac{3}{4}=\frac{14}{5}\times\frac{15}{4}=\frac{2\cdot 7}{5}\times\frac{3\cdot 5}{2\cdot 2}$
$=\frac{21}{2}$
$=10\frac{1}{2}$

16. $4\div\frac{8}{7}=\frac{4}{1}\times\frac{7}{2\cdot 4}=\frac{7}{2}=3\frac{1}{2}$

17. $2\frac{1}{3}\div 3\frac{1}{4}=\frac{7}{3}\div\frac{13}{4}=\frac{7}{3}\times\frac{4}{13}$
$=\frac{28}{39}$

18. $\frac{7}{8}=8\overline{)7.000}=0.875$

$$\begin{array}{r} 0.875 \\ 8\overline{)7.000} \\ \underline{64} \\ 60 \\ \underline{56} \\ 40 \\ \underline{40} \end{array}$$

19. $\frac{5}{9}=9\overline{)5.000}=0.\overline{5}$

$$\begin{array}{r} 0.555 \\ 9\overline{)5.000} \\ \underline{45} \\ 50 \\ \underline{45} \\ 50 \\ \underline{45} \\ 5 \end{array}$$

20. $\frac{3}{200}=200\overline{)3.000}=0.15$

$$\begin{array}{r} 0.15 \\ 200\overline{)3.000} \\ \underline{200} \\ 1000 \end{array}$$

21.
$$\begin{array}{r} 15.230 \\ 3.600 \\ +\quad 0.821 \\ \hline 19.651 \end{array}$$

22.
$$\begin{array}{r} 3.28 \\ \times\ 0.63 \\ \hline 984 \\ 1968 \\ \hline 2.0664 \end{array}$$

23. $\frac{3.015}{6.7} = 6.7\overline{)3.015} = 0.45$

$$\begin{array}{r} 0.45 \\ 6.7\overline{)3.015} \\ 268 \\ \hline 335 \\ 335 \\ \hline \end{array}$$

24.
$$\begin{array}{r} 12.130 \\ -9.884 \\ \hline 2.246 \end{array}$$

0.5 Exercises

1. Answers may vary.

   19% means 19 out of 100 parts or $\frac{19}{100}$

3. $0.28 = 28\%$

5. $0.568 = 56.8\%$

7. $0.076 = 7.6\%$

9. $2.39 = 239\%$

11. $3\% = 0.03$

13. $0.4\% = 0.004$

15. $250\% = 2.5$

17. $7.4\% = 0.074$

19. 0.8% of 65

    $0.008 \times 65 = 0.52$

21. 10% of 130

    $0.10 \times 130 = 13$

23. 112% of 65

    $1.12 \times 65 = 72.8$

25. 36 is what percent of 24?

    $\frac{36}{24} = 1.50 = 150\%$

27. What percent of 340 is 17?

    $\frac{17}{340} = \frac{1}{20} = 0.05 = 5\%$

29. 75 is what percent of 30?

    $\frac{75}{30} = 2.5 = 250\%$

31. $\frac{68}{80} = 0.85 = 85\%$

    His grade was 85%.

33. $0.15 \times 32.80 = 4.92$

    $32.80 + 4.92 = 37.72$

    \$4.92 tip, \$37.72 new total

35. $\frac{380}{1850} \approx 0.21 = 21\%$

    About 21% of the budget is for food.

37. $1.5\% \times 36{,}000 = 0.015 \times 36{,}000 = 540$ gifts

39. $0.08 \times 12.50 = 1$
$12.50 + 1 = 13.50$
\$1 per hour raise
\$13.50 per hour new rate

41. a. $(0.038)(780{,}000) = 29{,}640$
\$29,640
b. $29{,}640 + 12(450) = 35{,}040$
\$35,040

43. Discount: $33\% \times 45 = 0.33 \times 45 = 14.85$
Nancy: $20\% \times (45 - 14.85) = 0.2 \times 30.13 = \$6.03$
Fred: $20\% \times 45 = 0.2 \times 45 = \$9.00$
Increase: $\$9.00 - \$6.03 = \$2.97$

Cumulative Review

45.

45.50	1.50
1189.00	98.00
33.90	128.00
1268.40	56.89
	445.88
	730.27

Balance $1268.40 - 730.27 = \$538.13$

47. Distance: $69{,}229.5 - 68{,}459.5 = 770$
$\frac{770}{35} = 22$ miles per gallon

0.6 Exercises

1. $693 \times 307 \approx 700 \times 300 = 210{,}000$

3. $2862 \times 5986 \approx 3000 \times 6000 \approx 18{,}000{,}000$

5. $14 + 73 + 80 + 21 + 56 \approx 10 + 70 + 80 + 20 + 60 = 240$

7. $41\overline{)829{,}346} \approx 40\overline{)800{,}000}$, quotient $20{,}000$
$\approx 20{,}000$

9. $\frac{2714}{31{,}500} \approx \frac{3000}{30{,}000} = 0.1$

11. 317% of $21{,}365.85 \approx 0.20 \times 20{,}000 \approx \$4000.00$

13. $18.5 \times 22.50 \approx 20 \times 20 \approx \$400$

15. $4 \times 22 \times 82 \approx 4 \times 20 \times 80 = \$6400$

17. $\frac{117.7}{3.8} \approx \frac{100}{4} = 25$ miles per gallon

19. $\frac{491{,}549.00}{279} \approx \frac{500{,}000}{300} = \$1666.67$

21. $26\% \times 217{,}970{,}000{,}000 \approx 0.30 \times 200{,}000{,}000{,}000 \approx \$60{,}000{,}000{,}000$

23. $14\% \times 38{,}945.89 \approx 0.10 \times 40{,}000 \approx \$4000$

25. $\dfrac{3247.54}{26,533.67} \approx \dfrac{3000}{30,000} = 0.10 = 10\%$

27. $43 \times 3.24 \approx 40 \times 3 = 120$
$12 \times 120 \times 23\% \approx 10 \times 120 \times 0.2$
$\approx \$240$ saved

Cumulative Review

29. $0.6\% \times 350 = 0.006 \times 350 = 2.1$

31. Tip $= 15\% \times 5.85 = 0.88$
Total $= 5.85 + 0.88 = \$6.73$

0.7 Exercises

1. Area $= 12\frac{3}{4} \times 9\frac{1}{2} = \frac{51}{4} \times \frac{19}{2} = \frac{969}{8}$
Square yards $= \frac{969}{8} \div 9 = 13.4583$
Cost $= 20 \times 13.4583 \approx \$269.17$

3. a. Length $= 2 \times 15\frac{1}{2} + 2 \times 25\frac{2}{3}$
$= 2 \times \frac{31}{2} + 2 \times \frac{77}{3}$
$= 31 + \frac{154}{3}$
$= \frac{93}{3} + \frac{154}{3}$
$= \frac{247}{3} = 82\frac{1}{3}$ feet

b. Cost $= 2.10 \times \frac{247}{3} = \$172.90$

She should purchase the prepackaged 90 feet. She would save $\$172.90 - 155.00 = \$17.90$.

5.

Day	Jog	Walk	Rest	Walk
1	$1\frac{1}{5}$	$1\frac{3}{4}$	$2\frac{1}{2}$	1
3	$2\frac{2}{15}$	$3\frac{1}{9}$	$4\frac{4}{9}$	$1\frac{7}{9}$

Increase $1\frac{1}{3} \times 1\frac{1}{3} = \frac{16}{9}$

$\frac{16}{9} \times 1\frac{1}{5} = \frac{16}{9} \times \frac{6}{5} = \frac{96}{45} = 2\frac{2}{15}$
$\frac{16}{9} \times 1\frac{3}{4} = \frac{16}{9} \times \frac{7}{4} = \frac{28}{9} = 3\frac{1}{9}$
$\frac{16}{9} \times 2\frac{1}{2} = \frac{16}{9} \times \frac{5}{2} = \frac{40}{9} = 4\frac{4}{9}$
$\frac{16}{9} \times 1 = \frac{16}{9} = 1\frac{7}{9}$ miles

7. Betty will have a more demanding schedule on day 3 because Melinda increases each activity by $\frac{2}{3}$ by day 3 and Betty increases each activity by $\frac{7}{9}$ by day 3.

9. Increase: $1\frac{1}{2} \times \frac{1}{3} = \frac{3}{2} \times \frac{1}{3} = \frac{1}{2}$
Day 7: $1\frac{1}{2} + \left(6 \times \frac{1}{2}\right) = \frac{3}{2} + \frac{6}{2} = \frac{9}{2}$
$= 4\frac{1}{2}$ miles

11. 1st increase:

$30\% \times 350 = 66,200 = 19,860$

1990 price:

$66,200 + 19,860 = 86,060$

2nd increase:

$15\% \times 86,060 = 12,909$

2000 price:

$86,060 + 12,909 = \$98,969.00$

13. a. Percent available $= 100 - 28 = 72\%$

$0.72 \times 50,000 = \$36,000$

b. $0.31 \times 36,000 = \$11,160$

15. Federal: $\frac{139}{1150} \approx 0.12 = 12\%$

State: $\frac{68}{1150} \approx 0.06 = 6\%$

Local: $\frac{5}{1150} \approx 0.004 = 0\%$

Total: $12\% + 6\% = 18\%$

17. $\frac{790.47}{1150} \approx 0.69 = 69\%$

Putting Your Skills to Work

1. $26.6 - 13.9 = 12.7$

12,700 square miles

2. $31.8 - 22.4 = 9.4$

9400 square miles

3. $$\begin{array}{r} 4.2 \\ 7540\overline{)31,800.0} \\ \underline{30,160} \\ 16400 \\ \underline{15080} \\ 1320 \end{array}$$

About 4 times

4. $$\begin{array}{r} 319.4 \\ 72\overline{)23000.0} \\ \underline{216} \\ 140 \\ \underline{72} \\ 680 \\ \underline{648} \\ 320 \\ \underline{288} \\ 32 \end{array}$$

About 319 times

Chapter 0 Review Problems

1. $\frac{36}{48} = \frac{12 \times 3}{12 \times 4} = \frac{3}{4}$

2. $\frac{15}{30} = \frac{5 \times 3}{5 \times 10} = \frac{3}{10}$

3. $\frac{36}{82} = \frac{18 \times 2}{41 \times 2} = \frac{18}{41}$

4. $\frac{18}{30} = \frac{6 \times 3}{6 \times 5} = \frac{3}{5}$

5. $4\frac{3}{5} = \frac{4 \cdot 5 + 3}{5} = \frac{20 + 3}{5} = \frac{23}{5}$

6. $\frac{34}{5} = 6\frac{4}{5}$

7. $\frac{39}{6} = 6\overline{)39} = 6\frac{3}{6} = 6\frac{1}{2}$ (quotient 6; $39 - 36 = 3$)

8. $\frac{5}{8} = \frac{?}{24} \Rightarrow \frac{5\cdot 3}{8\cdot 3} = \frac{15}{24}$

9. $\frac{1}{7} = \frac{?}{35} \Rightarrow \frac{1\cdot 5}{7\cdot 5} = \frac{5}{35}$

10. $\frac{5}{9} = \frac{?}{72} \Rightarrow \frac{5\cdot 8}{9\cdot 8} = \frac{40}{72}$

11. $\frac{2}{5} = \frac{?}{55} \Rightarrow \frac{2\times 11}{5\times 11} = \frac{22}{55}$

12. $\frac{3}{5} + \frac{1}{4} = \frac{3\times 4}{5\times 4} + \frac{1\times 5}{4\times 5} = \frac{12}{20} + \frac{5}{20} = \frac{17}{20}$

13. $\frac{7}{12} + \frac{5}{8} = \frac{7\times 2}{12\times 2} + \frac{5\times 3}{8\times 3} = \frac{14}{24} + \frac{15}{24} = \frac{29}{24} = 1\frac{5}{24}$

14. $\frac{7}{20} - \frac{1}{12} = \frac{21}{60} - \frac{5}{60} = \frac{16}{60} = \frac{4}{15}$

15. $\frac{7}{10} - \frac{4}{15} = \frac{7\cdot 3}{10\cdot 3} - \frac{4\cdot 2}{5\cdot 2} = \frac{21}{30} - \frac{8}{30} = \frac{13}{30}$

16. $3\frac{1}{6} + 2\frac{3}{5} = 3\frac{5}{30} + 2\frac{18}{30} = 5\frac{23}{30}$

17. $1\frac{1}{4} + 2\frac{7}{10} = \frac{5}{4} + \frac{27}{10} = \frac{25}{20} + \frac{54}{20} = \frac{79}{20} = 3\frac{19}{20}$

18. $6\frac{2}{9} - 3\frac{5}{12} = \frac{56}{9} - \frac{41}{12} = \frac{56\cdot 4}{9\cdot 4} - \frac{41\cdot 3}{12\cdot 3}$
$= \frac{224}{36} - \frac{123}{36} = \frac{101}{36}$
$= 2\frac{29}{36}$

19. $3\frac{1}{15} - 1\frac{3}{20} = \frac{46}{15} - \frac{23}{20} = \frac{184}{60} - \frac{69}{60}$
$= \frac{115}{60} = 1\frac{55}{60} = 1\frac{11}{12}$

20. $6\times\frac{5}{11} = \frac{6}{1}\times\frac{5}{11} = \frac{30}{11}$ or $2\frac{8}{11}$

21. $2\frac{1}{3}\times 4\frac{1}{2} = \frac{7}{3}\times\frac{9}{2} = \frac{7}{3}\times\frac{3\cdot 3}{2} = \frac{21}{2} = 10\frac{1}{2}$

22. $1\frac{1}{8}\times 2\frac{1}{9} = \frac{9}{8}\times\frac{19}{9} = \frac{1}{8}\times\frac{19}{1} = \frac{19}{8}$ or $2\frac{3}{8}$

23. $\frac{4}{7}\times 5 = \frac{4}{7}\times\frac{5}{1} = \frac{20}{7} = 2\frac{6}{7}$

24. $\frac{3}{8}\div 6 = \frac{3}{8}\times\frac{1}{2\cdot 3} = \frac{1}{16}$

25. $\frac{\frac{8}{3}}{\frac{5}{9}} = \frac{8}{3}\times\frac{9}{5} = \frac{8}{3}\times\frac{3\cdot 3}{5} = \frac{24}{5} = 4\frac{4}{5}$

26. $\frac{15}{16} \div 6\frac{1}{4} = \frac{15}{16} \div \frac{25}{4} = \frac{15}{16}\times\frac{4}{25} = \frac{3}{4}\times\frac{1}{5} = \frac{3}{20}$

27. $2\frac{6}{7}\div\frac{10}{21} = \frac{20}{7}\times\frac{21}{10} = \frac{2\cdot 10}{7}\times\frac{3\cdot 7}{10} = 6$

28. $$\begin{array}{r} 1.634 \\ 3.007 \\ +\ 2.560 \\ \hline 7.201 \end{array}$$

29. $$\begin{array}{r} 24.831 \\ -\ 17.094 \\ \hline 7.737 \end{array}$$

30. $$\begin{array}{r} 47.251 \\ -\ 17.690 \\ \hline 29.561 \end{array}$$

31. $$\begin{array}{r} 1.900 \\ 2.530 \\ +\ 0.006 \\ \hline 4.436 \end{array}$$

32. $$\begin{array}{r} 5.35 \\ \times\ 0.007 \\ \hline 0.03745 \end{array}$$

33. $362.341 \times 1000 = 362,341$

34. $2.6 \times 0.03 \times 1.02 = 0.07956$

35. $1.08 \times 0.06 \times 160 = 10.368$

36. $0.186 \div 100 = 0.00186$

37. $71.32 \div 1000 = 0.07132$

38. $0.523 \div 0.4 = 1.3075$

39. $$0.015\overline{)1.350} \quad \text{quotient } 90$$
$$\begin{array}{r} 135 \\ \hline 00 \end{array}$$

40. $$\frac{4.186}{2.3} = 2.3\overline{)4.186} = 1.82 \quad \text{quotient } 1.82$$
$$\begin{array}{r} 23 \\ \hline 188 \\ 184 \\ \hline 46 \\ 46 \\ \hline \end{array}$$

41. $$0.38\overline{)0.190} \quad \text{quotient } 0.5$$
$$\begin{array}{r} 190 \\ \hline \end{array}$$

42. $$\frac{3}{8} = 8\overline{)3.000} = 0.375 \quad \text{quotient } 0.375$$
$$\begin{array}{r} 24 \\ \hline 60 \\ 56 \\ \hline 40 \end{array}$$

43. $0.36 = \frac{36}{100} = \frac{9}{25}$

44. $1.4\% = 0.014$

45. $36.1\% = 0.361$

46. $0.02\% = 0.0002$

47. $125.3\% = 1.253$

48. 85% of 600
$(0.85)(600) = 510$

49. 7.2% of 55

$0.072 \times 55 = 3.96$

50. 48 is what percent of 75?

$\frac{48}{75} = 0.64 = 64\%$

51. What percent of 120 is 15?

$\frac{15}{120} = 0.125 = 12.5\%$

52. $\frac{750}{1250} = 0.6 = 60\%$

53. 80.7% of 20,229,000

$0.807 \times 20,229,000 = 16,324,803$

54. $\frac{720}{960} = 0.75 = 75\%$

55. $234,897 \times 1,936,112$

$\approx 200,000 \times 2,000,000$

$= 400,000,000,000$

56. $400 + 900 + 800 + 400 = 2500$

57. $634,318 - 284,000 \approx 600,000 - 300,000$

$\approx 300,000$

58. $7 + 4 + 8 = 19$

59. 18% of 56,297 $\approx 0.2 \times 60,000 = \$12,000$

60. $12,000 \div 400 = 30$

61. $7.85 \times 32.5 \approx 8 \times 30 = \$240$

62. $\frac{900}{3} = 300$

\$300

63. Area $= 12\frac{1}{2} \times 9\frac{2}{3} = \frac{25}{2} \times \frac{29}{3}$

$= 120\frac{5}{6}$ sq. feet

$= \frac{120\frac{5}{6}}{9} = 13\frac{23}{54}$ sq. yards

Cost $= 26.00 \times 13\frac{23}{54} = 26 \times \frac{725}{54}$

$= \$349.07$

64. Increase $= 36,720 - 34,000 = 2720$

$\frac{2720}{34,000} = 0.08 = 8\%$

65. Maximum distance $= 7\frac{2}{3} \times 240$

$= \frac{23}{3} \times 240$

$= 1840$ miles

Longest trip $= 80\% \times 1840 = .8 \times 1840$

$= 1472$ miles

66. Maximum speed distance =

$6\frac{1}{4} \times 240 = \frac{25}{4} \times 240 = 1500$ miles

67. $\text{Volume} = 4 \times 2\frac{1}{2} \times \frac{3}{4}$
$= \frac{4}{1} \times \frac{5}{2} \times \frac{3}{4}$
$= \frac{15}{2}$
$= 7\frac{1}{2}$ cubic inches

68. $\text{Outer volume} = 16 \times 12\frac{1}{2} \times 7\frac{1}{2}$
$= \frac{16}{1} \times \frac{25}{2} \times \frac{15}{2}$
$= 1500$ cubic inches

$\frac{1500}{\frac{15}{2}} = \frac{1500}{1} \times \frac{2}{15} = 200$ boxes

69. 8% of 5785
$0.08 \times 5785 = \$462.80$

70. $\frac{34.5}{0.75} = 46$ packages

71. Amount paid $= 225 \times 4 \times 12 = \$10,800$
Interest $= 10,800 - 9214.50 = \$1585.50$

72. Regular pay $= 6.40 \times 40 = 256$
O/T hours $= 52 - 40 = 12$
O/T pay rate $= 1.5 \times 6.40 = 9.60$
O/T pay $= 9.60 \times 12 = 115.20$
Total pay $= 256 + 115.20 = \$371.20$

How Am I Doing? Chapter 0 Test

1. $\frac{16}{18} = \frac{2 \times 8}{2 \times 9} = \frac{8}{9}$

2. $\frac{48}{36} = \frac{4 \cdot 12}{3 \cdot 12} = \frac{4}{3}$

3. $6\frac{3}{7} = \frac{7 \times 6 + 3}{7} = \frac{45}{7}$

4. $\frac{105}{9} = 9\overline{)105}$ with quotient 11: $\frac{9}{15}$, $\frac{9}{6}$; $= 11\frac{6}{9} = 11\frac{2}{3}$

5. $\frac{2}{3} + \frac{5}{6} + \frac{3}{8} = \frac{2 \cdot 8}{3 \cdot 8} + \frac{5 \cdot 4}{6 \cdot 4} + \frac{3 \cdot 3}{8 \cdot 3}$
$= \frac{16}{24} + \frac{20}{24} + \frac{9}{24}$
$= \frac{45}{24} = \frac{15}{8}$
$= 1\frac{7}{8}$

6. $1\frac{1}{8} + 3\frac{3}{4} = \frac{9}{8} + \frac{15}{4} = \frac{9 + 30}{8} = 4\frac{7}{8}$

7. $3\frac{2}{3} - 2\frac{5}{6} = 3\frac{4}{6} - 2\frac{5}{6} = 2\frac{10}{6} - 2\frac{5}{6} = \frac{5}{6}$

8. $\frac{5}{7} \times \frac{28}{15} = \frac{1}{1} \times \frac{4}{3} = \frac{4}{3}$ or $1\frac{1}{3}$

9. $\frac{5}{18} \times \frac{3}{4} = \frac{5 \times 3}{6 \times 3 \times 4} = \frac{5}{24}$

10. $\frac{7}{4} \div \frac{1}{2} = \frac{7}{4} \times \frac{2}{1} = \frac{7}{2} \times \frac{1}{1}$
$= \frac{7}{2}$ or $3\frac{1}{2}$

11. $2\frac{1}{2} \times 3\frac{1}{4} = \frac{5}{2} \times \frac{13}{4} = \frac{65}{8} = 8\frac{1}{8}$

12. $5\frac{3}{8} \div 2\frac{3}{4} = \frac{43}{8} \div \frac{11}{4}$
$= \frac{43}{2 \cdot 4} \times \frac{4}{11}$
$= \frac{43}{22}$
$= 1\frac{21}{22}$

13. $\begin{array}{r} 1.60 \\ 3.24 \\ \underline{9.80} \\ 14.64 \end{array}$

14. $\begin{array}{r} 7.0046 \\ \underline{-\ 3.0149} \\ 3.9897 \end{array}$

15. $\begin{array}{r} 32.8 \\ \underline{\times\ 0.04} \\ 1.312 \end{array}$

16. $0.07385 \times 1000 = 73.85$

17. $\begin{array}{r} 230 \\ 0.056\overline{)12.880} \\ \underline{112} \\ 168 \\ \underline{168} \\ 0 \end{array}$

18. $26,325.9 \div 100 = 263.259$

19. $0.073 = 7.3\%$

20. $196.5\% = 1.965$

21. What is 3.5% of 180?
$0.035 \times 180 = 6.3$

22. $(0.02)(16.8) = 0.336$

23. 39 is what percent of 650?
$\frac{39}{650} = 0.06 = 6\%$

24. $\frac{138}{460} = 0.3 = 30\%$

25. $4 \div \frac{2}{9} = \frac{4}{1} \times \frac{9}{2} = \frac{2}{1} \times \frac{9}{1} = 18$
18 chips

26. $\begin{array}{r} 100 \\ 50000\overline{)5000000} \end{array}$

27. $285.36 + 311.85 + 113.6 \approx 300 + 300 + 100$
≈ 700

28. Commission: $(0.03)(870,000) = 26,100$

    Total income: $26,100 + 14,000 = 40,100$

    Percent: $\frac{26,100}{40,100} \approx 0.65 = 65\%$

29. $210 \div 3\frac{1}{2} = 210 \div \frac{7}{2} = 210 \times \frac{2}{7} = 60$ tiles

Chapter 1

1.1 Exercises

	Number	WN	RaN	IR	ReN
1.	23	x	x		x
3.	π				x
5.	$-6.666...$		x		x
7.	$-2.3434...$		x		x
9.	$\sqrt{2}$			x	x

11. $-20,000$

13. Lost $37\frac{1}{2} \Rightarrow -37\frac{1}{2}$

15. Rise 7^0 F $\Rightarrow +7$

17. Additive inverse of 8 is -8

19. Opposite of -2.73 is 2.73

21. $|-1.3| = 1.3$

23. $\left|\frac{5}{6}\right| = \frac{5}{6}$

25. $-6+(-5)=-11$

27. $-17+(-14)=-31$

29. $-\frac{5}{16}+\frac{9}{16}=\frac{4}{16}=\frac{1}{4}$

31. $\left(-\frac{2}{13}\right)+\left(-\frac{5}{13}\right)=-\frac{7}{13}$

33. $-\frac{2}{5}+\frac{3}{7}=\frac{1}{35}$

35. $(-1.5)+(-2.3)=-3.8$

37. $(+0.6)+(-0.2)=0.4$

39. $-5.26+(-8.9)=-14.16$

41. $-8+5+(-3)=-3+(-3)=-6$

43. $-3+5+(-7)=2+(-7)=-5$

45. $-\frac{4}{5}+\frac{8}{15}=-\frac{12}{15}+\frac{8}{15}=-\frac{4}{15}$

47. $-7+(-9)+8=-16+8=-8$

49. $8+(-11)=-3$

51. $-83+142=59$

53. $-\frac{4}{9}+\frac{5}{6}=-\frac{8}{18}+\frac{15}{18}=\frac{7}{18}$

55. $-\frac{1}{10}+\frac{1}{2}=-\frac{1}{10}+\frac{5}{10}=\frac{4}{10}=\frac{2}{5}$

57. $4.36+(-3.6)=0.76$

59. $4+(-8)+16=-4+16=12$

61. $34+(-18)+11+(-27)$
$=16+11+(-27)$
$=27+(-27)$
$=0$

63. $17.85+(-2.06)+0.15=15.79+0.15$
$=15.94$

65. Profit $=214-47=\$167$

67. $-2300+(-1500)=-3800$
He owed $-\$3800$.

69. $9+(-11)+5=-2+5=3$
3 yard gain

71. $8000+(-3000)+(-1500)=5000+(-1500)$
$=3500$

73. $30+14+(-12)=32$
$\$32,000,000$

75. $-13+?=5$
$?=18$

77. $10+(-8)+4=2+4=6$
They must make 6 yards.

Cumulative Review

79. $\frac{3}{7}+\frac{5}{21}=\frac{9}{21}+\frac{5}{21}=\frac{14}{21}=\frac{2}{3}$

81. $\frac{2}{15}-\frac{1}{20}=\frac{8}{60}-\frac{3}{60}=\frac{5}{60}=\frac{1}{12}$

83. $0.72+0.8=1.52$

85.
$$\begin{array}{r} 1.63 \\ \times \quad 0.7 \\ \hline 1.141 \end{array}$$

1.2 Exercises

1. $-8-(-3)=-8+3=-5$

3. $18-35=18+(-35)=-17$

5. $(+15)-(+20)=(+15)+(-20)=-5$

7. $(-14)-(-3)=(-14)+(3)=-11$

9. $(-52)-(-60)=(-52)+(+60)=8$

11. $(0)-(-5)=(0)+(+5)=5$

13. $(-18)-(-18)=(-18)+(+18)=0$

15. $-11-(-8)=-11+8=-3$

17. $\frac{2}{5}-\frac{4}{5}=\frac{2}{5}+\left(-\frac{4}{5}\right)=-\frac{2}{5}$

19. $\left(+\frac{3}{4}\right)-\left(-\frac{3}{5}\right)=\left(+\frac{15}{20}\right)+\left(+\frac{12}{20}\right)$
$=\frac{27}{20}=1\frac{7}{20}$

21. $\left(-\frac{3}{4}\right)-\left(+\frac{5}{6}\right)=\left(-\frac{9}{12}\right)+\left(-\frac{10}{12}\right)$
$=-\frac{19}{12}=-1\frac{7}{12}$

23. $(-0.6)-(+0.3)=(-0.6)+(-0.3)=-0.9$

25. $(+2.64)-(-1.83)=(+2.64)+(+1.83)$
$=4.47$

27. $\frac{3}{5}-4=\frac{3}{5}+\left(-\frac{20}{5}\right)=-\frac{17}{5}=-3\frac{2}{5}$

29. $-\frac{2}{6}+6=-\frac{2}{7}+\frac{42}{7}=\frac{40}{7}=5\frac{5}{7}$

31. $34-87=34+(-87)=-53$

33. $-25-48=-25+(-48)$

35. $2.3-(-4.8)=2.3+(+4.8)=7.1$

37. $8-\left(-\frac{3}{4}\right)=8+\frac{3}{4}=8\frac{3}{4}$

39. $\frac{5}{6}-7=\frac{5}{6}+\left(-\frac{42}{6}\right)=-\frac{37}{6}=-6\frac{1}{6}$

41. $-\frac{3}{10}-\frac{3}{4}=\left(-\frac{6}{20}\right)+\left(-\frac{15}{20}\right)$
$=-\frac{21}{20}=-1\frac{1}{20}$

43. $-135-(-126.5)=-135+(+126.5)=-8.5$

45. $\frac{1}{5}-6=\frac{1}{5}+\left(-\frac{30}{5}\right)=-\frac{29}{5}=-5\frac{4}{5}$

47. $5-(-3.162)=5+(+3.162)=8.162$

49. $-3-2.047=-3+(-2.047)=-5.047$

51. $-2-(-9)=-2+(+9)=7$

53. $-35-(13)=-35+(-13)=-48$

55. $7+(-6)-(+3)=7+(-6)+(-3)$
$=1+(-3)=-2$

57. $-10+6-(-15)=-10+6+(+15)$
$=-4+(+15)=11$

59. $7+(-42)-27=7+(-42)+(-27)$
$=-35+(-27)=-62$

61. $-6.4-(-2.7)+5.3$
$=-6.4+(2.7)+5.3$
$=-3.7+5.3$
$=1.6$

63. $42-(-30)-65-(-11)+20$
$=42+(+30)+(-65)+(+11)+20$
$=72+(-65)+(+11)+20$
$=7+(+11)+20$
$=18+20=38$

65. $112-(-37)=112+(+37)=\$149$

67. $5895-(-156)=5895+156$
$=6051$
The difference is 6051 m

69. $-37+16=-21$

71. $-3+(-6)+(-10)=-9+(-10)$
$=-19$

73. $\left(\frac{4}{5}\right)\left(8\frac{1}{3}\right)=\left(\frac{4}{5}\right)\left(\frac{25}{3}\right)=\frac{20}{3}$
$=6\frac{2}{3}$ miles

1.3 Exercises

1. To multiply two real numbers, multiply the absolute values. The sign of the result is positive if both numbers have the same sign, but negative if the two numbers have opposite signs.

3. $5(-4)=-20$

5. $0(-12)=0$

7. $16\times 1.5=24$

9. $(-1.32)(-0.2)=0.264$

11. $(0.7)(-2.5)=-1.75$

13. $\frac{3}{8}(-4)=\frac{3}{8}\left(-\frac{4}{1}\right)=-\frac{3}{2}=-1\frac{1}{2}$

15. $\left(-\frac{3}{5}\right)\left(-\frac{15}{11}\right)=\frac{9}{11}$

17. $\left(\frac{12}{13}\right)\left(\frac{-5}{24}\right)=-\frac{5}{26}$

19. $(-36)\div(-9)=4$

21. $-48\div(-8)=\frac{-48}{-8}=6$

23. $(-220)\div(-11)=20$

25. $156\div(-13)=-12$

27. $-9.1\div 0.07=-\frac{9.1}{0.07}=-130$

29. $0.54\div(-0.9)=\frac{0.54}{-0.9}=-0.6$

31. $-6.3\div 7=\frac{-6.3}{7}=-0.9$

33. $\left(-\frac{1}{5}\right)\div\left(\frac{2}{3}\right)=\left(-\frac{1}{5}\right)\left(\frac{3}{2}\right)=-\frac{3}{10}$

35. $\left(-\frac{5}{7}\right)\div\left(-\frac{3}{28}\right)=\frac{5}{7}\cdot\left(-\frac{28}{3}\right)=-\frac{20}{3}=-6\frac{2}{3}$

37. $-\frac{7}{12}\div\left(-\frac{5}{6}\right)=\left(\frac{7}{12}\right)\left(\frac{6}{5}\right)=\frac{7}{10}$

39. $\frac{-6}{-\frac{3}{7}}=\left(-\frac{6}{1}\right)\left(-\frac{7}{3}\right)=14$

41. $\frac{\frac{-2}{3}}{\frac{8}{15}}=\left(-\frac{2}{3}\right)\left(\frac{15}{8}\right)=-\frac{5}{4}=-1\frac{1}{4}$

43. $\frac{-\frac{7}{8}}{-\frac{14}{15}}=\left(-\frac{7}{8}\right)\left(\frac{15}{14}\right)=\frac{15}{16}$

45. $(-1)(-2)(-3)(4)=-(1)(2)(3)(4)=-24$

47. $-2(-2)(2)(-1)(-3)=24$

49. $(-3)(-2)\left(\frac{1}{3}\right)(-4)(2)$

$= -(3)(2)\left(\frac{1}{3}\right)(4)(2) = -16$

51. $-3(-0.03)(0.001)(-2)$

$-(0.09)(0.001)(-2)$

$= (0.00009)(-2)$

$= -0.00018$

53. $\left(-\frac{4}{5}\right)\left(-\frac{6}{7}\right)\left(-\frac{1}{3}\right) = -\left(\frac{24}{35}\right)\left(\frac{1}{3}\right) = -\frac{8}{35}$

55. $\left(-\frac{3}{4}\right)\left(-\frac{7}{15}\right)\left(-\frac{8}{21}\right)\left(-\frac{5}{9}\right)$

$= +\left(\frac{3}{4}\right)\left(\frac{7}{3\cdot 5}\right)\left(\frac{4\cdot 2}{3\cdot 7}\right)\left(\frac{5}{9}\right) = \frac{2}{27}$

57. $-36 \div (-4) = \frac{-36}{-4} = 9$

59. $5 + (-7) = -2$

61. $8 - (-9) = 8 + (9) = 17$

63. $6(-12) = -72$

65. $(-37) \div 37 = \frac{-37}{37} = -1$

67. $17.60 \div 4 = \frac{17.60}{4} = 4.40$

He gave $4.40 to each boy and to himself.

69. $\frac{15,768}{48} = \$328.50$

71. $5(4) = +20$, gained 20 yards

73. $-10(7) = -70$, lost 70 yards

75. Total $= -70 + 90 = 20$, gained 20 yards

77. Additional large-losses $= -15(3)$

$= -45$

Avoided medium-gains $= -(+15)(4)$

$= -60$

Fewer yards gained $= 45 + 60$

$= 105$ yards

The Panthers would have gained 105 fewer yards.

Cumulative Review

79. $-\frac{3}{4} + \left(-\frac{2}{3}\right) + \left(-\frac{5}{12}\right)$

$= -\frac{9}{12} + \left(-\frac{8}{12}\right) + \left(-\frac{5}{12}\right)$

$= -\frac{17}{12} + \left(-\frac{5}{12}\right)$

$= -\frac{22}{12} = -\frac{11}{6}$

$= -1\frac{5}{6}$

81. $(-37) - (51) = (-37) + (-51) = -88$

1.4 Exercises

1. $4^4 = (4)(4)(4)(4) = 256$

3. A negative number to an odd power is negative.

5. $(-2)^4 = (-2)(-2)(-2)(-2) = 64$
   $-2^4 = -(2)(2)(2)(2) = -64$

7. $(5)(5)(5)(5)(5)(5)(5) = 5^7$

9. $(w)(w) = w^2$

11. $(p)(p)(p)(p) = p^4$

13. $(3q)(3q)(3q) = (3q)^3$ or 3^3q^3

15. $3^3 = 27$

17. $3^4 = 81$

19. $6^3 = 216$

21. $(-3)^3 = -27$

23. $(-4)^2 = 16$

25. $-5^2 = -(5)(5) = -25$

27. $\left(\frac{1}{4}\right)^2 = \frac{1}{16}$

29. $\left(\frac{2}{5}\right)^3 = \frac{8}{125}$

31. $(1.1)^2 = 1.21$

23. $(0.2)^4 = 0.0016$

35. $(-16)^2 = 256$

37. $-16^2 = -256$

39. $5^3 + 6^2 = 125 + 36 = 161$

41. $5^3 - 3^2 = 125 - 9 = 116$

43. $(-3)^3 - (8)^2 = -27 - 64 = -91$

45. $2^5 - (-3)^2 = 32 - 9 = 23$

47. $(-4)^3(-3)^2 = (-64)(9) = -576$

49. $8^2(-2)^3 = 64(-8) = -512$

51. $4^{12} = 16,777,216$

53. $-343 = (-7)(-7)(-7) = (-7)^3$
    The number is -7.

Cumulative Review

55. $(-11) + (-13) + 6 + (-9) + 8$
    $= -24 + 6 + (-9) + 8$
    $= -18 + (-9) + 8$
    $= -27 + 8$
    $= -19$

57. $-17-(-9)=-17+9=-8$

59. Interest for 1 year:
$6\%(1600)=0.06(1600)=96$
Interest for 9 months:
$\frac{9}{12}(96)=72$
Total $=1600+72=\$1672$

1.5 Exercises

1. $3(4)+6(5)$

3. $3(4)+6(5)=12+6(5)=18(5)=90$
(b) $3(4)+6(5)=12+30=42$

5. $(2-5)^2\div 3\times 4=(-3)^2\div 3\times 4$
$=9\div 3\times 4$
$=3\times 4$
$=12$

7. $2(3-5+6)+5=2(4)+5$
$=8+5$
$=13$

9. $8-2^3\cdot 5+3=8-8\cdot 5+3$
$=8-40+3=-29$

11. $4+42\div 3\cdot 2-8$
$=4+14\cdot 2-8$
$=4+28-8$
$=32-8$
$=24$

13. $3\cdot 5+7\cdot 3-5\cdot 3=15+21-15$
$=21$

15. $8-5(2)^3\div(-8)=8-5(8)\div(-8)$
$=8-40\div(-8)$
$=8+5$
$=13$

17. $3(5-7)^2-6(3)=3(-2)^2-6(3)$
$=3(4)-6(3)=12-18=-6$

19. $5\cdot 6-(3-5)^2+8\cdot 2=5\cdot 6-(-2)^2+8\cdot 2$
$=5\cdot 6-4+8\cdot 2$
$=30-4+16=42$

21. $\frac{1}{2}\div\frac{2}{3}+6\cdot\frac{1}{4}=\frac{1}{2}\cdot\frac{3}{2}+\frac{6}{1}\cdot\frac{1}{4}$
$=\frac{3}{4}+\frac{6}{4}=\frac{9}{4}=2\frac{1}{4}$

23. $0.8+0.3(0.6-0.2)^2$
$=0.8+0.3(0.4)^2$
$=0.8+0.3(0.16)$
$=0.8+0.048=0.848$

25. $\frac{3}{4}\left(-\frac{2}{5}\right)-\left(-\frac{3}{5}\right)=-\frac{3}{10}+\frac{6}{10}=\frac{3}{10}$

27. $(3-7)^2\div 8+3=(-4)^2\div 8+3$
$=16\div 8+3$
$=2+3$
$=5$

29. $\left(\frac{3}{4}\right)^2(-16)+\frac{4}{5}\div\frac{-8}{25}$
$=\frac{9}{16}(-16)+\frac{4}{5}\div\frac{-8}{25}$
$=-9+\frac{4}{5}\div\frac{-8}{25}$
$=-9+\frac{4}{5}\left(\frac{25}{-8}\right)$
$=-9+\left(\frac{5}{-2}\right)$
$=-\frac{18}{2}+\left(-\frac{5}{2}\right)$
$=-\frac{23}{2}=-11\frac{1}{2}$

31. $-6.3-(-2.7)(1.1)+(3.3)^2$
$=-6.3+2.97+10.89=7.56$

33. $\left(\frac{1}{2}\right)^3+\left(\frac{1}{4}\right)-\left(\frac{1}{6}-\frac{1}{12}\right)-\frac{2}{3}\cdot\left(\frac{1}{4}\right)^2$
$=\left(\frac{1}{2}\right)^3+\frac{1}{4}-\left(\frac{1}{12}\right)-\frac{2}{3}\cdot\left(\frac{1}{4}\right)^2$
$=\frac{1}{8}+\frac{1}{4}-\frac{1}{12}-\frac{2}{3}\cdot\frac{1}{16}$
$=\frac{1}{8}+\frac{1}{4}-\frac{1}{12}-\frac{1}{24}$
$=\frac{6}{24}=\frac{1}{4}$

35. $3(-2)+9(-1)+5(0)+1(1)$

37. 1 above par

Cumulative Review

39. $(0.5)^3=0.125$

41. $-1^{20}=-1$

43. $15\div 2\times(6)=45$ ounces

45. $0.324\div 0.6$

$$\frac{0.324}{0.6}=0.6\overset{0.54}{\overline{)0.324}}=0.54$$

$$\begin{array}{r} 30 \\ \hline 24 \\ \underline{24} \end{array}$$

How Am I Doing? Sections 1.1-1.5

1. $3+(-12)=-9$

2. $-\frac{5}{6}+\left(-\frac{7}{8}\right)=-\frac{20}{24}+\left(-\frac{21}{24}\right)$
$=\frac{-41}{24}$
$=-1\frac{17}{24}$

3. $\begin{array}{r} 0.34 \\ +\ 0.90 \\ \hline 1.24 \end{array}$

4. $-14+3+(-2.5)+6.4$
$=-11+(-2.5)+6.4$
$=-13.5+6.4$
$=-7.1$

5. $-23-(-34)=-23+34=11$

6. $-\frac{4}{5}-\frac{1}{3}=-\frac{4}{5}+\left(-\frac{1}{3}\right)$
$=-\frac{12}{15}+\left(-\frac{5}{15}\right)$
$=-\frac{17}{15}$
$=-1\frac{7}{15}$

7. $4.5-(-7.8)=4.5+7.8=12.3$

8. $-4-(-5)+9=-4+5+9=10$

9. $(-3)(-8)(2)(-2)=24(2)(-2)$
$=48(-2)$
$=-96$

10. $\left(-\frac{6}{11}\right)\left(-\frac{5}{3}\right)=\frac{10}{11}$

11. $-0.072\div 0.08=\frac{-0.072}{0.08}=-0.9$

12. $\frac{5}{8}\div\left(-\frac{17}{16}\right)=\left(\frac{5}{8}\right)\cdot\left(-\frac{16}{17}\right)=-\frac{10}{17}$

13. $(0.7)^3=(0.7)(0.7)(0.7)$
$=(0.49)(0.7)$
$=0.343$

14. $(-4)^4=(-4)(-4)(-4)(-4)=256$

15. $0-2^8=-(2)(2)(2)(2)(2)(2)(2)(2)$
$=-256$

16. $\left(\frac{2}{3}\right)^3=\left(\frac{2}{3}\right)\left(\frac{2}{3}\right)\left(\frac{2}{3}\right)=\frac{8}{27}$

17. $5^3+(-2)^4=125+16=141$

18. $12\div 6(2)+3=2(2)+3=4+3=7$

19. $15+3-2+(-6)=18+(-2)+(-6)$
$=16+(-6)$
$=10$

20. $(9-13)^2+15\div(-3)$
$=(-4)^2+15\div(-3)$
$=16+15\div(-3)$
$=16+(-5)$
$=11$

21. $-0.12\div 0.6+(-3)(1.2)-(-0.5)$
$=-0.2+(-3)(1.2)+0.5$
$=-0.2+(-3.6)+0.5$
$=-3.8+0.5$
$=-3.3$

22. $\left(\frac{3}{4}\right)\left(-\frac{2}{5}\right)+\left(-\frac{1}{2}\right)\left(\frac{4}{5}\right)+\left(\frac{1}{2}\right)^2$

$=\left(\frac{3}{4}\right)\left(-\frac{2}{5}\right)+\left(-\frac{1}{2}\right)\left(\frac{4}{5}\right)+\frac{1}{4}$

$=-\frac{3}{10}+\left(-\frac{1}{2}\right)\left(\frac{4}{5}\right)+\frac{1}{4}$

$=-\frac{3}{10}+\left(-\frac{2}{5}\right)+\frac{1}{4}$

$=-\frac{6}{20}+\left(-\frac{8}{20}\right)+\frac{5}{20}$

$=-\frac{14}{20}+\frac{5}{20}$

$=-\frac{9}{20}$

1.6 Exercises

1. A var*iable* is a symbol used to represent an unknown number.

3. We are multiplying 4 by x by x. We know from the definition of exponents that x multiplied by x is x^2, this gives us an answer of $4x^2$.

5. Yes, $a(b-c)$ can be written as $a[b+(-c)]$.

$3(10-2)=(3\times10)-(3\times2)$

$3\times8=30-6$

$24=24$

7. $3(x-2y)=3(x)-3(2y)=3x-6y$

9. $-2(4a-3b)=-2(4a)-2(-3b)$

$=-8a+6b$

11. $3(3x+y)=3(3x)+3(y)=9x+3y$

13. $5(-2a-3b)=5(-2a)-5(3b)$

$=-10a-15b$

15. $-(x-3y)=(-1)(x)-(-1)(3y)$

$=-x+3y$

17. $-9(9x-5y+8)=-9(9x)-9(-5y)-9(8)$

$=-81x+45y-72$

19. $2(-5x+y-6)=2(-5x)+2(y)-2(6)$

$=-10x+2y-12$

21. $\frac{5}{6}(12x^2-24x+18)$

$=\left(\frac{5}{6}\right)(12x^2)+\left(\frac{5}{6}\right)(-24x)+\left(\frac{5}{6}\right)(18)$

$=10x^2-20x+15$

23. $\frac{x}{5}(x+10y-4)$

$=\frac{x}{5}(x)+\frac{x}{5}(10y)+\frac{x}{5}(-4)$

$=\frac{x^2}{5}+2xy-\frac{4x}{5}$

25. $5x(x+2y+z-1)$

$=5x(x)+5x(2y)+5x(z)+5x(-1)$

$=5x^2+10xy+5xz-5x$

27. $(2x-3)(-2)$

$=2x(-2)+(-3)(-2)$

$=-4x+6$

29. $(6x+y-1)(3x)=6x(3x)+y(3x)-1(3x)$
$=18x^2+3xy-3x$

31. $(3x+2y-1)(-xy)$
$=3x(-xy)+2y(-xy)-1(-xy)$
$=-3x^2y-2xy^2+xy$

33. $(2x+3y-2)(3xy)$
$=(2x)(3xy)+3y(3xy)-2(3xy)$
$=6x^2y+9xy^2-6xy$

35. $2.3(4.1x^2-2.3x+0.4)$
$=2.3(4.1x^2)-2.3(2.3x)+2.3(0.4)$
$=9.43x^2-5.29x+0.92$

37. $-0.3x(-1.2x^2-0.3x+0.5)$
$=-0.3x(-1.2x^2)+(-0.3x)(-0.3x)$
$+(-0.3x)(0.5)$
$=0.36x^3+0.09x^2-0.15x$

39. $0.5x(0.6x+0.8y-5)$
$=0.5x(0.6x)+0.5x(0.8y)+0.5x(-5)$
$=0.3x^2+0.4xy-2.5x$

41. $800(5x+14y)=800(5x)+800(14y)$
$=4000x+11{,}200y$ square feet

43. $4y(2x-5)=(8xy-20y)$ dollars

45. $4x(3000-2y)=4x(3000)+4x(-2y)$
$=12{,}000x-8xy$ square feet

Cumulative Review

47. $(-2)^6=(-2)(-2)(-2)(-2)(-2)(-2)$
$=64$

49. $25\div 5(2)+(-6)=5(2)+(-6)$
$=10+(-6)$
$=4$

51. $\frac{205}{365}\approx 0.56=56\%$

1.7 Exercises

1. A term is a number, a variable, or a product of numbers and variables.

3. The two terms $5x$ and $-8x$ are like terms because they both have the variable x with the exponent of one.

5. The only like terms are $7xy$ and $-14xy$ because the other two have different exponents even though they have the same variables.

7. $-14b^2-11b^2=(-14-11)b^2=-25b^2$

9. $10x^4+8x^4+7x^2=(10+8)x^4+7x^2$
$=18x^4+7x^2$

11. $3x+2y-8x-7y$
$=(3-8)x+(2-7)y$
$=-5x-5y$

13. $1.3x - 2.6y + 5.8x - 0.9y$
$= (1.3 + 5.8)x + (-2.6 - 0.9)y$
$= 7.1x - 3.5y$

15. $1.6x - 2.8y - 3.6x - 5.9y$
$= (1.6 - 3.6)x + (-2.8 - 5.9)y$
$= -2x - 8.7y$

17. $3p - 4q + 2p + 3 + 5q - 21$
$= (3 + 2)p + (-4 + 5)q + 3 - 21$
$= 5p + q - 18$

19. $2ab + 5bc - 6ac - 2ab$
$= (2 - 2)ab + 5bc - 6ac$
$= 5bc - 6ac$

21. $2x^2 - 3x - 5 - 7x + 8 - x^2$
$= (2 - 1)x^2 + (-3 - 7)x - 5 + 8$
$= x^2 - 10x + 3$

23. $2y^2 - 8y + 9 - 12y^2 - 8y + 3$
$= (2 - 12)y^2 + (-8 - 8)y + 9 + 3$
$= -10y^2 - 16y + 12$

25. $\frac{1}{3}x - \frac{2}{3}y - \frac{2}{5}x + \frac{4}{7}y$
$= \left(\frac{1}{3} - \frac{2}{5}\right)x + \left(-\frac{2}{3} + \frac{4}{7}\right)y$
$= \left(\frac{5}{15} - \frac{6}{15}\right)x + \left(-\frac{14}{21} + \frac{12}{21}\right)y$
$= -\frac{1}{15}x - \frac{2}{21}y$

27. $\frac{3}{4}a^2 - \frac{1}{3}b - \frac{1}{5}a^2 - \frac{1}{2}b$
$= \left(\frac{3}{4} - \frac{1}{5}\right)a^2 + \left(-\frac{1}{3} - \frac{1}{2}\right)b$
$= \left(\frac{15}{20} - \frac{4}{20}\right)a^2 + \left(-\frac{2}{6} - \frac{3}{6}\right)b$
$= -\frac{11}{20}a^2 - \frac{5}{6}b$

29. $ab - 7a - 9ab + 4a - 6b$
$= (1 - 9)ab + (-7 + 4)a - 6b$
$= -8ab - 3a - 6b$

31. $5(2a - b) - 3(5b - 6a)$
$= 10a - 5b - 15b + 18a$
$= 28a - 20b$

33. $-3b(5a - 3b) + 4(-3ab - 5b^2)$
$= -15ab + 9b^2 - 12ab - 20b^2$
$= -27ab - 11b^2$

35. $-3(x^2 + 3y) + 5(-6y - x^2)$
$= -3x^2 - 9y - 30y - 5x^2$
$= -8x^2 - 39y$

37. $4(2 - x) - 3(-5 - 12x)$
$= 8 - 4x + 15 + 36x$
$= 32x + 23$

39. $3a + 2b + 4a + 7b = 7a + 9b$

41. $2a + 7b + (5a + 3) = 7a + 7b + 3$ cm

43. $4[(9x - 2) - 3] = 4(9x - 5) = 36x - 20$ inches

Cumulative Review

45. $-\frac{1}{3}-\left(-\frac{1}{5}\right)=-\frac{5}{15}+\frac{3}{15}=-\frac{2}{15}$

47. $\frac{4}{5}+\left(-\frac{1}{25}\right)+\left(-\frac{3}{10}\right)$
$=\frac{40}{50}+\left(-\frac{2}{50}\right)+\left(-\frac{15}{50}\right)=\frac{23}{50}$

49. $\frac{1.9}{9.5}=0.2$ liters per serving

1.8 Exercises

1. If $x=3$, then $-2x+1$
$=-2(3)+1=-6+1=-5$

3. If $x=-9$, then $\frac{2}{3}x-5=\frac{2}{3}(-9)-5$
$=-6-5=-11$

5. If $x=\frac{1}{2}$, then $5x+10$
$=5\left(\frac{1}{2}\right)+10=\frac{5}{2}+10=\frac{25}{2}=12\frac{1}{2}$

7. If $x=7$, then $2-4x=2-4(7)$
$=2-28=-26$

9. If $x=2.4$, then $3.5-2x$
$=3.5-2(2.4)=3.5-4.8=-1.3$

11. If $x=-\frac{3}{4}$, then $9x+13$
$=9\left(-\frac{3}{4}\right)+13=-\frac{27}{4}+\frac{52}{4}=\frac{25}{4}$

13. If $x=-2$, then $x^2-3x=(-2)^2-3(-2)$
$=4+6=10$

15. If $x=-1$, then $3x^2=3(-1)^2=3$

17. If $x=2$, then $3x^3=-3(2)^3=-3(8)$
$=-24$

19. If $x=-2$, then $-5x^2=-5(-2)^2$
$=-5(4)=-20$

21. If $x=-3$, then $2x^2+3x=2(-3)^2+3(-3)$
$=2(9)-9=18-9=9$

23. If $x=3$, then $(2x)^2+x$
$=[2(3)]^2+3$
$=[6]^2+3=36+3=39$

25. If $x=-2$, then $2-(-x)^2$
$=2-[-(-2)]^2=2-(2)^2=2-4=-2$

27. If $x=-3$, then
$7x+(2x)^2=7(-3)+[2(-3)]^2$
$=-21+(-6)^2=-21+36=15$

29. If $x=-3$, $3x^2-5x=3(-3)^2-5(-3)$
$=3(9)+15=27+15=42$

31. If $x = 3$, then $x^2 - 7x + 3 = (3)^2 - 7(3) + 3$
$= 9 - 21 + 3 = -9$

33. If $x = -4$, then $\frac{1}{2}x^2 - 3x + 9$
$= \frac{1}{2}(-4)^2 - 3(-4) + 9$
$= \frac{1}{2}(16) + 12 + 9$
$= 8 + 12 + 9 = 29$

35. If $x = -3$ and $y = 4$, then $x^2 - 2y + 3y^2$
$= (-3)^2 - 2(4) + 3(4)^2$
$= 9 - 8 + 3(16)$
$= 9 - 8 + 48 = 49$

37. If $a = 5$, $b = 8$, and $c = -1$, then
$a^3 + 2abc - 3c^2 = 5^3 + 2(5)(9)(-1) - 3(-1)^2$
$= 125 - 90 - 3(1)$
$= 125 - 90 - 3 = 32$

39. If $a = -1$ and $b = -2$, then
$\frac{a^2 + ab}{3b} = \frac{(-1)^2 + (-1)(-2)}{3(-2)}$
$= \frac{1+2}{-6} = -\frac{1}{2}$

41. $A = ab$, $b = 22$, $a = 16$
$A = (22)(16) = 352$ square feet

43. $A = s^2$
Increase $= A_{\text{new}} - A_{\text{old}}$
$= (3.2)^2 - (3)^2$
$= 10.24 - 9$
$= 1.24$ square centimeters

45. $A = \frac{1}{2}a(b_1 + b_2)$, $a = 4$, $b_1 = 9$, $b_2 = 7$
$A = \frac{1}{2}(4)(9+7) = \frac{4(16)}{2}$
$= 32$ square inches

47. $A = \frac{1}{2}ab$, $a = 400$, $b = 280$
$A = \frac{1}{2}(400)(280) = 56,000$ square feet

49. $A = \pi r^2 = (3.14)(4)^2 = (3.14)(16)$
$= 50.24$ square cm

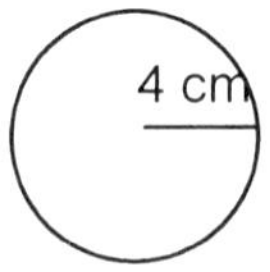

51. $C = \frac{5}{9}(F - 32)$
$C = \frac{5}{9}(-109.3 - 32)$
$C = \frac{5}{9}(-141.3)$
$C = 5(-15.7)$
$C = -78.5^\circ C$

53. $A = \frac{1}{2}ab,\ a = 20,\ b = 12$

$A = \frac{1}{2}(20)(12) = 120$ square feet

55. $F = \frac{9}{5}C + 32$

$F = \frac{9}{5}(60) + 32 = 108 + 32 = 140$

$F = \frac{9}{5}(-50) + 32 = -90 + 32 = -58$

140° F to -58° F

57. $k = 1.61r$

$2.3 = 1.61r$

$r = 1.4$

1.4 miles

The air is thin and lacks the usual amount of oxygen at this elevation.

Cumulative Review

59. $(-2)^4 - 4 \div 2 - (-2) = 16 - 2 + 2$

$= 16$

61. $\frac{93}{15} = 6.2$ minutes/song

1.9 Exercises

1. $-3x - 2y = -(3x + 2y)$

3. To simplify expressions with grouping symbols, we use the *distributive* property.

5. $6x - 3(x - 2y) = 6x - 3x + 6y$

$= 3x + 6y$

7. $2(a + 3b) - 3(b - a)$

$= 2a + 6b - 3b + 3a$

$= 5a + 3b$

9. $-3(x + 3y) + 2(2x + y)$

$= -3x - 9y + 4x + 2y$

$= x - 7y$

11. $2x[4x^2 - 2(x - 3)]$

$= 2x[4x^2 - 2x + 6]$

$= 8x^3 - 4x^2 + 12x$

13. $2[5(x + y) - 2(3x - 4y)]$

$= 2[5x + 5y - 6x + 8y]$

$= 2(-x + 13y)$

$= -2x + 26y$

15. $2(x - 2y) - [3 - 2(x - y)]$

$= 2x - 4y - (3 - 2x + 2y)$

$= 2x - 4y - 3 + 2x - 2y$

$= 4x - 6y - 3$

17. $5\left[3a-2a(3a+6b)+6a^2\right]$
$=5\left[3a-6a^2-12ab+6a^2\right]$
$=5[3a-12ab]$
$=15a-60ab$

19. $6a\left(2a^2-3a-4\right)-a(a-2)$
$=12a^3-18a^2-24a-a^2+2a$
$=12a^3-19a^2-22a$

21. $3a^2-4\left[2b-3b(b+2)\right]$
$=3a^2-4\left(2b-3b^2-6b\right)$
$=3a^2-4\left(-4b-3b^2\right)$
$=3a^2+16b+12b^2$

23. $6b-\left\{5a-2\left[a+(b-2a)\right]\right\}$
$=6b-\left\{5a-2[a+b-2a]\right\}$
$=6b-\left\{5a-2[-a+b]\right\}$
$=6b-\{5a+2a-2b\}$
$=6b-\{7a-2b\}$
$=6b-7a+2b$
$=-7a+8b$

25. $3\left\{3b^2+2\left[5b-(2-b)\right]\right\}$
$=3\left\{3b^2+2[5b-2+b]\right\}$
$=3\left\{3b^2+2[6b-2]\right\}$
$=3\left\{3b^2+12b-4\right\}$
$=9b^2+36b-12$

27. $-4\left\{3a^2-2\left[4a^2-\left(b+a^2\right)\right]\right\}$
$=-4\left[3a^2-2\left(4a^2-b-a^2\right)\right]$
$=-4\left[3a^2-2\left(3a^2-b\right)\right]$
$=-4\left(3a^2-6a^2+2b\right)$
$=-4\left(-3a^2+2b\right)$
$=12a^2-8b$

29. Attempts $=(4)(6)(365)=8760$
Successful $=2.5\%(8760)$
$=(.025)(8760)$
$=219$
Unsuccessful $=8760-219=8541$

Cumulative Review

31. If $C=36.4$, then
$F=1.8C+32$
$=1.8(36.4)+32=65.52+32=97.52^0\text{F}$

33. $A=\frac{1}{2}a(b_1+b_2)$,
$a=400,\ b_1=800, b_2=700$
$A=\frac{1}{2}(400)(800+700)=200(1500)$
$=300{,}000$ sq. ft.
Cost $=55(300{,}000)=\$16{,}500{,}000$

35. If $k=2.205p$
If $p=120,\ k=2.205(120)=264.6$
If $p=150,\ k=2.205(150)=330.75$
Great Danes weigh on average from 264.6 to 330.75 kg.

Putting Your Skills to Work

1. 66.8% of 31,723

   $0.668 \times 31,723 = 21,190.964$

   About 21,190,964 use e-mail.

2. $$\begin{array}{r} 31,712 \\ -\ 22,205 \\ \hline 9,518 \end{array}$$

   About 9,518,000 more use computers.

3. $(0.77)x = 65,190,000$

   $x = \dfrac{65,190,000}{0.77}$

   $x = 846,622,338$ people

4. 117.5% of 65,190,000

   $(1.175)(65,190,000) = 76,598,250$

   About 76,598,250 people.

5. $N = 65,190,000 + 4,500,000x$

   when $x = 6$

   $N = 65,190,000 + 4,500,000(6)$

   $N = 65,190,000 + 27,000,000$

   $N = 92,190,000$ people

6. $N = 65,190,000 + 4,500,000x$

   when $x = 9$

   $N = 65,190,000 + 4,500,000(9)$

   $N = 65,190,000 + 40,500,000$

   $N = 105,190,000$ people

Chapter 1 Review Problems

1. $(-6) + (-2) = -8$

2. $(-12) + (+7.8) = -4.2$

3. $(+5) + (-2) + (-12) = (+3) + (-12) = -9$

4. $(+3.7) + (-1.8) = 1.9$

5. $\left(+\dfrac{1}{2}\right) + \left(-\dfrac{5}{6}\right) = \left(+\dfrac{3}{6}\right) + \left(-\dfrac{5}{6}\right) = -\dfrac{2}{6} = -\dfrac{1}{3}$

6. $\left(-\dfrac{3}{11}\right) + \left(-\dfrac{1}{22}\right) = \left(-\dfrac{6}{22}\right) + \left(-\dfrac{1}{22}\right)$

   $= -\dfrac{7}{22}$

7. $\left(+\dfrac{3}{4}\right) + \left(-\dfrac{1}{12}\right) + \left(-\dfrac{1}{2}\right)$

   $= \left(+\dfrac{9}{12}\right) + \left(-\dfrac{1}{12}\right) + \left(-\dfrac{6}{12}\right) = \dfrac{2}{12} = \dfrac{1}{6}$

8. $\left(-\dfrac{4}{15}\right) + \left(+\dfrac{12}{5}\right) + \left(-\dfrac{2}{3}\right)$

   $= \left(-\dfrac{4}{15}\right) + \left(+\dfrac{36}{15}\right) + \left(-\dfrac{10}{15}\right) = \dfrac{22}{15}$

9. $(+5) - (-3) = (+5) + (+3) = 8$

10. $(-2) - (-15) = (-2) + (+15)$

    $= 13$

11. $(-30) - (+3) = (-30) + (-3)$

    $= -33$

12. $(+8)-(-1.2)=(+8)+(+1.2)$
$=9.2$

13. $\left(-\frac{7}{8}\right)+\left(-\frac{3}{4}\right)=\left(-\frac{7}{8}\right)+\left(-\frac{6}{8}\right)$
$=-\frac{13}{8}=-1\frac{5}{8}$

14. $\left(-\frac{3}{14}\right)+\left(+\frac{5}{7}\right)=\left(-\frac{3}{14}\right)+\left(+\frac{10}{14}\right)$
$=\frac{7}{14}=\frac{1}{2}$

15. $-20.8-1.9=-20.8+(-1.9)=-22.7$

16. $-151-(-63)=-151+63=-88$

17. $87\div(-29)=-3$

18. $-5(-6)+4(-3)=30+(-12)=18$

19. $\frac{-24}{-\frac{3}{4}}=\left(\frac{-24}{1}\right)\left(-\frac{4}{3}\right)=32$

20. $\left(-\frac{1}{2}\right)\div\frac{3}{4}=\left(-\frac{1}{2}\right)\left(\frac{2\cdot 2}{3}\right)=-\frac{2}{3}$

21. $\left(\frac{5}{7}\right)\div\left(-\frac{5}{25}\right)=\frac{5}{7}\cdot\left(-\frac{25}{5}\right)$
$=-\frac{25}{7}$ or $-3\frac{4}{7}$

22. $(-6)(3)(4)=(-18)(4)$
$=-72$

23. $(-1)(-2)(-3)(-5)=(1)(2)(3)(5)=30$

24. $(-5)\left(-\frac{1}{2}\right)(4)(-3)=\left(\frac{5}{2}\right)(-12)$
$=-30$

25. $-5+(-2)-(-3)=-5+(-2)+(+3)$
$=(-7)+(+3)=-4$

26. $6-(-4)+(-2)+(8)=6+4+(-2)+(8)$
$=18+(-2)$
$=16$

27. $(-16)+(-13)=-29$

28. $(-11)-(-12)=-11+12=1$

29. $-\frac{4}{3}+\frac{2}{3}+\frac{1}{6}=-\frac{2}{3}+\frac{1}{6}=-\frac{4}{6}+\frac{1}{6}$
$=-\frac{3}{6}=-\frac{1}{2}$

30. $-\frac{6}{7}+\frac{1}{2}+\left(-\frac{3}{14}\right)=-\frac{12}{14}+\frac{7}{14}+\left(-\frac{3}{14}\right)$
$=-\frac{5}{14}+\left(-\frac{3}{14}\right)$
$=-\frac{8}{14}$
$=-\frac{4}{7}$

31. $(-3)(-2)(-5)=-(3)(2)(5)=-30$

32. $-6+(-2)-(-3)=-6+(-2)+3$
$=-8+3$
$=-5$

33. $3.5(-2.6) = -9.1$

34. $(-5.4) \div 4(-6) = 0.9$

35. $5 - (-3.5) + 1.6 = 5 + 3.5 + 1.6 = 10.1$

36. $-8 + 2 - (-4.8) = -6 + 4.8$
$= -1.2$

37. $17 + 3.4 + (-16) + (-2.5) = 20.4 + (-18.5)$
$= 1.9$

38. $37 + (-44) + 12.5 + (-6.8)$
$= 49.5 + (-50.8)$
$= -1.3$

39. $(-8)(3) = -24$; 24 yards lost

40. $-34 + 12 = -22$
-22° F

41. $6895 - (-468) = 6895 + 468 = 7363$ feet

42. $+1\frac{1}{2} - 3\frac{1}{4} + 2 - 2\frac{1}{2} = 3\frac{1}{2} + \left(-5\frac{3}{4}\right)$
$= -2\frac{1}{4}$
$2\frac{1}{4}$ point loss

43. $(-3)^4 = (-3)(-3)(-3)(-3) = 81$

44. $(-2)^7 = -128$

45. $(-5)^4 = -625$

46. $\left(\frac{2}{3}\right)^3 = \frac{8}{27}$

47. $-9^2 = -81$

48. $(0.6)^2 = 0.36$

49. $\left(\frac{5}{6}\right)^2 = \left(\frac{5}{6}\right)\left(\frac{5}{6}\right) = \frac{25}{36}$

50. $\left(\frac{3}{4}\right)^3 = \left(\frac{3}{4}\right)\left(\frac{3}{4}\right)\left(\frac{3}{4}\right)$
$= \frac{27}{64}$

51. $(5)(-4) + (3)(-2)^3$
$= (5)(-4) + (3)(-8)$
$= -20 + (-24)$
$= -44$

52. $20 - (-10) - (-6) + (-5) - 1$
$= 20 + 10 + 6 + (-5) - 1$
$= 36 - 6$
$= 30$

53. $(7 - 9)^3 + (-6)(-2) + (-3)$
$= (-2)^3 + (-6)(-2) + (-3)$
$= -8 + (-6)(-2) + (-3)$
$= -8 + (12) + (-3) = 1$

54. $5(3x - 7y) = 5(3x) + 5(-7y) = 15x - 35y$

55. $2x(3x-7y+4)$
$=2x(3x)+2x(-7y)+2x(4)$
$=6x^2-14xy+8x$

56. $-(7x^2-3x+11)$
$=-1(7x^2)+(-1)(-3x)+(-1)(11)$
$=-7x^2+3x-11$

57. $(2xy+x-y)(-3y)$
$=(2xy)(-3y)+x(-3y)-y(-3y)$
$=-6xy^2-3xy+3y^2$

58. $3a^2b-2bc+6bc^2-8a^2b-6bc^2+5bc$
$=(3-8)a^2+(-2+5)bc+(6-6)bc^2$
$=-5a^2b+3bc$

59. $9x+11y-12x-15y=-3x-4y$

60. $4x^2-13x+7-9x^2-22x-16$
$=(4-9)x^2+(-13-22)x+7-16$
$=-5x^2-35x-9$

61. $-x+\frac{1}{2}+14x^2-7x-1-4x^2$
$=(14-4)x^2+(-7-1)x+\frac{1}{2}-1$
$=10x^2-8x-\frac{1}{2}$

62. If $x=-7$, then $7x-6=7(-7)-6$
$=-49-6=-55$

63. If $x=8$, then $7-\frac{3}{4}x=7-\frac{3}{4}(8)$
$=7-6=1$

64. $x^2+3x-4=(-3)^2+3(-3)-4$
$=9-9-4$
$=-4$

65. If $x=2$, then
$-3x^2-4x+5=-3(2)^2-4(2)+5$
$=-3(4)-4(2)+5$
$=-12-8+5=-15$

66. $-3x^3-4x^2+2x+6$
$=-3(-2)^3-4(-2)^2+2(-2)+6$
$=-3(-8)-4(4)-4+6$
$=24-16-4+6$
$=30-20$
$=10$

67. If $v=24$, $t=2$, and $a=32$, then
$vt-\frac{1}{2}at^2=24(2)-\frac{1}{2}(32)(2)^2$
$=48-16(4)=48-64=-16$

68. $\frac{nRT}{V}=\frac{16(-2)(4)}{-20}$
$=\frac{-32}{-5}=\frac{32}{5}$

69. If $p=6000$, $r=18\%$, and $t=\frac{3}{4}$, then
$I=prt=6000(0.18)\left(\frac{3}{4}\right)=\$810$

70. $F = \dfrac{9C+160}{5} = \dfrac{9(30)+160}{5}$

$= \dfrac{270+160}{5} = \dfrac{430}{5} = 86$

$86^\circ F$

71. If $r = 15$, $A = \pi r^2 = 3.14(15)^2$

$= 3.14(225) = 706.5$ square meters

Cost $= (706.5 \text{ sq m})(\$3/\text{sq m})$

$= \$2119.50$

72. $P = 180S - R - C$

$= 180(56) - 300 - 1200$

$= 10{,}080 - 300 - 1200$

$= 8580$

$\$8580$

73. $A = \dfrac{1}{2}a(b_1 + b_2)$,

$a = 200,\ b_1 = 700,\ b_2 = 300$

$A = \dfrac{1}{2}(200)(700+300)$

$= 100(1000)$

$= 100{,}000$ square feet

Cost $= 2(100{,}000) = \$200{,}000$

74. $A = \dfrac{1}{2}ab,\ a = 3.8,\ b = 5.5$

$A = \dfrac{1}{2}(3.8)(5.5) = 10.45$ square feet

Cost $= 66(10.45) = \$689.70$

75. $5x - 7(x-6) = 5x - 7x + 42$

$= -2x + 42$

76. $3(x-2) - 4(5x+3)$

$= 3x - 6 - 20x - 12$

$= -17x - 18$

77. $2[3-(4-5x)] = 2(3-4+5x)$

$= 2(-1+5x) - -2 + 10x$

78. $-3x[x+3(x-7)] = -3x(x+3x-21)$

$= -3x(4x-21)$

$= -12x^2 + 63x$

79. $2xy^3 - 6x^3y - 4x^2y^2 + 3(xy^3 - 2x^2y - 3x^2y^2)$

$= 2xy^3 - 6x^3y - 4x^2y^2 + 3xy^3 - 6x^2y - 9x^2y^2$

$= (2+3)xy^3 - 6x^3y + (-4-9)x^2y^2 - 6x^2y$

$= 5xy^3 - 6x^3y - 13x^2y^2 - 6x^2y$

80. $-5(x+2y-7) + 3x(2-5y)$

$= -5x - 10y + 35 + 6x - 15xy$

$= x - 10y + 35 - 15xy$

81. $2\{x - 3(y-2) + 4[x - 2(y+3)]\}$

$= 2[x - 3y + 6 + 4(x - 2y - 6)]$

$= 2(x - 3y + 6 + 4x - 8y - 24)$

$= 2(5x - 11y - 18)$

$= 10x - 22y - 36$

82. $-5\{2a - b[5a - b(3+2a)]\}$

$= -5[2a - b(5a - 3b - 2ab)]$

$= -5(2a - 5ab + 3b^2 + 2ab^2)$

$= -10a + 25ab - 15b^2 - 10ab^2$

83. $-3\{2x-[x-3y(x-2y)]\}$
$=-3[2x-(x-3xy+6y^2)]$
$=-3(2x-x+3xy-6y^2)$
$=-3(x+3xy-6y^2)$
$=-3x-9xy+18y^2$

84. $2\{3x+2[x+2y(x-4)]\}$
$=2[3x+2(x+2xy-8y)]$
$=2(3x+2x+4xy-16y)$
$=2(5x+4xy-16y)$
$=10x+8xy-32y$

85. $-6.3+4=-2.3$

86. $4+(-8)+12=-4+12=8$

87. $-\frac{2}{3}-\frac{4}{5}=-\frac{10}{15}+\left(-\frac{12}{15}\right)$
$=\frac{-22}{15}$
$=-1\frac{7}{15}$

88. $-\frac{7}{8}-\left(-\frac{3}{4}\right)=-\frac{7}{8}+\frac{6}{8}=-\frac{1}{8}$

89. $3-(-4)+(-8)=3+4+(-8)$
$=7+(-8)$
$=-1$

90. $-1.1-(-0.2)+0.4$
$=-1.1+0.2+0.4$
$=-0.9+0.4$
$=-0.5$

91. $\left(-\frac{3}{5}\right)\left(-2\frac{1}{2}\right)=\left(-\frac{3}{5}\right)\left(-\frac{5}{2}\right)$
$=\frac{3}{2}=1\frac{1}{2}$

92. $(-4.2)\div(-0.7)=6$

93. $-14.4\div(-0.06)=240$

94. $(-8.2)(3.1)=-25.42$

95. $400+1000-800=1400-800=600$
Her score was \$600.

96. $(-0.3)^4=(-0.3)(-0.3)(-0.3)(-0.3)$
$=0.0081$

97. $-0.5^4=-(0.5)(0.5)(0.5)(0.5)$
$=-0.0625$

98. $9(5)-5(2)^3+5=9(5)-5(8)+5$
$=45-5(8)+5$
$=45-40+5$
$=5+5$
$=10$

99. $3.8x-0.2y-8.7x+4.3y$
$=(3.8-8.7)x+(-0.2+4.3)y$
$=-4.9x+4.1y$

100. If $p=-2$ and $q=3$, then

$$\frac{2p+q}{3q}=\frac{2(-2)+3}{3(3)}$$
$$=\frac{-4+3}{9}$$
$$=-\frac{1}{9}$$

101. If $s=-3$ and $t=-2$, then

$$\frac{4s-7t}{s}=\frac{4(-3)-7(-2)}{-3}$$
$$=\frac{-12+14}{-3}$$
$$=-\frac{2}{3}$$

102. $F=\frac{9}{5}C+32,\ C=38.6$

$F=\frac{9}{5}(38.6)+32$

$F=69.48+32$

$F=101.48^\circ$

Your dog does not have a fever; in fact, its temperature is below normal.

103. $-7(x-3y^2+4)+3y(4-6y)$

$=-7x+21y^2-28+12y-18y^2$

$=-7x+3y^2+12y-28$

104. $-2\{6x-3[7y-2y(3-x)]\}$

$=-2\{6x-3[7y-6y+2xy]\}$

$=-2\{6x-3[y+2xy]\}$

$=-2\{6x-3y-6xy\}$

$=-12x+6y+12xy$

How Am I Doing? Chapter 1 Test

1. $-2.5+6.3+(-4.1)=3.8+(-4.1)=-0.3$

2. $-5-(-7)=-5+7=2$

3. $\left(-\frac{2}{3}\right)(7)=-\frac{14}{3}=-4\frac{2}{3}$

4. $(-5)(-2)(7)(-1)=-(10)(7)(1)$
   $=-(70)(1)=-70$

5. $(-12)\div(-3)=4$

6. $(-1.8)\div(0.6)=-3$

7. $(-4)^3=(-4)(-4)(-4)=-64$

8. $(1.6)^2=(1.6)(1.6)=2.56$

9. $\left(\frac{2}{3}\right)^4=\left(\frac{2}{3}\right)\left(\frac{2}{3}\right)\left(\frac{2}{3}\right)\left(\frac{2}{3}\right)=\frac{16}{81}$

10. $(0.2)^2-(2.1)(-3)+0.46$
   $=0.04-(2.1)(-3)+0.46$
   $=0.04-(-6.3)+0.46$
   $=0.04+6.3+0.46$
   $=6.34+0.46$
   $=6.8$

11. $3(4-6)^3 + 12 \div (-4) + 2$
$= 3(-2)^3 + 12 \div (-4) + 2$
$= 3(-8) + 12 \div (-4) + 2$
$= -24 - 3 + 2$
$= -25$

12. $-5x(x+2y-7)$
$= -5x(x) - 5x(2y) - 5x(-7)$
$= -5x^2 - 10xy + 35x$

13. $-2ab^2(-3a - 2b + 7ab)$
$= -2ab^2(-3a) - 2ab^2(-2b) - 2ab^2(7ab)$
$= 6a^2b^2 + 4ab^3 - 14a^2b^3$

14. $6ab - \frac{1}{2}a^2b + \frac{3}{2}ab + \frac{5}{2}a^2b$
$= \left(6 + \frac{3}{2}\right)ab + \left(-\frac{1}{2} + \frac{5}{2}\right)a^2b$
$= \left(\frac{12}{2} + \frac{3}{2}\right)ab + \frac{4}{2}a^2b$
$= \frac{15}{2}ab + 2a^2b$

15. $2.3x^2y - 8.1xy^2 + 3.4xy^2 - 4.1x^2y$
$= (2.4 - 4.1)x^2y + (-8.1 + 3.4)xy^2$
$= -1.8x^2y - 4.7xy^2$

16. $3(2-a) - 4(-6-2a) = 6 - 3a + 24 + 8a$
$= 5a + 30$

17. $5(3x-2y) - (x+6y)$
$= 15x - 10y - x - 6y$
$= (15-1)x + (-10-6)y$
$= 14x - 16y$

18. $x^3 - 3x^2y + 2y - 5$
$= 3^3 - 3(3)^2(-4) + 2(-4) - 5$
$= 27 - 3(9)(-4) - 8 - 5$
$= 27 + 108 - 8 - 5$
$= 122$

19. If $x = -3$, then $3x^2 - 7x - 11$
$= 3(-3)^2 - 7(-3) - 11$
$= 37$

20. $2a - 3b$
$= 2(-4) - 3(-3)$
$= -8 + 9$
$= 1$

21. (60 miles/hr)(1.61 km/mile)
= 96.6 kilometers per hour

22. $A = \frac{1}{2}a(b_1 + b_2)$,
$a = 120,\ b_1 = 200,\ b_2 = 180$
$A = \frac{1}{2}(120)(200+180) = 60(380)$
= 22,800 square feet

23. $A = \frac{1}{2}ab,\ a = 6.8,\ b = 8.5$
$A = \frac{1}{2}(6.8)(8.5) = 28.9$ square feet
Cost $= 0.80(28.9) = \$23.12$

24. $A = 60 \times 10 = 600$ sq. ft.
600 sq. ft. $\times \frac{1 \text{ can}}{200 \text{ sq. ft.}} = 3$ cans

25. $3\left[x-2y(x+2y)-3y^2\right]$
$=3\left[x-2xy-4y^2-3y^2\right]$
$=3\left[x-2xy-7y^2\right]$
$=3x-6xy-21y^2$

26. $-3\{a+b[3a-b(1-a)]\}$
$=-3\left[a+b(3a-b+ab)\right]$
$=-3\left(a+3ab-b^2+ab^2\right)$
$=-3a-9ab+3b^2-3ab^2$

Chapter 2

2.1 Exercises

1. When we use the equals sign, we indicate two expressions are equal in value.

3. The solution of an equation is a value of the variable that makes the equation true.

5. Answers may vary. A sample answer is to isolate the variable x.

7. $$\begin{aligned} x+11&=15 \\ x+11+(-11)&=15+(-11) \\ x&=4 \end{aligned}$$

Check: $4+11 \stackrel{?}{=} 15$

$15=15$

9. $$\begin{aligned} 17&=x+5 \\ 17+(-5)&=x+5+(-5) \\ 12&=x \end{aligned}$$

Check: $17 \stackrel{?}{=} 5+12$

$17=17$

11. $$\begin{aligned} x-3&=14 \\ x-3+3&=14+3 \\ x&=17 \end{aligned}$$

Check: $17-3 \stackrel{?}{=} 14$

$14=14$

13. $$\begin{aligned} 0&=x+5 \\ 0+(-5)&=x+5+(-5) \\ -5&=x \end{aligned}$$

Check: $0 \stackrel{?}{=} -5+5$

$0=0$

15. $$\begin{aligned} x-6&=-19 \\ x-6+6&=-19+6 \\ x&=-13 \end{aligned}$$

Check: $-13-6 \stackrel{?}{=} -19$

$-19=-19$

17. $$\begin{aligned} -12+x&=50 \\ -12+12+x&=50+12 \\ x&=62 \end{aligned}$$

Check: $-12+62 \stackrel{?}{=} 50$

$50=50$

19. $$\begin{aligned} 3+5&=x-7 \\ 8&=x-7 \\ 8+7&=x-7+7 \\ 15&=x \end{aligned}$$

Check: $3+5 \stackrel{?}{=} 15-7$

$8=8$

21. $$\begin{aligned} 32-17&=x-6 \\ 15&=x-6 \\ 15+6&=x-6+6 \\ 21&=x \end{aligned}$$

Check: $32\text{-}17 \stackrel{?}{=} 21-6$

$15=15$

23. $4+8+x=6+6$

$12+x=12$

$12+(-12)+x=12+(-12)$

$x=0$

Check: $4+8+0\stackrel{?}{=}6+6$

$12=12$

25. $8-23+7=1+x-2$

$-8=x-1$

$-8+1=x-1+1$

$-7=x$

Check: $8-23+7\stackrel{?}{=}1+(-7)-2$

$-8=-8$

27. $-12+x-3=15-18+9$

$-15+x=6$

$-15+15+x=6+15$

$x=21$

Check: $-12+21-3\stackrel{?}{=}15-18+9$

$6=6$

29. $-7+x=2,\ x\stackrel{?}{=}5$

$-7+5\stackrel{?}{=}2$

$-2=2$

$x=5$ is not a solution.

31. $-18-2\stackrel{?}{=}-3-7$

$-20\neq-10$

$x=-3$ is not the solution.

$-18-2=x-7$

$-20=x-7$

$-20+7=x-7+7$

$-13=x$

33. $x-23=-56,\quad x\stackrel{?}{=}-33$

$-33-23\stackrel{?}{=}-56$

$-56=-56$

$x=-33$ is a solution.

35. $15-3+20=x-3,\ \ x\stackrel{?}{=}35$

$15-3+20\stackrel{?}{=}35-3$

$32=32$

$x=35$ is the solution.

37. $2.5+x=0.7$

$2.5+(-2.5)+x=0.7+(-2.5)$

$x=-1.8$

39. $2.7+x-1.4=3.8$

$x+1.3=3.8$

$x+1.3+(-1.3)=3.8+(-1.3)$

$x=2.5$

41. $x-\frac{1}{4}=\frac{3}{4}$

$x-\frac{1}{4}+\frac{1}{4}=\frac{3}{4}+\frac{1}{4}$

$x=1$

43.
$$\frac{2}{3}+x=\frac{1}{6}+\frac{1}{4}$$
$$\frac{8}{12}+x=\frac{2}{12}+\frac{3}{12}$$
$$\frac{8}{12}+x=\frac{5}{12}$$
$$\frac{8}{12}+\left(-\frac{8}{12}\right)+x=\frac{5}{12}+\left(-\frac{8}{12}\right)$$
$$x=-\frac{3}{12}$$
$$x=-\frac{1}{4}$$

45.
$$3+x=-12+8$$
$$3+x=-4$$
$$3+(-3)+x=-4+(-3)$$
$$x=-7$$

47.
$$5\frac{1}{6}+x=8$$
$$\frac{31}{6}+x=\frac{48}{6}$$
$$\frac{31}{6}+\left(-\frac{31}{6}\right)+x=\frac{48}{6}+\left(-\frac{31}{6}\right)$$
$$x=\frac{17}{6}\text{ or }2\frac{5}{6}$$

49.
$$\frac{5}{12}-\frac{5}{6}=x-\frac{3}{2}$$
$$\frac{5}{12}-\frac{10}{12}=x-\frac{18}{12}$$
$$-\frac{5}{12}=x-\frac{18}{12}$$
$$-\frac{5}{12}+\frac{18}{12}=x-\frac{18}{12}+\frac{18}{12}$$
$$\frac{13}{12}=x$$
$$1\frac{1}{12}=x$$

51.
$$1.6+4x-3.2=-2x+5.6+5x$$
$$-1.6+4x=3x+5.6$$
$$-1.6+4x+(-3x)=-3x+(-3x)+5.6$$
$$-1.6+x=5.6$$
$$-1.6+1.6+x=5.6+1.6$$
$$x=7.2$$

53.
$$x+0.7513=2.2419$$
$$x+0.7513+(-0.7513)=2.2419+(-0.7513)$$
$$x=1.4906$$

Cumulative Review

55. $x+3y-5x-7y+2x$
$=(1-5+2)x+(3-7)y$
$=-2x-4y$

57. percent $=\frac{8}{413}=0.019=1.9\%$
yes, this was $1.9\% < 3\%$

59. Distance $=90-(-27)=117$ feet

2.2 Exercises

1. Divide each side of $6x = -24$ by $\underline{6}$.

3. Multiply each side of $\frac{1}{7}x = -2$ by $\underline{7}$.

5. $\frac{1}{9}x = 4$

$9\left(\frac{1}{9}x\right) = 9(4)$

$x = 36$

Check: $\frac{1}{9}(36) \stackrel{?}{=} 4$

$4 = 4$

7. $\frac{1}{3}x = -9$

$3\left(\frac{1}{3}x\right) = 3(-9)$

$x = -27$

Check: $\frac{1}{3}(-27) \stackrel{?}{=} -9$

$-9 = -9$

9. $\frac{x}{5} = 16$

$5\left(\frac{x}{5}\right) = 5(16)$

$x = 80$

Check: $\frac{80}{5} \stackrel{?}{=} 16$

$16 = 16$

11. $-3 = \frac{x}{5}$

$5(-3) = 5\left(\frac{x}{5}\right)$

$-15 = x$

Check: $-3 \stackrel{?}{=} \frac{-15}{5}$

$-3 = -3$

13. $13x = 52$

$\frac{13x}{13} = \frac{52}{13}$

$x = 4$

Check: $13(4) \stackrel{?}{=} 52$

$52 = 52$

15. $56 = 7x$

$\frac{56}{7} = \frac{7x}{7}$

$8 = x$

Check: $56 \stackrel{?}{=} 7(8)$

$56 = 56$

17. $-16 = 6x$

$\frac{-16}{6} = \frac{6x}{6}$

$-\frac{8}{3} = x$

Check: $-16 \stackrel{?}{=} 6\left(-\frac{8}{3}\right)$

$-16 = -16$

19. $1.5x = 75$

$\frac{1.5x}{1.5} = \frac{75}{1.5}$

$x = 50$

Check: $1.5(50) \stackrel{?}{=} 75$

$75 = 75$

21. $-15 = -x$

$\frac{-15}{-1} = \frac{-x}{-1}$

$15 = x$

Check: $-15 \stackrel{?}{=} (-1)(15)$

$-15 = -15$

23. $-112 = 16x$

$\frac{-112}{16} = \frac{16x}{16}$

$-7 = x$

Check: $-112 \stackrel{?}{=} 16(7)$

$-112 = -112$

25. $0.4x = 0.08$

$\frac{0.4x}{0.4} = \frac{0.08}{0.4}$

$x = 0.2$ or $\frac{1}{5}$

Check: $(0.4)(0.2) \stackrel{?}{=} 0.08$

$0.08 = 0.08$

27. $-3.9x = -15.6$

$\frac{-3.9x}{-3.9} = \frac{-15.6}{-3.9}$

$x = 4$

Check: $-3.9(4) \stackrel{?}{=} -15.6$

$-15.6 = -15.6$

29. $-3x = 21, \quad x \stackrel{?}{=} 7$

$-3(7) \stackrel{?}{=} 21$

$-21 \neq 21$

$x = 7$ is not the solution.

$-3x = 21$

$\frac{-3x}{-3} = \frac{21}{-3}$

$x = -7$

31. $-11x = 66, \quad x \stackrel{?}{=} -6$

$-11(-6) \stackrel{?}{=} 66$

$66 = 66$

$x = -6$ is the solution.

33. $7y = -0.21$

$\frac{7y}{7} = \frac{-0.21}{7}$

$y = -0.03$

35. $-56 = -21t$

$\frac{-56}{-21} = \frac{-21t}{-21}$

$\frac{8}{3} = t$

Check: $-56 \stackrel{?}{=} -21\left(\frac{8}{3}\right)$

$-56 = -56$

37. $4.6y = -3.22$
$$\frac{4.6y}{4.6} = \frac{-3.22}{4.6}$$
$$y = -0.7$$

39. $4x + 3x = 21$
$$7x = 21$$
$$\frac{7x}{7} = \frac{21}{7}$$
$$x = 3$$

41. $2x - 7x = 20$
$$-5x = 20$$
$$\frac{-5x}{-5} = \frac{20}{-5}$$
$$x = -4$$

43. $-6x - 3x = -7$
$$-9x = -7$$
$$\frac{-9x}{-9} = \frac{-7}{-9}$$
$$x = \frac{7}{9}$$

45. $12 - 19 = -7x$
$$-7 = -7x$$
$$\frac{-7}{-7} = \frac{-7x}{-7}$$
$$1 = x$$

47. $6x = -18 + 36$
$$6x = 18$$
$$\frac{6x}{6} = \frac{18}{6}$$
$$x = 3$$

49. $\frac{2}{3}x = 18$
$$\frac{3}{2}\left(\frac{2}{3}x\right) = \frac{3}{2}(18)$$
$$x = 27$$

51. $3.6172x = -19.026472$
$$\frac{3.6172x}{3.6172} = \frac{19.026472}{3.6172}$$
$$x = -5.26$$

53. To solve an equation, we are performing steps to get an equivalent equation that has the same solution. Now $a = b$ and $a(0) = b(0)$ are not equivalent equations because they do not have the same solution. So we must have the requirement that when we multiply both sides of the equation by c, it is essential that c is nonzero.

Cumulative Review

55. $(-6)(-8) + (-3)(2) = 48 - 6 = 42$

57. $5 + (2 - 6)^2 = 5 + (-4)^2$
$$= 5 + 16$$
$$= 21$$

59. $600 + 82 - 47 + 103 - 106$
$= 785 - 153$
$= \$632$

61. $100\% + 35\% = 135\%$
    135% of 20
    $1.35(20) = 27$
    Expect 27 earthquakes

2.3 Exercises

1. $4x + 13 = 21$
   $4x + 13 + (-13) = 21 + (-13)$
   $4x = 8$
   $\frac{4x}{4} = \frac{8}{4}$
   $x = 2$
   Check: $4(2) + 13 \stackrel{?}{=} 21$
   $21 = 21$

3. $4x - 11 = 13$
   $4x - 11 + 11 = 13 + 11$
   $4x = 24$
   $\frac{4x}{4} = \frac{24}{4}$
   $x = 6$
   Check: $4(6) - 11 \stackrel{?}{=} 13$
   $13 = 13$

5. $7x - 18 = -46$
   $7x - 18 + 18 = -46 + 18$
   $7x = -28$
   $\frac{7x}{7} = \frac{-28}{7}$
   $x = -4$
   Check: $7(-4) - 18 \stackrel{?}{=} -46$
   $-28 - 18 \stackrel{?}{=} -46$
   $-46 = -46$

7. $-4x + 17 = -35$
   $-4x + 17 + (-17) = -35 + (-17)$
   $-4x = -52$
   $\frac{-4x}{-4} = \frac{-52}{-4}$
   $x = 13$
   Check: $-4(13) + 17 \stackrel{?}{=} -35$
   $-52 + 17 \stackrel{?}{=} -35$
   $-35 = -35$

9. $2x + 3.2 = 9.4$
   $2x + 3.2 + (-3.2) = 9.4 + (-3.2)$
   $2x = 6.2$
   $\frac{2x}{2} = \frac{6.2}{2}$
   $x = 3.1$
   Check: $2(3.1) + 3.2 \stackrel{?}{=} 9.4$
   $6.2 + 3.2 \stackrel{?}{=} 9.4$
   $9.4 = 9.4$

11. $\frac{1}{5}x - 2 = 6$
    $\frac{1}{5}x - 2 + 2 = 6 + 2$
    $\frac{1}{5}x = 8$
    $5\left(\frac{1}{5}x\right) = 5(8)$
    $x = 40$
    Check: $\frac{1}{5}(40) - 2 \stackrel{?}{=} 6$
    $6 = 6$

13. $\frac{1}{3}x+5=-4$

$$\frac{1}{3}x+5+(-5)=-4+(-5)$$
$$\frac{1}{3}x=-9$$
$$3\left(\frac{1}{3}x\right)=3(-9)$$
$$x=-27$$

Check: $\frac{1}{3}(-27)+5\stackrel{?}{=}-4$

$$-9+5\stackrel{?}{=}-4$$
$$-4=-4$$

15. $8x=48+2x$

$$8x+(-2x)=48+2x+(-2x)$$
$$6x=48$$
$$\frac{6x}{6}=\frac{48}{6}$$
$$x=8$$

Check: $8(8)\stackrel{?}{=}48+2(8)$

$$64\stackrel{?}{=}48+16$$
$$64=64$$

17. $-6x=-27+3x$

$$-6x+(-3x)=-27+3x+(-3x)$$
$$-9x=-27$$
$$\frac{-9x}{-9}=\frac{-27}{-9}$$
$$x=3$$

Check: $-6(3)\stackrel{?}{=}-27+3(3)$

$$-18\stackrel{?}{=}-27+9$$
$$-18=-18$$

19. $63-x=8x$

$$63-x+x=8x+x$$
$$63=9x$$
$$\frac{63}{9}=\frac{9x}{9}$$
$$7=x$$

Check: $63-7\stackrel{?}{=}8(7)$

$$56=56$$

21. $54-2x=-8x$

$$54-2x+2x=-8x+2x$$
$$54=-6x$$
$$\frac{54}{-6}=\frac{-6x}{-6}$$
$$-9=x$$

Check: $54-2(-9)\stackrel{?}{=}-8(-9)$

$$54+18\stackrel{?}{=}72$$
$$72=72$$

23. $2y+3y=12-y$

$$2(2)+3(2)\stackrel{?}{=}12-2$$
$$4+6\stackrel{?}{=}10$$
$$10=10$$

$y=2$ is the solution.

25. $7x+6-3x=2x-5+x$

$$7(11)+6-3(11)\stackrel{?}{=}2(11)-5+11$$
$$77+6-33\stackrel{?}{=}22-5+11$$
$$50\neq 28$$

$x=11$ is not a solution.

$$7x+6-3x=2x-5+x$$
$$4x+6=3x-5$$

$$4x+(-3x)+6=3x+(-3x)-5$$
$$x+6=-5$$
$$x+6+(-6)=-5+(-6)$$
$$x=-11$$

27.
$$14-2x=-5x+11$$
$$14-2x+5x=-5x+5x+11$$
$$14+3x=11$$
$$3x+14+(-14)=11+(-14)$$
$$3x=-3$$
$$\frac{3x}{3}=\frac{-3}{3}$$
$$x=-1$$

29.
$$x-6=8-x$$
$$x+x-6=8-x+x$$
$$2x-6=8$$
$$2x-6+6=8+6$$
$$2x=14$$
$$\frac{2x}{2}=\frac{14}{2}$$
$$x=7$$

31.
$$0.8y-0.4=0.9-0.5y$$
$$0.8y+0.5y-0.4=0.9-0.5y+0.5y$$
$$1.3y-0.4=0.9$$
$$1.3y-0.4+0.4=0.9+0.4$$
$$1.3y=1.3$$
$$\frac{1.3y}{1.3}=\frac{1.3}{1.3}$$
$$y=1$$

33.
$$5x-9=3x+23$$
$$5x+(-3x)-9=3x+(-3x)+23$$
$$2x-9=23$$
$$2x-9+9=23+9$$
$$2x=32$$
$$\frac{2x}{2}=\frac{32}{2}$$
$$x=16$$

35.
$$-3+10y+6=15+12y-18$$

Left
$$10y+3=12y-3$$
$$10y+(-12y)+3=12y+(-12y)-3$$
$$-2y+3=-3$$
$$-2y+3+(-3)=-3+(-3)$$
$$-2y=-6$$
$$\frac{-2y}{-2}=\frac{-6}{-2}$$
$$y=3$$

Right
$$10y+3=12y-3$$
$$10y+(-10y)+3=12y+(-10y)-3$$
$$3=2y-3$$
$$3+3=2y-3+3$$
$$6=2y$$
$$\frac{6}{2}=\frac{2y}{2}$$
$$3=y$$

Neither approach is better.

37. $5(x+3)=35$

$5x+15=35$

$5x+15-15=35-15$

$5x=20$

$\frac{5x}{5}=\frac{20}{5}$

$x=4$

Check: $5(4+3)\stackrel{?}{=}35$

$5(7)\stackrel{?}{=}35$

$35=35$

39. $6(3x+2)-8=-2$

$18x+12-8=-2$

$18x+4=-2$

$18x=-6$

$x=-\frac{1}{3}$

Check: $6\left[3\left(-\frac{1}{3}\right)+2\right]-8\stackrel{?}{=}-2$

$6(-1+2)-8\stackrel{?}{=}-2$

$-2=-2$

41. $7x-3(5-x)=10$

$7x-15+3x=10$

$10x-15=10$

$10x=25$

$x=2.5$

Check: $7(2.5)-3[5-(2.5)]\stackrel{?}{=}10$

$17.5-3(2.5)\stackrel{?}{=}10$

$17.5-7.5\stackrel{?}{=}10$

$10=10$

43. $0.5x-0.3(2-x)=4.6$

$0.5x-0.6-0.3x=4.6$

$0.8x-0.6=4.6$

$0.8x-0.6+0.6=4.6+0.6$

$0.8x=5.2$

$\frac{0.8x}{0.8}=\frac{5.2}{0.8}$

$x=6.5$ or $6\frac{1}{2}$

Check: $0.5(6.5)-0.3(2-6.5)\stackrel{?}{=}4.6$

$0.5(6.5)-0.3(-4.5)\stackrel{?}{=}4.6$

$3.25+1.35\stackrel{?}{=}4.6$

$4.6=4.6$

45. 44. $5(x-3)+5=3(x+2)-4$

$5x-15+5=3x+6-4$

$5x-10=3x+2$

$5x-3x-10=3x-3x+2$

$2x-10=2$

$2x-10+10=2+10$

$2x=12$

$x=6$

Check: $5(6-3)+5 \stackrel{?}{=} 3(6+2)-4$

$5(3)+5 \stackrel{?}{=} 3(8)-4$

$15+5 \stackrel{?}{=} 24-4$

$20 = 20$

47. $-2(x+3)+4 = 3(x+4)+2$

$-2x-6+4 = 3x+12+2$

$-2x-2 = 3x+14$

$-2x+(-3x)-2 = 3x+(-3x)+14$

$-5x-2 = 14$

$-5x-2+2 = 14+2$

$-5x = 16$

$\frac{-5x}{-5} = \frac{16}{-5}$

$x = -3.2 \text{ or } -3\frac{1}{5}$

Check: $-2[(-3.2)+3]+4 \stackrel{?}{=} 3[(-3.2)+4]+2$

$-2(-0.2)+4 \stackrel{?}{=} 3(0.8)+2$

$0.4+4 \stackrel{?}{=} 2.4+2$

$4.4 = 4.4$

49. $-3(y-3y)+4 = -4(3y-y)+6+13y$

$-3(-2y)+4 = -4(2y)+6+13y$

$6y+4 = -8y+6+13y$

$6y+4 = 5y+6$

$6y-5y+4 = 5y-5y+6$

$y+4 = 6$

$y+4-4 = 6-4$

$y = 2$

Check: $-3[2-3(2)]+4$

$\stackrel{?}{=} -4[3(2)-2]+6+13(2)$

$-3(2-6)+4 \stackrel{?}{=} -4(6-2)+6+26$

$-3(-4)+4 \stackrel{?}{=} -4(4)+32$

$12+4 \stackrel{?}{=} 16+32$

$16 = 16$

51. $5.7x+3 = 4.2x-3$

$5.7x-4.2x+3 = 4.2x-4.2x-3$

$1.5x+3 = -3$

$1.5x+3-3 = -3-3$

$1.5x = -6$

$\frac{1.5x}{1.5} = \frac{-6}{1.5}$

$x = -4$

53. $5z+7-2z = 32-2z$

$3z+7 = 32-2z$

$3z+2z+7 = 32-2z+2z$

$5z = 25$

$\frac{5z}{5} = \frac{25}{5}$

$z = 5$

55. $-0.3a+1.4 = -1.2-0.7a$

$-0.3a+0.7a+1.4 = -1.2-0.7a+0.7a$

$0.4a+1.4 = -1.2$

$0.4a+1.4+(-1.4) = -1.2+(-1.4)$

$0.4a = -2.6$

$\frac{0.4a}{0.4} = \frac{-2.6}{0.4}$

$a = -6.5$

57.
$$\begin{aligned}6x+8-3x&=11-12x-13\\3x+8&=-12x-2\\3x+12x+8&=-12x+12x-2\\15x+8&=-2\\15x+8-8&=-2-8\\15x&=-10\\\frac{15x}{15}&=\frac{-10}{15}\\x&=-\frac{2}{3}\end{aligned}$$

59.
$$\begin{aligned}-3.5x+1.3&=-2.7x+1.5\\-3.5x+3.5x+1.3&=-2.7x+3.5x+1.5\\1.3&=0.8x+1.5\\1.3-1.5&=0.8x+1.5-1.5\\-0.2&=0.8x\\\frac{-0.2}{0.8}&=\frac{0.8x}{0.8}\\-0.25&=x\end{aligned}$$

61.
$$\begin{aligned}5(4+x)&=3(3x-1)-9\\20+5x&=9x-3-9\\20+5x&=9x-12\\20+5x-5x&=9x-5x-12\\20&=4x-12\\20+12&=4x-12+12\\32&=4x\\\frac{32}{4}&=\frac{4x}{4}\\8&=x\end{aligned}$$

63.
$$\begin{aligned}4x+3.2-1.9x&=0.3x-4.9\\2.1x+3.2&=0.3x-4.9\\2.1x+(-0.3x)+3.2&=0.3x+(-0.3x)-4.9\\1.8x+3.2&=-4.9\\1.8x+3.2+(-3.2)&=-4.9+(-3.2)\\1.8x&=-8.1\\\frac{1.8x}{1.8}&=\frac{-8.1}{1.8}\\x&=-4.5\end{aligned}$$

65.
$$\begin{aligned}3(x+4)-5(3x-2)&=8\\3x+12-15x+10&=8\\-12x+22&=8\\-12x+22-22&=8-22\\-12x&=-14\\\frac{-12x}{-12}&=\frac{-14}{-12}\\x&=\frac{7}{6}\end{aligned}$$

67.
$$\begin{aligned}1.63x-9.23&=5.71x+8.04\\1.63x-1.63x-9.23&=5.71x-1.63x+8.04\\-9.23&=4.08x+8.04\\-9.23-8.04&=4.08x+8.04-8.04\\-17.27&=4.08x\\\frac{-17.27}{4.08}&=\frac{4.08x}{4.08}\\-4.23&=x\end{aligned}$$

Cumulative Review

69.
$$\begin{aligned}&2x(3x-y)+4(2x^2-3xy)\\&=6x^2-2xy+8x^2-12xy\\&=(6+8)x^2+(-2-12)xy\\&=14x^2-14xy\end{aligned}$$

71. $4\times 52\frac{1}{8}=4\times\frac{417}{8}=208\frac{1}{2}$

$$3.2\left(81\frac{7}{8}\right)=\left(3\frac{1}{5}\right)\left(81\frac{7}{8}\right)=\left(\frac{16}{5}\right)\left(\frac{655}{8}\right)$$
$$=262$$

$$5.2\left(71\frac{7}{8}\right)=\left(5\frac{1}{5}\right)\left(71\frac{7}{8}\right)=\left(\frac{26}{5}\right)\left(\frac{575}{8}\right)$$
$$=373\frac{3}{4}$$

$$\text{Total}=208\frac{1}{2}+262+373\frac{3}{4}=\$844.25$$

2.4 Exercises

1. $\frac{1}{2}x+\frac{2}{3}=\frac{1}{6}$

$$6\left(\frac{1}{2}x\right)+6\left(\frac{2}{3}\right)=6\left(\frac{1}{6}\right)$$
$$3x+4=1$$
$$3x+4+(-4)=1+(-4)$$
$$3x=-3$$
$$\frac{3x}{3}=\frac{-3}{3}$$
$$x=-1$$

Check: $\frac{1}{2}(-1)+\frac{2}{3}\overset{?}{=}\frac{1}{6}$

$$-\frac{1}{2}+\frac{2}{3}\overset{?}{=}\frac{1}{6}$$
$$-\frac{3}{6}+\frac{4}{6}\overset{?}{=}\frac{1}{6}$$
$$\frac{1}{6}=\frac{1}{6}$$

3. $\frac{2}{3}x=\frac{1}{15}x+\frac{3}{5}$

$$15\left(\frac{2}{3}x\right)=15\left(\frac{1}{15}x\right)+15\left(\frac{3}{5}\right)$$
$$10x=x+9$$
$$10x+(-x)=x+(-x)+9$$
$$9x=9$$
$$\frac{9x}{9}=\frac{9}{9}$$
$$x=1$$

Check: $\frac{2}{3}(1)\overset{?}{=}\frac{1}{15}(1)+\frac{3}{5}$

$$\frac{2}{3}\overset{?}{=}\frac{1}{15}+\frac{9}{15}$$
$$\frac{2}{3}\overset{?}{=}\frac{10}{15}$$
$$\frac{2}{3}=\frac{2}{3}$$

5. $\frac{x}{2}+\frac{x}{5}=\frac{7}{10}$

$$10\left(\frac{x}{2}\right)+10\left(\frac{x}{5}\right)=10\left(\frac{7}{10}\right)$$
$$5x+2x=7$$
$$7x=7$$
$$\frac{7x}{7}=\frac{7}{7}$$
$$x=1$$

Check: $\frac{1}{2}+\frac{1}{5}\stackrel{?}{=}\frac{7}{10}$

$\frac{5}{10}+\frac{2}{10}\stackrel{?}{=}\frac{7}{10}$

$\frac{7}{10}=\frac{7}{10}$

7. $20-\frac{1}{3}x=\frac{1}{2}x$

$6(20)-6\left(\frac{1}{3}x\right)=6\left(\frac{1}{2}x\right)$

$120-2x=3x$

$120=5x$

$24=x$

Check: $20-\frac{1}{3}(24)\stackrel{?}{=}\frac{1}{2}(24)$

$20-8\stackrel{?}{=}12$

$12=12$

9. $2+\frac{y}{2}=\frac{3y}{4}-3$

$4(2)+4\left(\frac{y}{2}\right)=4\left(\frac{3y}{4}\right)-4(3)$

$8+2y=3y-12$

$8=y-12$

$20=y$

Check: $2+\left(\frac{20}{2}\right)\stackrel{?}{=}\frac{3(20)}{4}-3$

$2+10\stackrel{?}{=}15-3$

$12=12$

11. $\frac{x-3}{5}=1-\frac{x}{3}$

$15\left(\frac{x-3}{5}\right)=15(1)-15\left(\frac{x}{3}\right)$

$3(x-3)=15-5x$

$3x-9=15-5x$

$3x+5x-9=15-5x+5x$

$8x-9=15$

$8x-9+9=15+9$

$8x=24$

$\frac{8x}{8}=\frac{24}{8}$

$x=3$

Check: $\frac{3-3}{5}\stackrel{?}{=}1-\frac{3}{3}$

$0=0$

13. $\frac{x+3}{4}=\frac{x}{2}+\frac{1}{6}$

$12\left(\frac{x+3}{4}\right)=12\left(\frac{x}{2}\right)+12\left(\frac{1}{6}\right)$

$3(x+3)=6x+2$

$3x+9=6x+2$

$3x+(-6x)+9=6x+(-6x)+2$

$-3x+9=2$

$-3x+9+(-9)=2+(-9)$

$-3x=-7$

$\frac{-3x}{-3}=\frac{-7}{-3}$

$x=\frac{7}{3}$ or $2\frac{1}{3}$

Check: $\dfrac{\frac{7}{3}+3}{4} \stackrel{?}{=} \dfrac{\frac{7}{3}}{2}+\dfrac{1}{6}$

$\dfrac{\frac{7}{3}+\frac{9}{3}}{4} \stackrel{?}{=} \dfrac{7}{3}\cdot\dfrac{1}{2}+\dfrac{1}{6}$

$\dfrac{16}{3}\cdot\dfrac{1}{4} \stackrel{?}{=} \dfrac{7}{6}+\dfrac{1}{6}$

$\dfrac{4}{3} \stackrel{?}{=} \dfrac{8}{6}$

$\dfrac{4}{3}=\dfrac{4}{3}$

15. $$0.6x+5.9=3.8$$
$$10(0.6x)+10(5.9)=10(3.8)$$
$$6x+59=38$$
$$6x+59-59=38-59$$
$$6x=-21$$
$$\frac{6x}{6}=\frac{-21}{6}$$
$$x=-\frac{7}{2}=-3.5$$

Check: $0.6(-3.5)+5.9 \stackrel{?}{=} 3.8$

$-2.1+5.9 \stackrel{?}{=} 3.8$

$3.8=3.8$

17. $\dfrac{1}{2}(y-2)+2=\dfrac{3}{8}(3y-4),\quad y\stackrel{?}{=}4$

$\dfrac{1}{2}(4-2)+2\stackrel{?}{=}\dfrac{3}{8}(3\cdot 4-4)$

$1+2\stackrel{?}{=}\dfrac{3}{8}(8)$

$3=3$

$y=4$ is a solution.

19. $\dfrac{1}{2}\left(y-\dfrac{1}{5}\right)=\dfrac{1}{5}(y+2),\quad y\stackrel{?}{=}\dfrac{5}{8}$

$\dfrac{1}{2}\left(\dfrac{5}{8}-\dfrac{1}{5}\right)\stackrel{?}{=}\dfrac{1}{5}\left(\dfrac{5}{8}+2\right)$

$\dfrac{1}{2}\left(\dfrac{17}{40}\right)\stackrel{?}{=}\dfrac{1}{5}\left(\dfrac{21}{8}\right)$

$\dfrac{17}{80}\neq\dfrac{42}{80}$

$y=\dfrac{5}{8}$ is not a solution.

21. $$\frac{3}{4}(3x+1)=2(3-2x)+1$$
$$4\left(\frac{3}{4}\right)(3x+1)=4(2)(3-2x)+4(1)$$
$$3(3x+1)=8(3-2x)+4$$
$$9x+3=24-16x+4$$
$$9x+3=-16x+28$$
$$9x+16x+3=-16x+16x+28$$
$$25x+3=28$$
$$25x+3-3=28-3$$
$$25x=25$$
$$\frac{25x}{25}=\frac{25}{25}$$
$$x=1$$

23. $$2(x-2)=\frac{2}{5}(3x+1)+2$$
$$5(2)(x-2)=5\left(\frac{2}{5}\right)(3x+1)+5(2)$$
$$10(x-2)=2(3x+1)+10$$
$$10x-20=6x+2+10$$
$$10x-20=6x+12$$
$$10x-6x-20=6x-6x+12$$
$$4x-20=12$$
$$4x-20+20=12+20$$
$$4x=32$$
$$\frac{4x}{4}=\frac{32}{4}$$
$$x=8$$

25. $$0.3x-0.2(3-5x)=-0.5(x-6)$$
$$10(0.3x)-10(0.2)(3-5x)=10(-0.5)(x-6)$$
$$3x-2(3-5x)=-5(x-6)$$
$$3x-6+10x=-5x+30$$
$$13x-6=-5x+30$$
$$13x+5x-6=-5x+5x+30$$
$$18x-6=30$$
$$18x-6+6=30+6$$
$$18x=36$$
$$\frac{18x}{18}=\frac{36}{18}$$
$$x=2$$

27. $$-5(0.2x+0.1)-0.6=1.9$$
$$-x-0.5-0.6=1.9$$
$$-x-1.1=1.9$$
$$-x=3$$
$$x=-3$$

29. $$\frac{1}{3}(y+2)=3y-5(y-2)$$
$$3\left(\frac{1}{3}\right)(y+2)=3(3y)-3(5)(y-2)$$
$$y+2=9y-15y+30$$
$$y+2=-6y+30$$
$$7y=28$$
$$y=4$$

31. $$\frac{1+2x}{5}+\frac{4-x}{3}=\frac{1}{15}$$
$$15\left(\frac{1+2x}{5}\right)+15\left(\frac{4-x}{3}\right)=15\left(\frac{1}{15}\right)$$
$$3+6x+20-5x=1$$
$$x+23=1$$
$$x=-22$$

33. $$\frac{1}{5}(x+3)=2x-3(2-x)-3$$
$$5\left(\frac{1}{5}\right)(x+3)=5(2x)-5(3)(2-x)-5(3)$$
$$x+3=10x-30+15x-15$$
$$x+3=25x-45$$
$$x-x+3=25x-x-45$$
$$3=24x-45$$
$$3+45=24x-45+45$$
$$48=24x$$
$$\frac{48}{24}=\frac{24x}{24}$$
$$2=x$$

35. $\frac{1}{3}(x-2) = 3x - 2(x-1) + \frac{16}{3}$

$3\left(\frac{1}{3}\right)(x-2) = 3(3x) - 3(2)(x-1) + 3\left(\frac{16}{3}\right)$

$x - 2 = 9x - 6x + 6 + 16$

$x - 2 = 3x + 22$

$-2x = 24$

$x = -12$

37. $\frac{4}{5}x - \frac{2}{3} = \frac{3x+1}{2}$

$30\left(\frac{4}{5}x\right) - 30\left(\frac{2}{3}\right) = 30\left(\frac{3x+1}{2}\right)$

$24x - 20 = 15(3x+1)$

$24x - 20 = 45x + 15$

$24x + (-45x) - 20 = 45x + (-30x) + 15$

$-21x - 20 = 15$

$-21x - 20 + 20 = 15 + 20$

$-21x = 35$

$\frac{-21x}{-21} = \frac{35}{-21}$

$x = -\frac{5}{3}$ or $-1\frac{2}{3}$

39. $0.2(x+3) = 4(0.5x - 0.03)$

$0.2x + 0.6 = 2xx - 0.12$

$0.2x + (-0.2x) + 0.6 = 2x + (-0.2x) - 0.12$

$0.6 = 1.8x - 0.12 + 0.12$

$0.72 = 1.8x$

$\frac{0.72}{1.8} = \frac{1.8x}{1.8}$

$0.4 = x$

41. $-1 + 5(x-2) = 12x + 3 - 7x$

$-1 + 5x - 10 = 5x + 3$

$5x - 11 = 5x + 3$

$5x - 5x - 11 = 5x - 5x + 3$

$-11 = 3$, No solution

43. $9(x+3) - 6 = 24 - 2x - 3 + 11x$

$9x + 27 - 6 = 9x + 21$

$9x + 21 = 9x + 21$

$9x - 9x + 21 = 9x - 9x + 21$

$21 = 21$

Infinite number of solutions

45. $7x + 6 = 2(3x-1) + 8$

$7x + 6 = 6x - 2 + 8$

$7x + 6 = 6x + 6$

$7x + (-6x) + 6 = 6x + (-6x) + 6$

$x + 6 = 6$

$x + 6 + (-6) = 6 + (-6)$

$x = 0$

47. $3(4x+1) - 2x = 2(5x-3)$

$12x + 3 - 2x = 10x - 6$

$10x + 3 = 10x - 6$

$10x + (-10x) + 3 = 10x + (-10x) - 6$

$3 = -6$

No solution

Cumulative Review

49. $\frac{3}{7}+1\frac{5}{10}=\frac{3}{7}+\frac{15}{10}$

$$=\frac{30}{70}+\frac{105}{70}$$

$$=\frac{135}{70}=1\frac{65}{70}=1\frac{13}{14}$$

51. $100\%+20\%=120\%$

$120\% \text{ of } 40=1.20(40)=48$

$100\%+450\%=550\%$

$550\% \text{ of } 48=5.50(48)=264$

264 pairs were thriving in 2000.

53. Population in 1985

$40+20\%(40)=48$

Population in 2000

$48+450\%(48)=264$ pairs

55. $A=\pi r^2,\ r=6$

$A\approx 3.14(6)^2=113.04$ square inches

Cost $=2(113.04)=\$226.08$

How Am I Doing? Sections 2.1-2.4

1. $5-8+x=-12$

$$-3+x=-12$$
$$-3+3+x=-12+3$$
$$x=-9$$

2. $3.6+x=-7.x$

$$3.6+(-3.6)+x=-7.3+(-3.6)$$
$$x=-10.9$$

3. $-45=-5x$

$$\frac{-45}{-5}=\frac{-5x}{-5}$$
$$9=x$$

4. $12x-6x=-48$

$$6x=-48$$
$$\frac{6x}{6}=\frac{-48}{6}$$
$$x=-8$$

5. $-1.2x+3.5=2.7$

$$-1.2x+3.5+(-3.5)=2.7+(-3.5)$$
$$-1.2x=-0.8$$
$$\frac{-1.2x}{-1.2}=\frac{-0.8}{-1.2}$$
$$x=\frac{2}{3}$$

6.
$$\begin{aligned}9x-3&=-17x+4\\9x+17x-3&=-17x+17x+4\\26x-3&=4\\26x&=7\\\frac{26x}{26}&=\frac{7}{26}\\x&=\frac{7}{26}\end{aligned}$$

7.
$$\begin{aligned}14x+2(7x-2)&=20\\14x+14-4x&=20\\10x+14&=20\\10x+14+(-14)&=20+(-14)\\10x&=6\\\frac{10x}{10}&=\frac{6}{10}\\x&=\frac{3}{5}\end{aligned}$$

8.
$$\begin{aligned}0.5(1.2x-3.4)&=-1.4x+5.8\\0.6x-1.7&=-1.4x+5.8\\0.6x+1.4x-1.7&=-1.4x+1.4x+5.8\\2x-1.7&=5.8\\2x-1.7+1.7&=5.8+1.7\\2x&=7.5\\\frac{2x}{2}&=\frac{7.5}{2}\\x&=3.75\end{aligned}$$

9.
$$\begin{aligned}3(x+6)&=-2(4x-1)+x\\3x+18&=-8x+2+x\\3x+18&=-7x+2\\3x+7x+18&=-7x+7x+2\\10x+18&=2\\10x+18+(-18)&=2+(-18)\\10x&=-16\\\frac{10x}{10}&=-\frac{16}{10}\\x&=-\frac{8}{5}\text{ or }-1\frac{3}{5}\end{aligned}$$

10.
$$\begin{aligned}\frac{x}{5}+\frac{x}{4}&=\frac{2}{5}\\20\left(\frac{x}{5}\right)+20\left(\frac{x}{4}\right)&=20\left(\frac{2}{5}\right)\\4x+5x&=8\\9x&=8\\\frac{9x}{9}&=\frac{8}{9}\\x&=\frac{8}{9}\end{aligned}$$

11.
$$\frac{1}{4}(x+3)=4x-2(x-3)$$
$$4\left(\frac{1}{4}\right)(x+3)=4(4x)-4(2)(x-3)$$
$$x+3=16x-8x+24$$
$$x+3=8x+24$$
$$x+(-x)+3=8x+(-x)+24$$
$$3=7x+24$$
$$3+(-24)=7x+24+(-24)$$
$$-21=7x$$
$$\frac{-21}{7}=\frac{7x}{7}$$
$$-3=x$$

12.
$$\frac{1}{2}(x-1)+2=3(2x-1)$$
$$2\left(\frac{1}{2}\right)(x-1)+2(2)=2(3)(2x-1)$$
$$x-1+4=12x-6$$
$$x+3=12x-6$$
$$x+(-x)+3=12x+(-x)-6$$
$$3=11x-6$$
$$3+6-11x-6+6$$
$$9=11x$$
$$\frac{9}{11}=\frac{11x}{11}$$
$$\frac{9}{11}=x$$

13.
$$\frac{1}{7}(7x-14)-2=\frac{1}{3}(x-2)$$
$$21\left(\frac{1}{7}\right)(7x-14)-21(2)=21\left(\frac{1}{3}\right)(x-2)$$
$$3(7x-14)-42=7(x-2)$$
$$21x-42-42=7x-14$$
$$21x-84=7x-14$$
$$21x+(-7x)-84=7x+(-7x)-14$$
$$14x-84=-14$$
$$14x-84+84=-14+84$$
$$14x=70$$
$$\frac{14x}{14}=\frac{70}{14}$$
$$x=5$$

14.
$$0.2(x-3)=4(0.2x-0.1)$$
$$0.2x-0.6=0.8x-0.4$$
$$0.2x+(-0.2x)-0.6=0.8x+(-0.2x)-0.4$$
$$-0.6=0.6x-0.4$$
$$-0.6+0.4=0.6x-0.4+0.4$$
$$-0.2=0.6x$$
$$\frac{-0.2}{0.6}=\frac{0.6x}{0.6}$$
$$-\frac{1}{3}=x$$

2.5 Exercises

1. Multiply each term by 5. Then add −160 to each side. Then divide each side by 9. We would obtain $\frac{5F-160}{9}=C$.

3. (a) $A=\frac{1}{2}bh$

$$\frac{2A}{h}=\frac{2\left(\frac{1}{2}bh\right)}{h}=b$$

$A=60\text{ m}^2,\quad h=12\text{ m}$

$$b=\frac{2(60)}{12}=10$$

base $=10$ m

(b) $A=\frac{1}{2}bh$

$$\frac{2A}{b}=\frac{2\left(\frac{1}{2}bh\right)}{b}$$

$$\frac{2A}{b}=h$$

$A=88\ m^2,\ b=11\ m$

$$h=\frac{2(88)}{11}=16$$

height $=16\ m$

5. $4x+3y=18$

$$4x+(-4x)+3y=18+(-4x)$$

$$3y=18-4x$$

$$\frac{3y}{3}=\frac{18-4x}{3}$$

$$y=\frac{18-4x}{3}\text{ or }y=6-\frac{4}{3}x$$

7. $A=\frac{1}{2}bh$

$$2A=bh$$

$$\frac{2A}{h}=b$$

9. $I=\text{P}tr$

$$\frac{I}{rt}=P$$

11. $y=mx+b$

$$y-b=mx$$

$$\frac{y-b}{x}=m$$

13. $8x-12y=24$

$$8x+(-8x)-12y=24+(-8x)$$

$$-12y=24-8x$$

$$\frac{-12y}{-12}=\frac{24-8x}{12}$$

$$y=-\frac{8}{12}x+\frac{24}{12}$$

$$y=-\frac{2}{3}x+2$$

15. $$y = -\frac{2}{3}x + 4$$
$$3y = 3\left(-\frac{2}{3}x\right) + 3(4)$$
$$3y = -2x + 12$$
$$3y + (-12) = -2x + 12 + (-12)$$
$$3y - 12 = -2x$$
$$\frac{3y - 12}{-2} = \frac{-2x}{-2}$$
$$-\frac{3}{2}y + 6 = x$$

17. $$ax + by = c$$
$$by = c - ax$$
$$y = \frac{c - ax}{b}$$

19. $$A = \pi r^2$$
$$\frac{A}{\pi} = r^2$$

21. $$S = \frac{1}{2}gt^2$$
$$2S = 2gt^2$$
$$\frac{2S}{t^2} = g$$

23. $$A = P(1 + rt)$$
$$A = P + Ptr$$
$$A - P = Ptr$$
$$\frac{A - P}{\Pr} = t$$

25. $$S = 2\pi rh + 2\pi r^2$$
$$S - 2\pi r^2 = 2\pi rh + 2\pi r^2 - 2\pi r^2$$
$$S - 2\pi r^2 = 2\pi rh$$
$$\frac{S - 2\pi r^2}{2\pi r} = \frac{2\pi rh}{2\pi r}$$
$$\frac{S - 2\pi r^2}{2\pi r} = h$$

27. $$V = \pi r^2 h$$
$$\frac{V}{\pi r^2} = \frac{\pi r^2 h}{\pi r^2}$$
$$\frac{V}{\pi r^2} = h$$

29. $$V = LWH$$
$$\frac{V}{WH} = L$$

31. $$V = \frac{1}{3}\pi r^2 h$$
$$3V = \pi r^2 h$$
$$\frac{3V}{\pi h} = r^2$$

33. $$P = 2L + 2W$$
$$P - 2L = 2W$$
$$\frac{P - 2L}{2} = W$$

35. $$c^2 = a^2 + b^2$$
$$c^2 - b^2 = a^2$$

37. $F = \frac{9}{5}C + 32$

$F - 32 = \frac{9}{5}C$

$\frac{5}{9}(F - 32) = C$

39. $P = \frac{E^2}{R}$

$RP = R\left(\frac{E^2}{R}\right)$

$RP = E^2$

$\frac{RP}{P} = \frac{E^2}{P}$

$R = \frac{E^2}{P}$

41. $A = \frac{\pi r^2 S}{360}$

(a) $360A = \pi r^2 S$

$\frac{360A}{\pi r^2} = r^2$

43. $W = \frac{P - 2L}{2}$, $P = 5.8$, $L = 2.1$

$W = \frac{5.8 - 2(2.1)}{2} = 0.8$ miles

45. $L = \frac{V}{WH}$, $V = 5940$, $W = 22$, $H = 9$

$L = \frac{5940}{(22)(9)} = 30$ feet

47. (a) $V = 1100x + 7050$

$V - 7050 = 1100x$

$\frac{V - 7050}{1100} = x$

(b) $V = 25,750$

$x = \frac{25,750 - 7050}{1100} = 17$ years

Year $= 1985 + 17 = 2002$

49. $A = \frac{1}{2}ab$

If $b \to 2b$, $A = \frac{1}{2}a(2b) = ab$

A doubles

51. $A = \pi r^2$

If $r \Rightarrow 2r$, $A \Rightarrow \pi(2r)^2 = 4\pi r^2$

A is quadrupled if r doubles.

Cumulative Review

53. 12% of $260 = 0.12(260) = 31.2$

55. 40% of what is 500?

$0.40x = 500$

$\frac{0.40x}{0.40} = \frac{500}{0.40}$

$x = 1250$ feet of fencing

57. $\left(3\frac{1}{4}\right)(12,000) = \frac{13}{4}(12,000)$

$= 39,000$ square feet

2.6 Exercises

1. $5 > -6$ is equivalent to $-6 < 5$.
   Both statements imply that 5 is to the right of -6 on the number line.

3. $9 > -3$

5. $-4 < -2$

7. $\frac{3}{5} ? \frac{4}{7}$

   $\frac{21}{35} ? \frac{20}{35}$

   $\frac{3}{5} > \frac{4}{7}$

9. $-4 < -2$

11. $-\frac{13}{3} < -4$

13. $-\frac{5}{8} < -\frac{3}{5}$

15. $\frac{123}{4986} = 0.024669$

    0.0247 is greater

17. $x > 7$

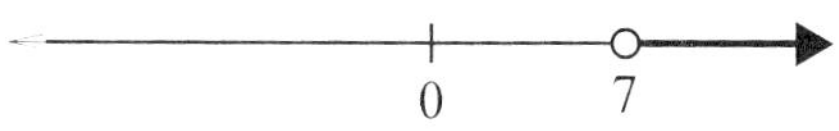

19. $x \geq -6$

21. $x > \frac{3}{4}$

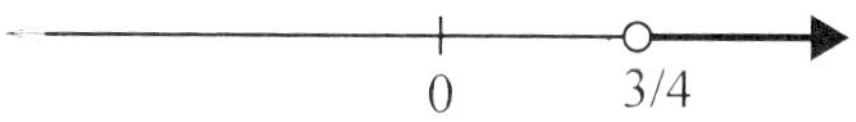

23. $x \leq -5.3$

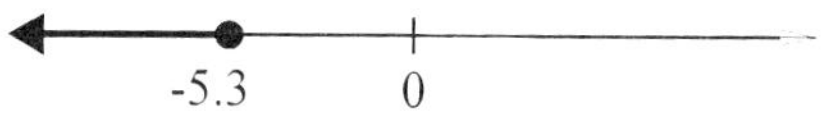

25. $25 < x$

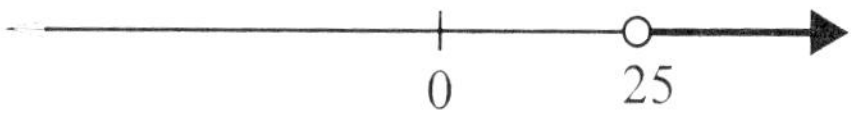

27. $x \geq -\frac{2}{3}$

29. $x < -20$

31. $x > 2.8$

33. $W > 175$

35. $h \geq 37$

37. $h \geq 48$

39. $x \leq 2$, $x > -3$, $x < \frac{5}{2}$, $x \geq -\frac{5}{2}$

    $-\frac{5}{2} \leq x \leq 2$

Cumulative Review

41. 16% of $38 = 0.16(38) = 6.08$

43. $\dfrac{16}{800} = 0.02 = 2\%$

45. $\text{Increase} = 10 - 8 = 2$

$\text{Percent Increase} = \dfrac{2}{8} = 0.25 = 25\%$

$100\% + 25\% = 125\%$

$\text{New price} = 125\% \text{ of } 12$

$= 1.25(12)$

$= \$15$

2.7 Exercises

1. $x + 7 \le 4$

$x \le -3$

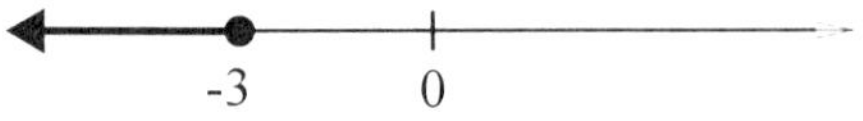

3. $5x \le 25$

$x \le 5$

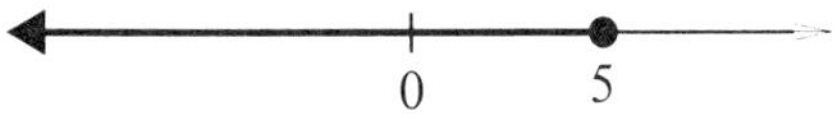

5. $-2x < 18$

$x > -9$

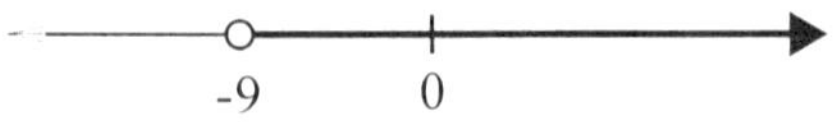

7. $\dfrac{1}{2}x \ge 4$

$x \ge 8$

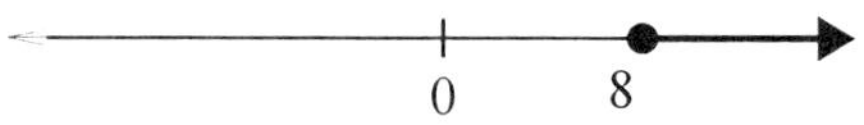

9. $-\dfrac{1}{4}x > 3$

$x < -12$

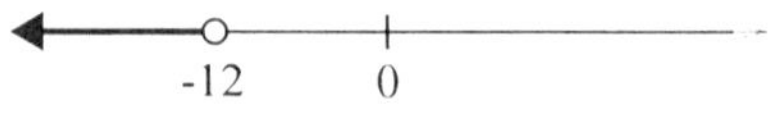

11. $2x - 3 < 4$

$2x < 7$

$x < \dfrac{7}{2}$

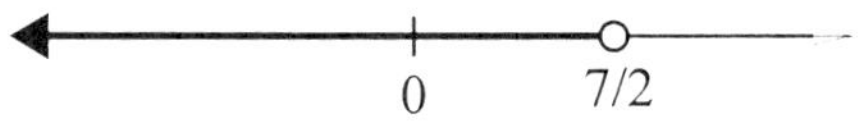

13. $5 - 2x \le 9$

$5 + (-5) - 2x \le 9 + (-5)$

$-2x \le 4$

$\dfrac{-2x}{-2} \ge \dfrac{4}{-2}$

$x \ge -2$

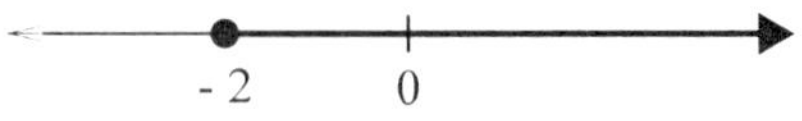

15.
$$-4+5x<-3x+8$$
$$-4+5x+3x<-3x+3x+8$$
$$-4+8x<8$$
$$-4+4+8x<8+4$$
$$8x<12$$
$$\frac{8x}{8}<\frac{12}{8}$$
$$x<\frac{3}{2}$$

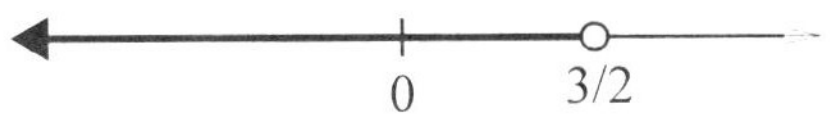

17.
$$\frac{5x}{6}-5>\frac{x}{6}-9$$
$$5x-30>x-54$$
$$4x>-24$$
$$x>-6$$

19.
$$2(3x+4)>3(x+3)$$
$$6x+8>3x+9$$
$$3x+8>9$$
$$3x>1$$
$$x>\frac{1}{3}$$

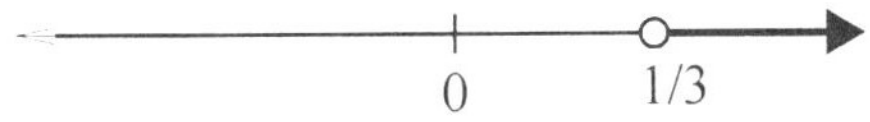

21.
$$5>3$$
$$5+(-2)>3+(-2)$$
$$3>1$$

Adding any number to both sides of an inequality does not reverse the direction.

23.
$$3x+8<7x-4$$
$$3x+(-7x)+8<7x+(-7x)-4$$
$$-4x+8<-4$$
$$-4x+8+(-8)<-4+(-8)$$
$$-4x<-12$$
$$\frac{-4x}{-4}>\frac{-12}{-4}$$
$$x>3$$

25.
$$6x-2\geq 4x+6$$
$$2x\geq 8$$
$$x\geq 4$$

27.
$$0.3(x-1)<0.1x-0.5$$
$$0.3x-0.3<0.1x-0.5$$
$$3x-3<x-5$$
$$2x<-2$$
$$x<-1$$

29.
$$3+5(2-x)\geq -3(x+5)$$
$$3+10-5x\geq -3x-15$$
$$13-5x\geq -3x-15$$
$$-2x+13\geq -15$$
$$-2x\geq -28$$
$$x\leq 14$$

31. $$\frac{x+6}{7}-\frac{6}{14}>\frac{x+3}{2}$$
$$14\left(\frac{x+6}{7}\right)-14\left(\frac{6}{14}\right)>14\left(\frac{x+3}{2}\right)$$
$$2(x+6)-6>7(x+3)$$
$$2x+12-6>7x+21$$
$$2x+6>7x+21$$

$$-5x>15$$
$$x<-\frac{15}{5}$$
$$x<-3$$

33. $$\frac{1}{6}-\frac{1}{2}(3x+2)<\frac{1}{3}\left(x-\frac{1}{2}\right)$$
$$1-3(3x+2)<2\left(x-\frac{1}{2}\right)$$
$$1-9x-6<2x-1$$
$$-9x-5<2x-1$$
$$-11x<4$$
$$x>-\frac{4}{11}$$

35. $$\frac{75+83+86+x}{4}\geq 80$$
$$\frac{244+x}{4}\geq 80$$
$$244+x\geq 320$$
$$x\geq 76$$
Must get a 76 or higher.

37. $$268+4x\geq 300$$
$$4x\geq 32$$
$$x\geq 8$$
8 days or more.

Cumulative Review

39. Width $=36+8=44$
Length $=78+8=86$
$P=2L+2W=2(86)+2(44)=260$ feet

41. $A=\pi r^2$, $r=\frac{9}{2}=4.5$
$A\approx 3.14(4.5)^2\approx 63.6$ square inches

Putting Your Skills to Work

1. $28.3-19.9=8.4$ gallons per hour

2. $30(8.4)=252$ gallons

3. $y=0.8x-3.7$, $x=45$
$y=0.8(45)-3.7$
$y=36-3.7$
$y=32.3$ gallon per hour

4. $y = 0.8x - 3.7,\ y = 40.3$

$$40.3 = 0.8x - 3.7$$
$$40.3 + 3.7 = 0.8x - 3.7 + 3.7$$
$$44 = 0.8x$$
$$\frac{44}{0.8} = \frac{0.8x}{0.8}$$
$$55 = x$$

It travels 55 miles per hour.

5. $y = 0.5x - 0.1$

$$y + 0.1 = 0.5x - 0.1 + 0.1$$
$$y + 0.1 = 0.5x$$
$$\frac{y+0.1}{0.5} = \frac{0.5x}{0.5}$$
$$\frac{y+0.1}{0.5} = x \text{ or } x = 2y + 0.2$$

6. $x = 2y + 0.2,\ y = 17.4$

$$x = 2(17.4) + 0.2$$
$$x = 34.8 + 0.2$$
$$x = 35 \text{ miles per hour}$$

Chapter 2 Review Problems

1. $5x = -35$

$$\frac{5x}{5} = \frac{-35}{5}$$
$$x = -7$$

2. $x - 19 = -22$

$$x - 19 + 19 = -22 + 19$$
$$x = -3$$

3. $6 - 18x = 4 - 17x$

$$6 - 18x + 18x = 4 - 17x + 18x$$
$$6 = 4 + x$$
$$6 - 4 = 4 - 4 + x$$
$$2 = x$$

4. $18 - 10x = 63 + 5x$

$$18 - 10x + 10x = 63 + 5x + 10x$$
$$18 = 63 + 15x$$
$$18 - 63 = 63 - 63 + 15x$$
$$-45 = 15x$$
$$\frac{-45}{15} = \frac{15x}{15}$$
$$-3 = x$$

5. $$6x-2(x+3)=5$$
$$6x-2x-6=5$$
$$4x-6=5$$
$$4x-6+6=5+6$$
$$4x=11$$
$$\frac{4x}{4}=\frac{11}{4}$$
$$x=\frac{11}{4}=2.75$$

6. $$1-2(6-x)=3x+2$$
$$1-12+2x=3x+2$$
$$2x-11=3x+2$$
$$2x-11+11=3x+2+11$$
$$2x=3x+13$$
$$-3x+2x=-3x+3x+13$$

7. $$x-(0.5x+2.6)=17.6$$
$$x-0.5x-2.6=17.6$$
$$0.5x-2.6=17.6$$
$$10(0.5x)-10(2.6)=10(17.6)$$
$$5x-26=176$$
$$5x-26+26=176+26$$

$$5x=202$$
$$\frac{5x}{5}=\frac{202}{5}$$
$$x=40.4$$

8. $$-0.2(x+1)=0.3(x+11)$$
$$10[-0.2(x+1)]-10[0.3(x+11)]$$
$$-2(x+1)=3(x+11)$$
$$-2x-2=3x+33$$
$$-2x-2-33=3x+33-33$$
$$-2x-35=3x$$
$$2x-2x-35=2x+3x$$
$$-35=5x$$
$$\frac{-35}{5}=\frac{5x}{5}$$
$$-7=x$$

9. $$3(x-2)=-4(5+x)$$
$$3x-6=-20-4x$$
$$3x+4x-6=-20-4x+4x$$
$$7x-6=-20$$
$$7x-6+6=-20+6$$
$$7x=-14$$
$$\frac{7x}{7}=\frac{-14}{7}$$
$$x=-2$$

10. $$\frac{2}{3}x=-18$$
$$\frac{3}{2}\left(\frac{2}{3}x\right)=\frac{3}{2}(-18)$$
$$x=-27$$

11. $$\frac{3}{4}x=15$$
$$\frac{4}{3}\left(\frac{3}{4}x\right)=\frac{4}{3}(15)$$
$$x=20$$

12.
$$\begin{aligned} 4(2x+3) &= 5(x-3) \\ 8x+12 &= 5x-15 \\ 8x+(-5x)+12 &= 5x+(-5x)-15 \\ 3x+12 &= -15 \\ 3x+12+(-12) &= -15+(-12) \\ 3x &= -27 \\ \frac{3x}{3} &= \frac{-27}{3} \\ x &= -9 \end{aligned}$$

13.
$$\begin{aligned} 3(x-3) &= 13x+21 \\ 3x-9 &= 13x+21 \\ 3x+(-3x)-9 &= 13x+(-3x)+21 \\ -9 &= 10x+21 \\ -21-9 &= 10x+21+(-21) \\ -30 &= 10x \\ \frac{-30}{10} &= \frac{10x}{10} \\ -3 &= x \end{aligned}$$

14.
$$\begin{aligned} 0.9x+1.0 &= 0.3x+0.4 \\ 0.9x+(-0.3x)+1.0 &= 0.3x+(-0.3x)+0.4 \\ 0.6x+1.0 &= 0.4 \\ 0.6x+1.0+(-1.0) &= 0.4+(-1.0) \\ 0.6x &= -0.6 \\ \frac{0.6x}{0.6} &= \frac{-0.6}{0.6} \\ x &= -1 \end{aligned}$$

15.
$$\begin{aligned} 2.4-0.3x &= 0.4(x-1) \\ 2.4-0.3x &= 0.4x-0.4 \\ 2.4-0.3x+0.3x &= 0.4x+0.3x-0.4 \\ 2.4 &= 0.7x-0.4 \\ 2.4+0.4 &= 0.7x-0.4+0.4 \\ 2.8 &= 0.7x \\ \frac{2.8}{0.7} &= \frac{0.7x}{0.7} \\ 4 &= x \end{aligned}$$

16.
$$\begin{aligned} 12-x+2 &= 3x-10+4x \\ -x+14 &= 7x-10 \\ -x+14+10 &= 7x-10+10 \\ -x+24 &= 7x \\ x-x+24 &= x+7x \\ 24 &= 8x \\ \frac{24}{8} &= \frac{8x}{8} \\ 3 &= x \end{aligned}$$

17.
$$\begin{aligned} 36 &= 9x-(3x-18) \\ 36 &= 9x-3x+18 \\ 36 &= 6x+18 \\ 36+(-18) &= 6x+18+(-18) \\ 18 &= 6x \\ \frac{18}{6} &= \frac{6x}{6} \\ 3 &= x \end{aligned}$$

18.
$$\begin{aligned}
12-5x &= -7x-2\\
12-5x+7x &= -7x+7x-2\\
12+2x &= -2\\
12-12+2x &= -2-12\\
2x &= -14\\
\frac{2x}{2} &= \frac{-14}{2}\\
x &= -7
\end{aligned}$$

19.
$$\begin{aligned}
2(3-x) &= 1-(x-2)\\
6-2x &= 1-x+2\\
6-2x+x &= 3-x+x\\
6-x &= 3\\
6+(-6)-x &= 3+(-6)\\
-x &= -3\\
\frac{-x}{-1} &= \frac{-3}{-1}\\
x &= 3
\end{aligned}$$

20.
$$\begin{aligned}
4(x+5)-7 &= 2(x+3)\\
4x+2x-7 &= 2x+6\\
4x+13 &= 2x+6\\
4x+13-13 &= 2x+6-13\\
4x &= 2x-7\\
-2x+4x &= -2x+2x-7\\
2x &= -7\\
\frac{2x}{2} &= \frac{-7}{2}\\
x &= -\frac{7}{2} \text{ or } -3\frac{1}{2}
\end{aligned}$$

21.
$$\begin{aligned}
0.9y+3 &= 0.4y+1.5\\
10(0.9y)+10(3) &= 10(0.4y)+10(1.5)\\
9y+30 &= 4y+15\\
9y+(-4y)+30 &= 4y+(-4y)+15\\
5y+30 &= 15\\
54y+30+(-30) &= 15+(-30)\\
5y &= -15\\
\frac{5y}{5} &= \frac{-15}{5}\\
y &= -3
\end{aligned}$$

22.
$$\begin{aligned}
7y-3.4 &= 1.3\\
7y-3.4+3.4 &= 1.3+3.4\\
7y &= 14.7\\
\frac{7y}{7} &= \frac{14.7}{7}\\
y &= 2.1
\end{aligned}$$

23.
$$\begin{aligned}
3 &= 2x+5-3(x-1)\\
3 &= 2x+5-3x+3\\
3 &= -x+8\\
3+(-8) &= -x+8+(-8)\\
-5 &= -x\\
\frac{-5}{-1} &= \frac{-x}{-1}\\
5 &= x
\end{aligned}$$

24. $2(5x-1)-7=3(x-1)+5-4x$

$$10x-2-7=3x-3+5-4x$$
$$10x-9=-x+2$$
$$10x+x-9=-x+x+2$$
$$11x-9=2$$
$$11x-9+9=2+9$$
$$11x=11$$
$$\frac{11x}{11}=\frac{11}{11}$$
$$x=1$$

25. $\frac{3}{4}x-3=\frac{1}{2}x+2$

$$4\left(\frac{3}{4}x-3\right)=4\left(\frac{1}{2}x+2\right)$$
$$3x-12=2x+8$$
$$3x-12+12=2x+8+12$$
$$3x=2x+20$$
$$-2x+3x=-2x+2x+20$$
$$x=20$$

26. $1=\frac{5x}{6}+\frac{2x}{3}$

$$6=5x+4x$$
$$6=9x$$
$$\frac{2}{3}=x$$

27. $\frac{7x}{5}=5+\frac{2x}{5}$

$$5\left(\frac{7x}{5}\right)=5(5)+5\left(+\frac{2x}{5}\right)$$
$$7x=25+2x$$
$$7x-2x=25+2x-2x$$
$$5x=25$$
$$\frac{5x}{5}=\frac{25}{5}$$
$$x=5$$

28. $\frac{7x-3}{2}-4=\frac{5x+1}{3}$

$$6\left(\frac{7x-3}{2}\right)-6(4)=6\left(\frac{5x+1}{3}\right)$$
$$3(7x-3)-24=2(5x+1)$$
$$21x-9-24=10x+2$$
$$21x-33=10x+2$$
$$21x+(-10x)-33=10x+(-10x)+2$$
$$11x-33=2$$
$$11x-33+33=2+33$$
$$11x=35$$
$$\frac{11x}{11}=\frac{35}{11}$$
$$x=\frac{35}{11}$$

29. $$\frac{3x-2}{2}+\frac{x}{4}=2+x$$
$$4\left(\frac{3x-2}{2}+\frac{x}{4}\right)=4(2+x)$$
$$2(3x-2)+x=8+4x$$
$$6x-4+x=8+4x$$
$$7x-4=4x+8$$
$$7x-4+4=4x+8+4$$
$$7x=4x+12$$
$$-4x+7x=-4x+4x+12$$
$$3x=12$$
$$\frac{3x}{3}=\frac{12}{3}$$
$$x=4$$

30. $$\frac{-3}{2}(x+5)=1-x$$
$$2\left(\frac{-3}{2}\right)(x+5)=2(1-x)$$
$$-3x-15=2-2x$$
$$-3x+3x-15=2-2x+3x$$
$$-15=2+x$$
$$-15+(-2)=2+(-2)+x$$
$$-17=x$$

31. $$\frac{-4}{3}(2x+1)=-x-2$$
$$3\left[\frac{-4}{3}(2x+1)\right]=3(-x-2)$$
$$-4(2x+1)=-3x-6$$
$$-8x-4=-3x-6$$
$$-8x-4+4=-3x-6+4$$
$$-8x=-3x-2$$
$$3x-8x=3x-3x-2$$
$$-5x=-2$$

32. $$\frac{1}{3}(x-2)=\frac{x}{4}+2$$
$$12\left(\frac{1}{3}\right)(x-2)=12\left(\frac{x}{4}\right)+12(2)$$
$$4(x-2)=3x+24$$
$$4x-8=3x+24$$
$$4x+(-3x)-8=3x+(-3x)+24$$
$$x-8=24$$
$$x-8+8=24+8$$
$$x=32$$

33.
$$\begin{aligned}
\frac{1}{5}(x-3) &= 20-\frac{x}{2}\\
10\left[\frac{1}{5}(x-3)\right] &= 10\left(2-\frac{x}{2}\right)\\
2(x-3) &= 20-5x\\
2x-6 &= 20-5x\\
2x+5x-6 &= 20-5x+5x\\
7x-6 &= 20\\
7x-6+6 &= 20+6\\
7x &= 26\\
\frac{7x}{7} &= \frac{26}{7}\\
x &= \frac{26}{7}
\end{aligned}$$

34.
$$\begin{aligned}
\frac{4}{5}+\frac{1}{2}x &= \frac{1}{5}x+\frac{1}{2}\\
10\left(\frac{4}{5}\right)+10\left(\frac{1}{2}x\right) &= 10\left(\frac{1}{5}x\right)+10\left(\frac{1}{2}\right)\\
8+5x &= 2x+5\\
5x+(-2x)+8 &= 2x+(-2x)+5\\
3x+8 &= 5\\
3x+8+(-8) &= 5+(-8)\\
3x &= -3\\
\frac{3x}{3} &= \frac{-3}{3}\\
x &= -1
\end{aligned}$$

35.
$$\begin{aligned}
2x-\frac{3}{4}+\frac{7}{2}x &= \frac{1}{2}x+\frac{1}{4}\\
4(2x)-4\left(\frac{3}{4}\right)+4\left(\frac{7}{2}x\right) &= 4\left(\frac{1}{2}x\right)+4\left(\frac{1}{4}\right)\\
8x-3+14x &= 2x+1\\
22x-3 &= 2x+1\\
22x+(-2x)-3 &= 2x+(-2x)+1\\
20x-3 &= 1\\
20x-3+3 &= 1+3\\
20x &= 4\\
\frac{20x}{20} &= \frac{4}{20}\\
x &= \frac{1}{5}
\end{aligned}$$

36.
$$\begin{aligned}
\frac{3}{2}x-\frac{5}{6}+x &= \frac{1}{2}x+\frac{2}{3}\\
6\left(\frac{3}{2}x\right)-6\left(\frac{5}{6}\right)+6x &= 6\left(\frac{1}{2}x\right)+6\left(\frac{2}{3}\right)\\
9x-5+6x &= 3x+4\\
15x-5 &= 3x+4\\
15x+(-3x)-5 &= 3x+(-3x)+4\\
12x-5 &= 4\\
12x-5+5 &= 4+5\\
12x &= 9\\
\frac{12x}{12} &= \frac{9}{12}\\
x &= \frac{3}{4}
\end{aligned}$$

37.
$$-\frac{8}{3}x - 8 + 2x - 5 = -\frac{5}{3}$$
$$-\frac{8}{3}x - 13 + 2x = -\frac{5}{3}$$
$$3\left(-\frac{8}{3}x - 13 + 2x\right) = 3\left(-\frac{5}{3}\right)$$
$$-8x - 39 + 6x - -5$$
$$-2x - 39 + 6x = -5$$
$$-2x - 39 + 39 = -5 + 39$$
$$-2x = 34$$
$$\frac{-2x}{-2} = \frac{34}{-2}$$
$$x = -17$$

38.
$$3.5(2x + 3) = 2.4x - 1$$
$$7x + 10.5 = 2.4x - 1$$
$$7x + (-2.4x) + 10.5 = 2.4x + (-2.4x) - 1$$
$$4.6x + 10.5 = -1$$
$$4.6x + 10.5 + (-10.5) = -1 + (-10.5)$$
$$4.6x = -11.5$$
$$\frac{4.6x}{4.6} = \frac{-11.5}{4.6}$$
$$x = -2.5$$

39.
$$\frac{7}{12}(x - 3) = -\frac{1}{3}x + 1$$
$$12\left(\frac{7}{12}\right)(x - 3) = -12\left(\frac{1}{3}x\right) + 12(1)$$
$$7x - 21 = -4x + 12$$
$$7x + 4x - 21 = -4x + 4x + 12$$
$$11x - 21 = 12$$
$$11x - 21 + 21 = 12 + 21$$
$$11x = 33$$
$$\frac{11x}{11} = \frac{33}{11}$$
$$x = 3$$

40.
$$\frac{1}{6} + \frac{1}{3}(x - 3) = \frac{1}{2}(x + 9)$$
$$6\left(\frac{1}{6}\right) + 6\left(\frac{1}{3}\right)(x - 3) = 6\left(\frac{1}{2}\right)(x + 9)$$
$$1 + 2(x - 3) = 3(x + 9)$$
$$1 + 2x - 6 = 3x + 27$$
$$2x - 5 = 3x + 27$$
$$2x - 2x - 5 = 3x - 2x + 27$$
$$-5 = x + 27$$
$$-5 - 27 = x + 27 - 27$$
$$-32 = x$$

41.
$$\frac{1}{7}(x+5)-\frac{6}{14}=\frac{1}{2}(x+3)$$
$$14\left(\frac{1}{7}\right)(x+5)-14\left(\frac{6}{14}\right)=14\left(\frac{1}{2}\right)(x+3)$$
$$2x+10-6=7x+21$$
$$2x+4=7x+21$$
$$2x+(-2x)+4=7x+(-2x)+21$$
$$4=5x+21$$
$$4+(-21)=5x+21+(-21)$$
$$-17=5x$$
$$\frac{-17}{5}=\frac{5x}{5}$$
$$-\frac{17}{5}=x$$

42.
$$\frac{1}{6}(8x+3)=\frac{1}{2}(2x+7)$$
$$6\left[\frac{1}{6}(8x+3)\right]=6\left[\frac{1}{2}(2x+7)\right]$$
$$8x+3=3(2x+7)$$
$$8x+3=3(2x+7)$$
$$8x+3-3=6x+21-3$$
$$8x=6x+18$$
$$-6x+8x=-6x+6x+18$$
$$2x=18$$
$$\frac{2x}{2}=\frac{18}{2}$$
$$x=9$$

43.
$$-\frac{2}{5}(3x+1)=\frac{1}{3}(2-x)$$
$$15\left(-\frac{2}{5}\right)(3x+1)=15\left(\frac{1}{3}\right)(2-x)$$
$$-6(3x+1)=5(2-x)$$
$$-18x-6=10-5x$$
$$-18x+5x-6=10-5x+5x$$
$$13x-6=10$$
$$-13x-6+6=10+6$$
$$-13x=16$$
$$\frac{-13x}{-13}=\frac{16}{-13}$$
$$x=-\frac{16}{13} \text{ or } -1\frac{3}{13}$$

44.
$$\frac{7}{9}x+\frac{2}{3}=2+\frac{1}{3}x$$
$$9\left(\frac{7}{9}x+\frac{2}{3}\right)=9\left(2+\frac{1}{3}x\right)$$
$$7x+6=18+3x$$
$$7x+6-6=18-6+3x$$
$$7x=12+3x$$
$$7x-3x=12+3x-3x$$
$$4x=12$$
$$\frac{4x}{4}=\frac{12}{4}$$
$$x=3$$

45.
$$3x-y=10$$
$$3x+(-3x)-y=-3x+10$$
$$-y=-3x+10$$
$$\frac{-y}{-1}=\frac{-3x+10}{-1}$$
$$y=3x-10$$

46. $5x+2y+7=0$
$$5x+2y=-7$$
$$2y=-5x-7$$
$$y=\frac{-5x-7}{2}$$

47. $A=P(1+rt)$
$$A=P+\Pr t$$
$$A+(-P)=P+(-P)+\Pr t$$
$$A-P=\Pr t$$
$$\frac{A-P}{Pt}=\frac{\Pr t}{Pt}$$
$$\frac{A-P}{Pt}=r$$

48. $A=4\pi r^2+2\pi rh$
$$A-4\pi r^2=2\pi rh$$
$$\frac{A-4\pi r^2}{2\pi r}=h$$

49. $H=\frac{1}{3}(a+2p+3)$
$$3H=3\left(\frac{1}{3}\right)(a+2p+3)$$
$$3H=a+2p+3$$
$$3H+(-a)+(-3)=a+(-a)+2p+3+(-3)$$
$$3H-a-3=2p$$
$$\frac{3H-a-3}{2}=\frac{2p}{2}$$
$$\frac{3H-a-3}{2}=p$$

50. $ax+by=c$
$$ax-ax+by=c-ax$$
$$by=c-ax$$
$$\frac{by}{b}=\frac{c-ax}{b}$$
$$y=\frac{c-ax}{b}$$

51. a. $C=\frac{WRT}{1000}$
$$1000C=WRT$$
$$\frac{1000C}{WR}=T$$

b. $T=\frac{1000(0.36)}{(30)(0.002)}=6000$

52. a. $5x-3y=12$
$$5x+(-5x)-3y=-5x+12$$
$$-3y=-5x+12$$
$$\frac{-3y}{-3}=\frac{-5x+12}{-3}$$
$$y=\frac{5}{3}x-4$$

b. $x=9$
$$y=\frac{5}{3}(9)-4=15-4=11$$

53. a. $V = lwh$

$$\frac{V}{lw} = \frac{lwh}{lw}$$

$$\frac{V}{lw} = h$$

b. $V = 48,\ l = 2,\ w = 4$

$$h = \frac{48}{2(4)} = 6$$

54. $7 - 2x \geq 4x - 5$

$$-6x \geq -12$$

$$x \leq 2$$

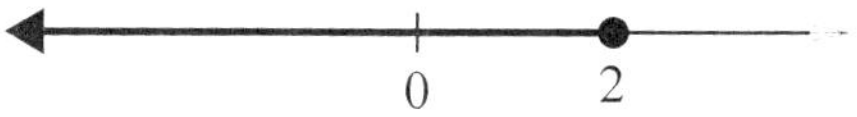

55.

$$\begin{aligned} 2 - 3x &\leq -5 + 4x \\ -2 + 2 - 3x &\leq -2 - 5 + 4x \\ -3x &\leq -7 + 4x \\ -3x - 4x &\leq -7 + 4x - 4x \\ -7x &\leq -7 \\ \frac{-7x}{-7} &\geq \frac{-7}{-7} \\ x &\geq 1 \end{aligned}$$

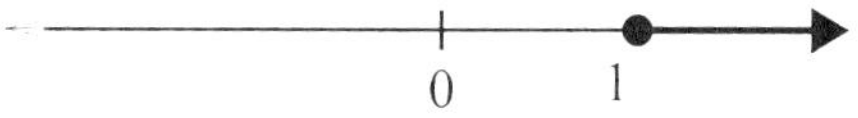

56. $2x - 3 + x > 5(x + 1)$

$$3x - 3 > 5x + 5$$

$$-2x > 8$$

$$x < -4$$

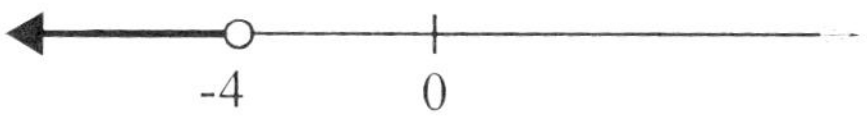

57.

$$\begin{aligned} -x + 4 &< 3x + 16 \\ -x + 4 - 4 &< 3x + 16 - 4 \\ -x &< 3x + 12 \\ -3x - x &< -3x + 3x + 12 \\ -4x &< 12 \\ \frac{-4x}{4} &> \frac{12}{-4} \\ x &> -3 \end{aligned}$$

58.

$$\begin{aligned} 8 - \frac{1}{3}x &\leq x \\ 3(8) - 3\left(\frac{1}{3}x\right) &\leq 3x \\ 24 - x &\leq 3x \\ 24 - x + (-3x) &\leq 3x + (-3x) \\ 24 - 4x &\leq 0 \\ 24 + (-24) - 4x &\leq 0 + (-24) \\ -4x &\leq -24 \\ \frac{-4x}{-4} &\geq \frac{-24}{-4} \\ x &\geq 6 \end{aligned}$$

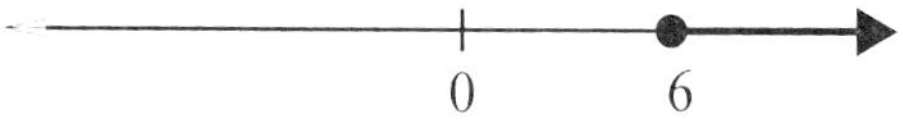

59. $$7-\frac{3}{5}x>4$$
$$5(7)-5\left(\frac{3}{5}x\right)>5(4)$$
$$35-3x>20$$
$$35+(-35)-3x>20+(-35)$$
$$-3x>-15$$
$$\frac{-3x}{-3}<\frac{-15}{-3}$$
$$x<5$$

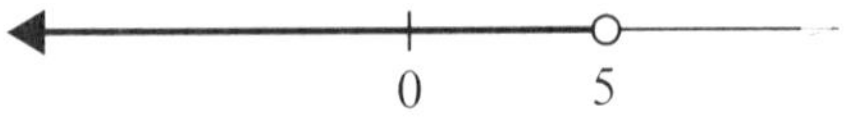

60. $$-4x-14<4-2(3x-1)$$
$$-4x-14<4-6x+2$$
$$-4x-14<6-6x$$
$$2x<20$$
$$x<10$$

61. $$3(x-2)+8<7x+14$$
$$3x-6+8<7x+14$$
$$3x+2<7x+14$$
$$3x-2+2<7x+14-2$$
$$3x<7x+12$$
$$-7x+3x<-7x+7x+12$$
$$-4x<12$$
$$\frac{-4x}{-4}>\frac{12}{-4}$$
$$x>-3$$

62. $$\frac{1}{2}(2x+3)>10$$
$$2x+3>20$$
$$2x>17$$
$$x>\frac{17}{2}$$

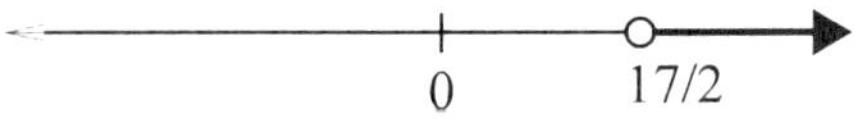

63. $$5(1-x)<3(x-1)-2(3-x)$$
$$5-5x<3x-3-6+2x$$
$$5-5x<5x-9$$
$$5+9-5x<5x-9+9$$
$$14-5x<5x$$
$$14-5x+5x<5x+5x$$
$$14<10x$$
$$\frac{14}{10}<\frac{10x}{10}$$
$$\frac{7}{5}<x \text{ or } x>\frac{7}{5}$$

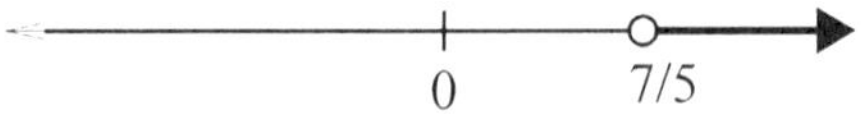

64. $$15h\le 480$$
$$\frac{15h}{15}\le\frac{480}{15}$$
$$h\le 32 \text{ hours}$$

65. $$85n\le 1445$$
$$n\le\frac{1445}{85}$$
$$n\le 17$$

66.
$$
\begin{aligned}
8(3x+5)-10 &= 9(x-2)+13\\
24x+40-10 &= 9x-18+13\\
24x+30 &= 9x-5\\
24x+(-9x)+30 &= 9x+(-9x)-5\\
15x+30 &= -5\\
15x+30+(-30) &= -5+(-30)\\
15x &= -35\\
\frac{15x}{15} &= \frac{-35}{15}\\
x &= \frac{-7}{3} \text{ or } -2\frac{1}{3}
\end{aligned}
$$

67.
$$
\begin{aligned}
8-3x+5 &= 13+4x+2\\
-3x+13 &= 4x+15\\
-3x+13-13 &= 4x+15-13\\
-3x &= 4x+2\\
-4x-3x &= -4x+4x+2\\
-7x &= 2\\
\frac{-7x}{-7} &= \frac{2}{-7}\\
x &= -\frac{2}{7}
\end{aligned}
$$

68.
$$
\begin{aligned}
-2(x-3) &= -4x+3(3x+2)\\
-2x+6 &= -4x+9x+6\\
-2x+6 &= 5x+6\\
-2x+6-6 &= 5x+6-6\\
-2x &= 5x\\
2x-2x &= 2x+5x\\
0 &= 7x\\
\frac{0}{7} &= \frac{7x}{7}\\
0 &= x
\end{aligned}
$$

69.
$$
\begin{aligned}
\frac{1}{2}+\frac{5}{4}x &= \frac{2}{5}x-\frac{1}{10}+4\\
20\left(\frac{1}{2}\right)+20\left(\frac{5}{4}x\right) &= 20\left(\frac{2}{5}x\right)-20\left(\frac{1}{10}\right)+20(4)\\
10+25x &= 8x-2+80\\
10+25x &= 8x+78\\
10+25x+(-8x) &= 8x+(-8x)+78\\
10+17x &= 78\\
10+(-10)+17x &= 78+(-10)\\
17x &= 68\\
\frac{17x}{17} &= \frac{68}{17}\\
x &= 4
\end{aligned}
$$

70.
$$
\begin{aligned}
\frac{1}{6}x-\frac{2}{3} &= \frac{1}{3}(x-4)\\
6\left(\frac{1}{6}x\right)-6\left(\frac{2}{3}\right) &= 6\left(\frac{1}{3}\right)(x-4)\\
x-4 &= 2x-8\\
x+(-x)-4 &= 2x+(-x)-8\\
-4 &= x-8\\
-4+8 &= x-8+8\\
4 &= x
\end{aligned}
$$

71.
$$
\begin{aligned}
\frac{1}{2}(x-3) &= \frac{1}{4}(3x-1)\\
4\left[\frac{1}{2}(x-3)\right] &= 4\left[\frac{1}{4}(3x-1)\right]\\
2(x-3) &= 3x-1\\
2x-6 &= 3x-1\\
2x-6+1 &= 3x-1+1\\
2x-5 &= 3x\\
-2x+2x-5 &= -2x+3x\\
-5 &= x
\end{aligned}
$$

72.
$$\begin{aligned}
H + 2d &= 6c - 3d \\
H + 2d + 3d &= 6c - 3d + 3d \\
H + 5d &= 6c \\
H - H + 5d &= 6c - H \\
5d &= 6c - H \\
\frac{5d}{5} &= \frac{6c - H}{5} \\
d &= \frac{6c - H}{5}
\end{aligned}$$

73.
$$\begin{aligned}
H &= \frac{3c + 2b}{4} \\
4H &= 4\left(\frac{3c + 2b}{4}\right) \\
4H &= 3c + 2b \\
4H - 3c &= 3c - 3c + 2b \\
4H - 3c &= 2b \\
\frac{4H - 3c}{2} &= \frac{2b}{2} \\
\frac{4H - 3c}{2} &= b
\end{aligned}$$

74.
$$\begin{aligned}
5 - \frac{1}{2}x &> 4 \\
2\left(5 - \frac{1}{2}x\right) &> 2(4) \\
10 - x &> 8 \\
-10 + 10 - x &> -10 + 8 \\
-x &> -2 \\
\frac{-x}{-1} &< \frac{-2}{-1} \\
x &< 2
\end{aligned}$$

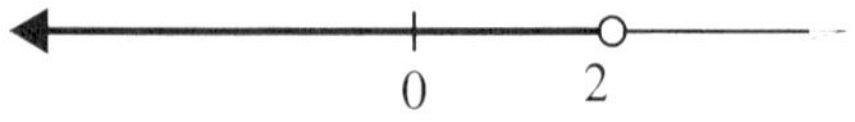

75.
$$\begin{aligned}
2(x - 1) &\geq 3(2 + x) \\
2x - 2 &\geq 6 + 3x \\
-x &\geq 8 \\
x &\leq -8
\end{aligned}$$

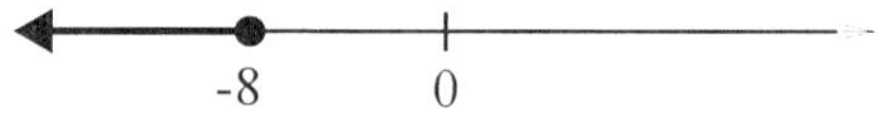

76.
$$\begin{aligned}
\frac{1}{3}(x + 2) &\leq \frac{1}{2}(3x - 5) \\
6\left[\frac{1}{3}(x + 2)\right] &\leq 6\left[\frac{1}{2}(3x - 5)\right] \\
2(x + 2) &\leq 3(3x - 5) \\
2x + 4 &\leq 9x - 15 \\
2x + 4 + 15 &\leq 9x - 15 + 15 \\
2x + 19 &\leq 9x \\
-2x + 2x + 19 &\leq -2x + 9x \\
19 &\leq 7x \\
\frac{19}{7} &\leq \frac{7x}{7} \\
\frac{19}{7} \leq x &\text{ or } x \geq \frac{19}{7}
\end{aligned}$$

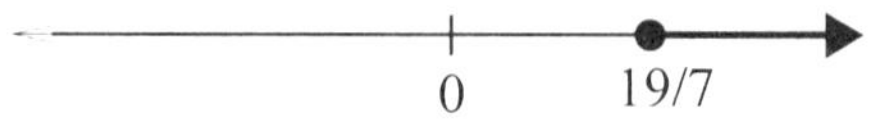

77.
$$\begin{aligned}
4(2 - x) - (-5x + 1) &\geq -8 \\
8 - 4x + 5x - 1 &\geq -8 \\
x + 7 &\geq -8 \\
x &\geq -15
\end{aligned}$$

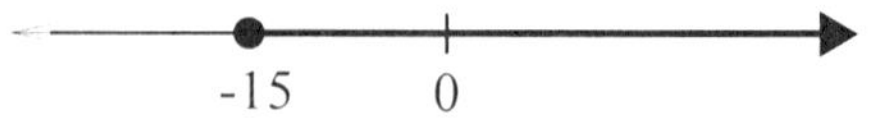

How Am I Doing? Chapter 2 Test

1.
$$3x + 5.6 = 11.6$$
$$3x + 5.6 - 5.6 = 11.6 - 5.6$$
$$3x = 6$$
$$\frac{3x}{3} = \frac{6}{3}$$
$$x = 2$$

2.
$$9x - 8 = -6x - 3$$
$$9x + 6x - 8 = -6x + 6x - 3$$
$$15x - 8 = -3$$
$$15x - 8 + 8 = -3 + 8$$
$$15x = 5$$
$$\frac{15x}{15} = \frac{5}{15}$$
$$x = \frac{1}{3}$$

3.
$$2(2y - 3) = 4(2y + 2)$$
$$4y - 6 = 8y + 8$$
$$4y - 6 + 6 = 8y + 8 + 6$$
$$4y = 8y + 14$$
$$-8y + 4y = -8y + 8y + 14$$
$$-4y = 14$$
$$\frac{-4y}{-4} = \frac{14}{-4}$$
$$y = -\frac{7}{2}$$

4.
$$\frac{1}{7}y + 3 = \frac{1}{2}y$$
$$14\left(\frac{1}{7}y\right) + 14(3) = 14\left(\frac{1}{2}y\right)$$
$$2y + 42 = 7y$$
$$2y - 2y + 42 = 7y - 2y$$
$$42 = 5y$$
$$\frac{42}{5} = \frac{5y}{5}$$
$$8\frac{2}{5} = y$$

5.
$$4(7 - 4x) = 3(6 - 2x)$$
$$28 - 16x = 18 - 6x$$
$$28 - 16x + 6x = 18 - 6x + 6x$$
$$28 - 10x = 18$$
$$28 + (-28) - 10x = 18 + (-28)$$
$$-10x = -10$$
$$\frac{-10x}{-10} = \frac{-10}{-10}$$
$$x = 1$$

6.
$$0.8x + 0.18 - 0.4x = 0.3(x + 0.2)$$
$$100(0.8x + 0.18 - 0.4x) = 100[0.3(x + 0.2)]$$
$$80x + 18 - 40x = 30(x + 0.2)$$
$$40x + 18 = 30x + 6$$
$$40x + 18 - 18 = 30x + 6 - 18$$
$$40x = 30x - 12$$
$$-30x + 40x = -30x + 30x - 12$$
$$10x = -12$$
$$\frac{10x}{10} = \frac{-12}{10}$$
$$x = -\frac{6}{5} \text{ or } -1.2$$

7.
$$\frac{2y}{3}+\frac{1}{5}-\frac{3y}{5}+\frac{1}{3}=1$$
$$15\left(\frac{2y}{3}\right)+15\left(\frac{1}{5}\right)-15\left(\frac{3y}{5}\right)+15\left(\frac{1}{3}\right)=15(1)$$
$$10y+3-9y+5=15$$
$$y+8=15$$
$$y+8-8=15-8$$
$$y=7$$

8.
$$3-2y=2(3y-2)-5y$$
$$3-2y=6y-4-5y$$
$$3-2y=y-4$$
$$3-2y+2y=y+2y-4$$
$$3=3y-4$$
$$3+3=3y-4+4$$
$$7=3y$$
$$\frac{7}{3}=\frac{3y}{3}$$
$$\frac{7}{3}=y$$

9.
$$5(20-x)+10x=165$$
$$100-5x+10x=165$$
$$100+5x=165$$
$$-100+100+5x=-100+165$$
$$5x=65$$
$$\frac{5x}{5}=\frac{65}{5}$$
$$x=13$$

10.
$$5(x+40)-6x=9x$$
$$5x+200-6x=9x$$
$$200-x=9x$$
$$200=10x$$
$$\frac{200}{10}=\frac{10x}{10}$$
$$20=x$$

11.
$$-2(2-3x)=76-2x$$
$$-4+6x=76-2x$$
$$-76-4+6x=-76+76-2x$$
$$-80+6x=-2x$$
$$-80+6x-6x=-2x-6x$$
$$-80=-8x$$
$$\frac{-80}{-8}=\frac{-8x}{-8}$$
$$10=x$$

12.
$$20-(2x+6)=5(2-x)+2x$$
$$20-2x-6=10-5x+2x$$
$$-2x+14=-3x+10$$
$$3x-2x+14=3x-3x+10$$
$$x+14=10$$
$$x+14-14=10-14$$
$$x=-4$$

13.
$$2x-3=12-6x+3(2x+3)$$
$$2x-3=12-6x+6x+9$$
$$2x-3=21$$
$$2x-3+3=21+3$$
$$2x=24$$
$$\frac{2x}{2}=\frac{24}{2}$$
$$x=12$$

14. $\frac{1}{3}x - \frac{3}{4}x = \frac{1}{12}$

$$12\left(\frac{1}{3}x - \frac{3}{4}x\right) = 12\left(\frac{1}{12}\right)$$
$$4x - 9x = 1$$
$$-5x = 1$$
$$\frac{-5x}{-5} = \frac{1}{-5}$$
$$x = -\frac{1}{5}$$

15. $\frac{3}{5}x + \frac{7}{10} = \frac{1}{3}x + \frac{3}{2}$

$$30\left(\frac{3}{5}x + \frac{7}{10}\right) = 30\left(\frac{1}{3}x + \frac{3}{2}\right)$$
$$18x + 21 = 10x + 45$$
$$18x + 21 - 21 = 10x + 45 - 21$$
$$18x = 10x + 24$$
$$-10x + 18x = -10x + 10x + 24$$
$$8x = 24$$
$$\frac{8x}{8} = \frac{24}{8}$$
$$x = 3$$

16. $\frac{15x - 2}{28} = \frac{5x - 3}{7}$

$$28\left(\frac{15x - 2}{28}\right) = 28\left(\frac{5x - 3}{7}\right)$$
$$15x - 2 = 4(5x - 3)$$
$$15x - 2 = 20x - 12$$
$$15x - 2 + 12 = 20x - 12 + 12$$
$$15x + 10 = 20x$$
$$-15x + 15x + 10 = -15x + 20x$$
$$10 = 5x$$
$$\frac{10}{5} = \frac{5x}{5}$$
$$2 = x$$

17. $\frac{2}{3}(x + 8) + \frac{3}{5} = \frac{1}{5}(11 - 6x)$

$$15\left(\frac{2}{3}\right)(x + 8) + 15\left(\frac{3}{5}\right) = 15\left(\frac{1}{5}\right)(11 - 6x)$$
$$10(x + 8) + 9 = 3(11 - 6x)$$
$$10x + 80 + 9 = 33 - 18x$$
$$10x + 89 = 33 - 18x$$
$$10x + 18x + 89 = 33 - 18x + 18x$$
$$28x + 89 = 33$$
$$28x + 89 + (-89) = 33 + (-89)$$
$$28x = -56$$
$$\frac{28x}{28} = \frac{-56}{28}$$
$$x = -2$$

18. $$A = 3w + 2P$$
$$A - 2P = 3w$$
$$\frac{A - 2P}{3} = w$$

19. $$\frac{2w}{3} = 4 - \frac{1}{2}(x + 6)$$
$$\frac{2w}{3} = 4 - \frac{1}{2}x - 3$$
$$6\left(\frac{2w}{3}\right) = 6(4) - 6\left(\frac{1}{2}x\right) - 6(3)$$
$$4w = 24 - 3x - 18$$
$$4w = 3x + 6$$
$$w = \frac{-3x + 6}{4}$$

20. $$A = \frac{1}{2}h(a + b)$$
$$2A = h(a + b)$$
$$2A = ha + hb$$
$$2A - hb = ha$$
$$\frac{2A - hb}{h} = a$$

21. $$5ax(2 - y) = 3axy + 5$$
$$10ax - 5axy = 3axy + 5$$
$$10ax - 5 = 8axy$$
$$\frac{10ax - 5}{8ax} = y$$

22. $$V = \frac{1}{3}Bh$$
$$3V = 3\left(\frac{1}{3}Bh\right)$$
$$3V = Bh$$
$$\frac{3V}{h} = \frac{Bh}{h}$$
$$\frac{3V}{h} = B$$

23. $B = \frac{3V}{h}$, $V = 140$, $h = 14$

$$B = \frac{3(140)}{14} = 30 \text{ square inches}$$

24. $$3(x - 2) \geq 5x$$
$$3x - 6 \geq 5x$$
$$3x + (-5x) - 6 \geq 5x + (-5x)$$
$$-2x - 6 \geq 0$$
$$-2x \geq 6$$
$$\frac{-2x}{-2} \leq \frac{6}{-2}$$
$$x \leq -3$$

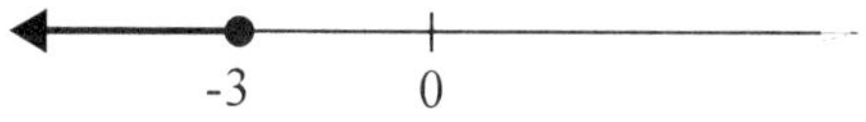

25. $2-7(x+1)-5(x+2)<0$

$2-7x-7-5x-10<0$

$-12x-15<0$

$-12x-15+15<0+15$

$-12x<15$

$\frac{-12x}{-12}>\frac{15}{-12}$

$x>-\frac{5}{4}$

26. $5+8x-4<2x+13$

$8x+1<2x+13$

$8x+1-1<2x+13-1$

$8x<2x+12$

$-2x+8x<-2x+2x+12$

$6x<12$

$\frac{6x}{6}<\frac{12}{6}$

$x<2$

27. $\frac{1}{4}x+\frac{1}{16}\le\frac{1}{8}(7x-2)$

$16\left[\frac{1}{4}x+\frac{1}{16}\right]\le16\left[\frac{1}{8}(7x-2)\right]$

$4x+1\le2(7x-2)$

$4x+1\le14x-4$

$4x+1+4\le14x-4+4$

$4x+5\le14x$

$-4x+4x+5\le-4x+14x$

$5\le10x$

$\frac{5}{10}\le\frac{10x}{10}$

$\frac{1}{2}\le x$

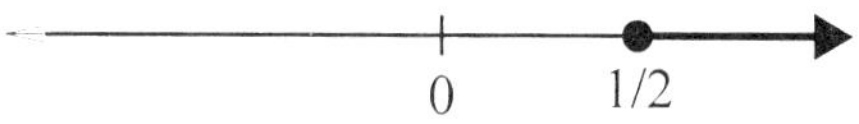

Cumulative Test Chapters 0-2

1. $\frac{6}{7}-\frac{2}{3}=\frac{18}{21}-\frac{14}{21}=\frac{4}{21}$

2. $1\frac{3}{4}+2\frac{1}{5}=1\frac{15}{20}+2\frac{4}{20}$

$=\frac{35}{20}+\frac{44}{20}=\frac{79}{20}$

$=3\frac{19}{20}$

3. $3\frac{1}{5}\div1\frac{1}{2}=\frac{16}{5}\cdot\frac{2}{3}=\frac{32}{15}=2\frac{2}{15}$

4. $$\begin{array}{r} 1.23 \\ \times\ 0.56 \\ \hline 738 \\ 615\ \\ \hline 0.6888 \end{array}$$

5. $$\begin{array}{r} 0.12 \\ 1.2\overline{)0.144} \\ \underline{12}\ \\ 24 \\ \underline{24} \end{array}$$

6. What is 4.3% of 830?
$0.043(830) = 35.69$

7. $(-3)(-5)(-1)(2)(-1) = (3)(5)(1)(2)(1)$
$= 30$

8. $5ab - 7ab^2 - 3ab - 12ab^2 + 10ab - 9ab^2$
$= (5 - 3 + 10)ab + (-7 - 12 - 9)ab^2$
$= 12ab - 28ab^2$

9. $(5x)^2 = 5^2(x)^2 = 25x^2$

10. $2\{3x - 4[5 - 3y(2 - x)]\}$
$= 2[3x - 4(5 - 6y + 3xy)]$
$= 2(3x - 20 + 24y - 12xy)$
$= 6x - 40 + 48y - 24xy$

11. $$\begin{aligned} 4(7 - 2x) &= 3x - 12 \\ 28 - 8x &= 3x - 12 \\ 28 - 8x + 8x &= 3x + 8x - 12 \\ 28 &= 11x - 12 \\ 28 + 12 &= 11x - 12 + 12 \\ 40 &= 11x \\ \frac{40}{11} &= \frac{11x}{11} \\ \frac{40}{11} &= x \text{ or } 3\frac{7}{11} \end{aligned}$$

12. $$\begin{aligned} \frac{1}{3}(x + 5) &= 2x - 5 \\ 3\left[\frac{1}{3}(x + 5)\right] &= 3(2x - 5) \\ x + 5 &= 6x - 15 \\ x + 5 + 15 &= 6x - 15 + 15 \\ x + 20 &= 6x \\ -x + x + 20 &= -x + 6x \\ 20 &= 5x \\ \frac{20}{5} &= \frac{5x}{5} \\ 4 &= x \end{aligned}$$

13. $$\begin{aligned} \frac{2y}{3} - \frac{1}{4} &= \frac{1}{6} + \frac{y}{4} \\ 8y - 3 &= 2 + 3y \\ 5y &= 5 \\ y &= 1 \end{aligned}$$

14.
$$5a-8b=6+2a$$
$$5a+(-2a)-8b=6+2a+(-2a)$$
$$3a-8b=6$$
$$3a-8b+8b=8b+6$$
$$3a=8b+b$$
$$\frac{3a}{3}=\frac{8b+6}{3}$$
$$a=\frac{8b+6}{3}$$

15.
$$H=\frac{2}{3}(b+4a)$$
$$\frac{3H}{2}=b+4a$$
$$\frac{3H}{2}-4a=b \text{ or } b=\frac{3H-8a}{2}$$

16. $I=\mathrm{P}tr$
$$\frac{I}{\mathrm{Pr}}=t$$

17.
$$A=\frac{ha}{2}+\frac{hb}{2}$$
$$2A=ha+hb$$
$$2A-hb=ha$$
$$\frac{2A-hb}{h}=a \text{ or } a=\frac{2A}{h}-b$$

18.
$$-6x-3<2x-10x+7$$
$$-6x-3<-8x+7$$
$$8x-6x-3<8x-8x+7$$
$$2x-3<7$$
$$2x-3+3<7+3$$
$$2x<10$$
$$\frac{2x}{2}<\frac{10}{2}$$
$$x<5$$

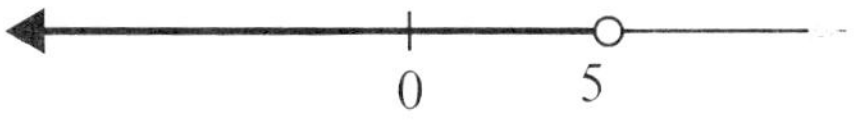

19. $\frac{1}{2}(x-5)\geq x-4$
$$x-5\geq 2x-8$$
$$-x\geq -3$$
$$x\leq 3$$

20.
$$4(2-x)>1-5x-8$$
$$8-4x>-7-5x$$
$$-8+8-4x>-8-7-5x$$
$$-4x>-15-5x$$
$$-4x+5x>-15-5x+5x$$
$$x>-15$$

21. $x+\frac{5}{9}\leq\frac{1}{3}+\frac{7}{9}x$
$$9x+5\leq 3+7x$$
$$2x\leq -2$$
$$x\leq -1$$

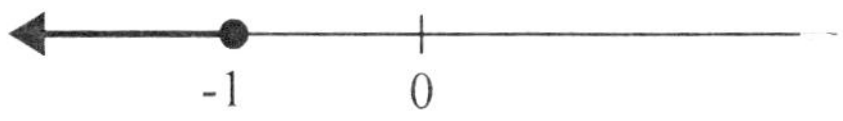

22. $$7x-13 \le 3(4-5x)$$
$$7x-13 \le 12-15x-3$$
$$7x-13 \le 9-15x$$
$$7x+15x-13 \le 9-15x+15x$$
$$22x-13 \le 9$$
$$22x-13+13 \le 9+13$$
$$22x \le 22$$
$$\frac{22x}{22} = \frac{22}{22}$$
$$x \le 1$$

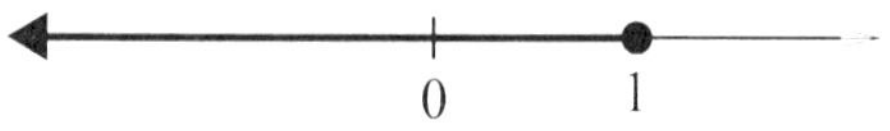

23. $$\frac{0+82+89+87+x}{5} \ge 70$$
$$258 + x \ge 5(70)$$
$$x \ge 350 - 258$$
$$x \ge 92$$
He must receive a 92 or better.

Chapter 3

3.1 Exercises

1. $x+5$

3. $x-6$

5. $\frac{x}{8}$

7. $2x$

9. $3+\frac{1}{2}x$

11. $2x+9$

13. $\frac{1}{3}(x+7)$

15. $\frac{1}{3}x-2x$

17. $3x-7$

19. $x =$ value of a share of AT&T stock
$x+74.50 =$ value of a share of IBM stock

21. $w =$ width of rectangle
$2w+7 =$ length of rectangle

23. $x =$ number of boxes of cookies sold by Keiko
$x-43 =$ number of boxes of cookies sold by Sarah
$x+53 =$ number of boxes of cookes sold by Imelda

25. $s =$ measure of 2nd angle
$s-16 =$ measure of 1st angle
$2s =$ measure of 3rd angle

27. $v =$ value of exports of Canada
$2v =$ value of exports of Japan

29. $x =$ 2nd angle
$3x =$ 1st angle
$x-14 =$ 3rd angle

31. $A =$ area of Minnesota
$\frac{1}{2}A =$ area of Kentucky
$\frac{2}{5}A =$ area of Maine

33. $x =$ points for an arrow in the blue ring
$3x-6 =$ points awarded for an arrow in the gold ring

35. $x =$ number of men aged 16 to 24
$x+51 =$ number of men aged 25 to 34
$x-60 =$ number of men aged 35 to 44
$x-132 =$ number of men aged 45 and above

Cumulative Review

37. $$x+\frac{1}{2}(x-3)=9$$
$$2x+2\left(\frac{1}{2}\right)(x-3)=2(9)$$
$$2x+(x-3)=18$$
$$2x+x-3=18$$
$$3x-3=18$$
$$3x-3+3=18+3$$
$$3x=21$$
$$\frac{3x}{3}=\frac{21}{3}$$
$$x=7$$

39. $$5(x-8)=13+x-5$$
$$5x-40=13+x-5$$
$$5x-40=x+8$$
$$4x=48$$
$$x=12$$

3.2 Exercises

1. x = the number
$$x-543=718$$
$$x=1261$$
Check:
$$1261-543\stackrel{?}{=}718$$
$$718=718$$
The number is 1261.

3. x = the number
$$\frac{x}{8}=296$$
$$x=2368$$
Check:
$$\frac{2368}{8}\stackrel{?}{=}296$$
$$296=296$$
The number is 2368.

5. x = the number
$$x+17=199$$
$$x=182$$
Check:
$$182+17\stackrel{?}{=}199$$
$$199=199$$
The number is 182.

7. x = the number
$$2x+7=93$$
$$2x=86$$
$$x=43$$
Check:
$$2(43)+7\stackrel{?}{=}93$$
$$86+7\stackrel{?}{=}93$$
$$93=93$$
The number is 43.

9. $x =$ the number

$$18-\frac{2}{3}x=12$$

$$54-2x=36$$

$$-2x=-18$$

$$x-9$$

Check:

$$18-\frac{2}{3}(9)\overset{?}{=}12$$

$$18-6\overset{?}{=}12$$

$$12=12$$

The number is 9.

11. $x =$ the number

$$2x-5=3x$$

$$-5=x$$

Check:

$$2(-5)-5\overset{?}{=}3(-5)$$

$$-10-5\overset{?}{=}-15$$

$$-15=-15$$

The number is -5.

13. $x =$ the number

$$x+\frac{1}{2}x+\frac{1}{3}x=22$$

$$6x+3x+2x=132$$

$$11x=132$$

$$x=12$$

Check:

$$12+\frac{1}{2}(12)+\frac{1}{3}(12)\overset{?}{=}22$$

$$12+6+4\overset{?}{=}22$$

$$22=22$$

The number is 12.

15. $x =$ number of used bikes

$4x =$ number of new bikes

$$4x=60$$

$$x=15$$

Used bikes in stock: 15

17. $x =$ wildfires between 1/1/03 and 7/27/03

$$2x-16,723=52,107$$

$$2x=68,830$$

$$x=34,415$$

34,415 wildfires between 1/1/03 and 7/27/03

19. $x =$ number of CDs

$$218+11x=284$$

$$11x=66$$

$$x=6$$

Kyle purchased 6 CDs.

21. x = number of items of jewelry

$$2(38)+49+11.50x=171$$
$$76+49+11.50x=171$$
$$125+11.50x=171$$
$$11.50x=46$$
$$x=4$$

She could buy 4 items of jewelry.

23. $r=\frac{d}{t}$, $d=6404$, $t=20$

$$r=\frac{6404}{20}=320.2$$

It traveled at 320.2 mi/hr

25. $J=2.5E$, $E=220$

$$J=2.5(220)=550 \text{ lb}$$

27. Nell: $r=12$, $t=2.5$

$$d=rt=12(2.5)=30$$

Kristin: $r=14$, $t=2.5$

$$d=rt=14(2.5)=35$$

Distance apart $=35-30=5$ miles

29. r = rate

Highway route: $5r=320$
$$r=64$$

Highway route at 64 mph.

Mountain route: $6r=312$
$$r=52$$

Mountain route at 52 mph.

Difference: $64-52=12$ mph

31. x = temperature on the 9th day

$$101+100+102.5+103+108+109$$
$$+105.5+102+x$$
$$=831+x$$
$$\frac{831+x}{9}=104$$
$$831+x=936$$
$$x=105$$

It needs to be 105°F.

33. x = old station wagon mileage

$$\frac{38+21+x}{3}=22\frac{2}{3}$$
$$\frac{59+x}{3}=\frac{68}{3}$$
$$59+x=68$$
$$x=9 \text{ mpg}$$

35. x = number of cricket chirps

F = Fahrenheit temperature

(a) $F-40=\frac{x}{4}$

(b) $90-40=\frac{x}{4}$
$$50=\frac{x}{4}$$
$$x=200 \text{ chirps}$$

(c) $F-40=\frac{148}{4}$
$$F-40=37$$
$$F=77^0$$

Cumulative Review

37. $-2a(ab-3b+5a)=-2a^2b+6ab-10a^2$

39. $5x^2y-7xy^2-8xy-9x^2y$
$=(5-9)x^2y-7xy^2-8xy$
$=-4x^2y-7xy^2-8xy$

41. $x=$ number of people invited
$23.75x=1425$
$x=60$
She can invite 60 people.

3.3 Exercises

1. $x=$ long length
$x-17=$ short length
$x+x-17=47$
$2x=64$
$x=32$
$x-17=15$
The long piece is 32 m long.
The short piece is 15 m long.

3. $x=$ Dave's salary
$x+2600=$ Elsie's salary
$x+x+2600=71,200$
$2x+2600=71,200$
$2x=68,600$
$x=34,300$
$x+2600-34,300+2600-36,900$
Dave earned \$34,300
Elsie earned \$36,900

5. $x=$ hours David worked
$x+15=$ hours Sarah worked
$x-5=$ hours Kate worked
$x+x+15+x-5=100$
$3x+10=100$
$3x=90$
$x=30$
$x+15=45$
$x-5=25$

David worked 30 hours
Sarah worked 45 hours
Kate worked 25 hours

7. x = number born in Oklahoma

$x + 8$ = number born in Texas

$x - 9$ = number born in Arizona

$$x + x + 8 + x - 9 = 32$$
$$3x - 1 = 32$$
$$3x = 33$$
$$x = 11$$
$$x + 8 = 11 + 8 = 19$$
$$x - 9 = 11 - 9 = 2$$

11 were born in Oklahoma

19 were born in Texas

2 were born in Arizona

9. x = width

$2x - 3$ = length

$$P = 2L + 2W$$
$$42 = 2(2x - 3) + 2x$$
$$42 = 4x - 6 + 2x$$
$$42 = 6x - 6$$
$$48 = 6x$$
$$8 = x$$
$$2x - 3 = 2(8) - 3 = 16 - 3 = 13$$

width = 8 meters

length = 13 meters

11. x = width

$3x + 20$ = length

$$P = 2L + 2W$$
$$96 = 2(3x + 20) + 2x$$
$$96 = 6x + 40 + 2x$$
$$96 = 8x + 40$$
$$56 = 8x$$
$$7 = x$$
$$3x + 20 = 3(7) + 20 = 41$$

width = 7 cm

length = 41 cm

13. w = width

$3w - 35$ = length

$$2w + 2(3w - 35) = 190$$
$$2w + 6w - 70 = 190$$
$$8w = 260$$
$$w = 32.5$$
$$3w - 35 = 62.5$$

width = 32.5 cm

length = 62.5 cm

15. $x =$ speed of jackal

$2x =$ speed of cheetah

$x+10 =$ speed of elk

$$x+2x+x+10=150$$
$$4x+10=150$$
$$4x=140$$
$$x=35$$
$$2x=70$$
$$x+10=45$$

35 mi/hr is the speed of the jackal.

70 mi/hr is the speed of the cheetah.

45 mi/hr is the speed of the elk.

17. $x =$ length C

$2x =$ length A

$1.5x =$ length B

$x+3 =$ length D

$$x+2x+1.5x+x+3=58$$
$$5.5x+3=58$$
$$x=10$$

$x =$ length C $= 10''$

$2x =$ length A $= 20''$

$1.5x =$ length B $= 15''$

$x+3 =$ length D $= 13''$

$$x+2x+1.5x+x+3=58$$
$$x=10''$$

19. $x =$ time over land

$x+3 =$ time over water

$$x+x+3=39$$
$$2x+3=39$$
$$2x=36$$
$$x-18$$
$$x+3=18+3=21$$

a. 18 hours over land

b. 21 hours over water

c. rate over land $= r = \frac{d}{t} = \frac{684}{18} = 38$ mi/hr

rate over water $= r = \frac{d}{t} = \frac{1197}{21} = 57$ mi/hr

$57-38=19$ mi/hr faster over water

21. $x =$ length of side

$x+2 =$ length of new side

$$4(x+2)=5x-3$$
$$4x+8=5x-3$$
$$11=x$$

The original square was 11 m by 11 m.

Cumulative Review

23. $-4x(2x^2-3x+8)=-8x^3+12x^2-32x$

25. $-7x+10y-12x-8y-2$

$=(-7-12)x+(10-8)y-2$

$=-19x+2y-2$

27. Professional L & C: $\frac{117}{650}=0.18=18\%$

Not professional L & C: $100\%-18\%=82\%$

How Am I Doing? Sections 3.1-3.3

1. $3x-40$

2. $\frac{1}{2}x+12$

3. $\frac{5x}{12}$

4. $\frac{1}{4}(x+10)$

5. x = the number
$2x+5=-47$
$2x=-52$
$x=-26$
The number is -26.

6. x = the number
$\frac{x}{2}-9=43$
$\frac{x}{2}=52$
$x=104$
The number is 104.

7. x = number of months
$310x=12{,}400$
$x=40$
In 40 months

8. x = hours of overtime
$24(40)+\frac{1}{2}(24)x=1140$
$960+12x=1140$
$12x=180$
$x=15$

9. x = 4th person's salary
$x+2500$ = 5th person's salary
$40{,}000+13{,}000$ = 3rd person's salary
$40{,}000+35{,}000+53{,}500+x+x+2500$
$=2x+131{,}000$
$\frac{2x+131{,}000}{5}=50{,}200$
$2x+131{,}000=251{,}000$
$2x=120{,}000$
$x=60{,}000$
$x+2500=60{,}000+2500=62{,}500$
The 4th person's salary was \$60,000.
The 5th person's salary was \$62,500.

10. x = length of short piece
$2x+18$ = length of long piece
$x+2x+18=84$
$3x+18=84$
$3x=66$
$x=22$
$2x+18=2(22)+18=62$
The long piece is 62 feet.
The short piece is 22 feet.

11. x = number of pages on Sunday

$2x-25$ = number of pages on Monday

$3x-15$ = number of pages on Tuesday

$$x+2x-25+3x-15=260$$
$$6x-40=260$$
$$6x=300$$
$$x=50$$
$$2x-25=2(50)-25=75$$
$$3x-15=3(50)-15=135$$

50 pages on Sunday

75 pages on Monday

135 pages on Tuesday

12. x = hours to plow 2nd field

$3x$ = hours to plow 1st field

$\frac{x}{4}$ = hours to plow 3rd field

$$x+3x+\frac{x}{4}=510$$
$$4x+12x+x=2040$$
$$17x=120$$
$$x=120$$
$$3x=3(120)=360$$
$$\frac{x}{4}=\frac{120}{4}=30$$

2nd field takes 120 hours

1st field takes 360 hours

3rd field takes 30 hours

3.4 Exercises

1. x = number of bags

$$0.75x+3(6.50)=27.75$$
$$0.75x+19.50=27.75$$
$$0.75x=8.25$$
$$x=11$$

He filled 11 bags.

3. x = overtime hours

$$6(40)+9x=303$$
$$240+9x=303$$
$$9x=63$$
$$x=7$$

He needs 7 hours of overtime per week.

5. x = time to pay off the uniforms

$$600+105x=1817.75$$
$$105x=1217.75$$
$$x\approx 11.6$$

It will take about 12 weeks.

7. x = original price

$$0.28x=100.80$$
$$x=360$$

The original price was $360.

9. x = last year's salary

$0.03x$ = raise

$$x+0.03x=22,660$$
$$1.03x=22,660$$
$$x=22,000$$

Last year's salary was $22,000

11. $x = \text{investment}$

$0.09x = \text{interest}$

$$x + 0.09x = 6540$$
$$1.09x = 6540$$
$$x = 6000$$

He invested \$6000.

13. $x = \text{amount earning } 7\%$

$5000 - x = \text{amount earning } 5\%$

$$0.07x + 0.05(5000 - x) = 310$$
$$0.07x + 250 - 0.05x = 310$$
$$0.02x + 250 = 310$$
$$0.02x = 60$$
$$x = 3000$$
$$5000 - x = 2000$$

They invested \$3000 at 7% and \$2000 at 5%.

15. $x = \text{amount invested at } 12\%$

$400,000 - x = \text{amount invested at } 8\%$

$$0.12x + 0.08(400,000 - x) = 38,000$$
$$0.12x + 32,000 - 0.08x = 38,000$$
$$0.04x + 32,000 = 38,000$$
$$0.04x = 6,000$$
$$x = 150,000$$
$$400,000 - x = 250,000$$

They invested \$150,000 at 12% and \$250,000 at 8%.

17. $x = \text{amount invested}$

$\frac{x}{2} = \text{amount invested at } 5\%$

$\frac{x}{3} = \text{amount invested at } 4\%$

$$x - \frac{x}{2} - \frac{x}{3} = \frac{6x - 3x - 2x}{6} = \frac{x}{6} =$$

amount invested at 3.5%

$$0.05\left(\frac{x}{2}\right) + 0.04\left(\frac{x}{3}\right) + (0.035)\left(\frac{x}{6}\right) = 530$$
$$\left(\frac{0.05}{2} + \frac{0.04}{3} + \frac{0.035}{6}\right)x = 530$$
$$\frac{53}{1200}x = 530$$
$$x = 12,000$$

He invested \$12,000.

19. $x = \text{number of quarters}$

$x - 4 = \text{number of nickels}$

$$0.25x + 0.05(x - 4) = 3.70$$
$$25x + 5(x - 4) = 370$$
$$25x + 5x - 20 = 370$$
$$30x = 390$$
$$x = 13$$
$$x - 4 = 9$$

She has 13 quarters and 9 nickels.

21. $x = \text{number of dimes}$

$x + 3 = \text{number of quarters}$

$2(x + 3) = 2x + 6 = \text{number of nickels}$

$$0.25(x + 3) + 0.10x + 0.05(2x + 6) = 3.75$$
$$25(x + 3) + 10x + 5(2x + 6) = 375$$
$$25x + +75 + 10x + 10x + 30 = 375$$
$$45x + 105 = 375$$

$45x = 270$

$x = 6$

$x + 3 = 6 + 3 = 9$

$2x + 6 = 12 + 6 = 18$

He has 9 quarters, 6 dimes, and 18 nickels

23. x = number of regular stamps

$x - 8$ = number of breast cancer stamps

$0.37x + 0.4(x - 8) = 19.90$

$0.37x + 0.4x - 3.2 = 19.90$

$0.77x = 23.1$

$x = 30$

$x - 8 = 16$

$x + 3 = 30 - 8 = 22$

30 regular stamps

22 breast cancer stamps

25. x = number of \$10 bills

$2x$ = number of \$20 bills

$x + 3$ = number of \$100 bills

$10x + 20(2x) + 100(x + 3) = 1500$

$10x + 40x + 100x + 300 = 1500$

$150x = 1200$

$x = 8$

$2x = 16$

$x + 3 = 11$

Eight \$10 bills, sixteen \$20 bills, eleven \$100 bills.

27. x = amount of sales

$18,000 + 0.04x = 55,000$

$0.04x = 37,000$

$x = 925,000$

She must sell \$925,000 worth of furniture.

29. x = miles driven

Golden Gate has the better bargain when the miles driven exceeds the number which makes their total charge equal to that of the West Suburban Agency.

$0.24x + 3(35) = 0.16x + 3(41)$

$24x + 10,500 = 16x + 12,300$

$8x = 1800$

$x = 225$

The salesperson must drive more than 225 miles.

Cumulative Review

31. $5(3) + 6 \div (-2) = 15 + (-3) = 12$

33. If $x = -2$ and $y = 3$, then

$2x^2 + 3xy - 2y^2$

$= 2(-2)^2 + 3(-2)(3) - 2(3)^2$

$= 2(4) - 18 - 2(9)$

$= 8 - 18 - 18 = -28$

35. Discount $= 0.15(22,400) = 3360$

Discounted cost $= 22,400 - 3360$

$= \$19,040$

Average cost $= \dfrac{19,040}{200} = \$95.20/\text{part}$

3.5 Exercises

1. Perimeter is the distance around a plane figure.

3. Area is a measure of the amount of surface in a region.

5. The sum of the interior angles of any triangle is 180^0.

9. $a = 14, \quad b = 7$

$A = ba = (14)(7) = 98 \text{ in.}^2$

11. $A = 80, \quad b = 16$

$A = \dfrac{1}{2}ab$

$80 = \dfrac{1}{2}a(16)$

$80 = 8a$

$10 = a$

$a = 10$ feet

13. x = length of adjacent side

$2(14) + 2x = 46$

$28 + 2x = 46$

$28 + 2x = 46$

$2x = 18$

$x = 9$

The length is 9 inches.

15. $A = \pi r^2, \; r = 7.00$

$A \approx (3.14)(7.00)^2 = 153.86$ square feet

17. $C = \pi d, \; d = 3032$

$C = (3.14)(3032)$

$C = 9520.48$ miles

19. $A = 600, \quad b_1 = 20, \quad b_2 = 30$

$A = \dfrac{1}{2}a(b_1 + b_2)$

$600 = \dfrac{1}{2}a(20 + 30)$

$1200 = a(50)$

$a = 24$ in.

21. length $= 3(80) = 240$

width $= 2(280) = 560$

$A = LW = 240(560) = 134,400 \text{ m}^2$

23. $C = 2\pi r, \; C = 31.4$

$31.4 = 2(3.14)r$

$31.4 = 6.28r$

$5 = r$

Radius is 5 centimeters

25. $r = \frac{d}{2}$, $d = 64$

$r = \frac{64}{2} = 32$

$C = 2\pi r$, $r = 32$

$C = 2(3.14)(32) = 200.96$ centimeters

27. x = measure of 3rd angle

47 = measure of 1st angle

59 = measure of 2nd angle

Sum of the angles is 180°

$x + 47 + 59 = 180$

$x + 106 = 180$

$x = 74$

The measure of the 3rd angle is 74°

29. x = measure of 3rd angle

$4x$ = measure of the equal angles

Sum of the angles is 180°

$x + 4x + 4x = 180$

$9x = 180$

$x = 20$

$4x = 80$

The measure of the 3rd angle is 20°

The measure of the equal angles is 80°

31. x = measure of the equal angles

38 = measure of smallest angle

Sum of the angles is 180°

$x + x + 38 = 180$

$2x + 38 = 180$

$2x = 142$

$x = 71$

The measure of the equal angles is 71°

33. x = measure of 2nd angle

$2x$ = measure of 1st angle

$2(2x) + 19$ = measure of 3rd angle

$x + 2x + 4x + 19 = 180$

$7x + 19 = 180$

$7x = 161$

$x = 23$

$2x = 46$

$4x + 19 = 111$

1st angle measures 46^0

2nd angle measures 23^0

3rd angle measures 111^0

35. x = the angle

$x + 38 + 90 = 180$

$x + 128 = 180$

$x = 52$

The angle is 52°.

37. Number of cans of paint $= \frac{40}{20} = 2$

Area painted $= 2(102) = 204$

Area of one side $= \frac{204}{4} = 51$

$A = \frac{1}{2}a(b_1 + b_2)$

$51 = \frac{1}{2}a(8+9)$

$102 = a(17)$

$6 = a$

Altitude is 6 feet.

39. $V = 235.5, \quad r = 15$

$V = \pi r^2 h, \ r = 10, \ h = 8$

$V = (3.14)(10)^2(8)$

$= (3.14)(100)(8)$

$= 2512$

The volume is 2512 cubic inches

41. $V = \frac{4}{3}\pi r^3, \ r = \frac{d}{2} = \frac{8}{2} = 4$

$V = \frac{4}{3}(3.14)(4)^3 = 267.95$

$267.95 > 175$

Yes, it will stay aloft.

43. $r = 6, \ h = 4$

(a) $V = 4\pi r^2 h$

$= 4(3.14)(6)^2(4)$

$= 452.16$

The volume is 452.16 cm^3

(b) $S = 2\pi rh + 2\pi r^2$

$= 2(3.14)(6)(4) + 2(3.14)(6)^2$

$= 150.72 + 226.08$

$= 376.8$

The total surface area is 376.8 cm^2

45. Area of the $\frac{3}{4}$ circle:

$A = \frac{3}{4}\pi r^2, \ r = 8$

$A = \frac{3}{4}(3.14)(8)^2 = 150.72$

Area of the $\frac{1}{4}$ circle:

$A = \frac{1}{4}\pi r^2, \ r = 4$

$A = \frac{1}{4}(3.14)(4)^2 = 12.56$

Total area $= 150.72 + 12.56$

$= 163.28$

Goat eats 163.28 square feet

47. (a) $A = l_1 w_1 + \frac{1}{4}\pi r^2 + l_2 w_2$

$l_1 = 1.5,\ w_1 = 9.5,\ r = 1.5,$

$l_2 = 1.5,\ w_2 = 4.5$

$A = (1.5)(9.5) + \frac{1}{4}(3.14)(1.5)^2 + (1.5)(4.5)$

$= 14.25 + 1.77 + 6.75$

$= 22.77 \text{ yards}^2$

(b) $\text{Cost} = (2.50)(22.77) = \$56.93$

49. a. x = length of straight edge

$P = 2\pi r + 2x,\ r = 7,\ x = 15$

$P = 2(3.14)(7) + 2(15) = 43.96 + 30 = 73.96$

The perimeter is 73.96 inches.

b. $V = (\pi r^2 + 2rx)h,\ h = 0.5$

$V = \left[3.14(7)^2 + 2(7)(15)\right](0.5)$

$V = 363.86(0.5)$

$V = 181.93$ cubic inches of plastic

51. $\text{Perimeter} = \frac{120.93}{1.39} = 87$ feet

w = width, $4w - 3$ = length

$2w + 2(4w - 3) = 87$

$2w + 8w - 6 = 87$

$10w = 93$

$w = 9.3$

$4w - 3 = 34.2$

width is 9.3 feet, length is 34.2 feet

53. 1st possibility: 2 sides are 4 and one is 7.5

Perimeter $= 4 + 4 + 7.5 = 15.5$ feet

2nd possibility: one side is 4 and 2 are 7.5

Perimeter $= 4 + 7.5 + 7.5 = 19$ feet

55. Moon's radius $= 1080(5280) = 5,702,400$

4-ft poles: $C = 2\pi r,\ r = 5,702,404$

$C = 2(3.14)(5,702,404) = 35,811,097.12$

3-ft poles: $C = 2\pi r,\ r = 5,702,404$

$C = 2(3.14)(5,702,403) = 35,811,090.84$

$C_4 - C_3 = 35,811,097.12 - 35,811,090.84$

$= 6.28$ feet

Cumulative Review

57. $-5(x+3) - 2[4 + 3(x-1)]$

$= -5x - 15 - 2(4 + 3x - 3)$

$= -5x - 15 - 2(3x + 1)$

$= -5x - 15 - 6x - 2$

$= -11x - 17$

59. $-2\{x + 3[2 - 4(x-3)]\}$

$= -2\{x + 3[2 - 4x + 12]\}$

$= -2\{x + 3[-4x + 14]\}$

$= -2\{x - 12x + 42\}$

$= -2\{-11x + 42\}$

$= 22x - 84$

3.6 Exercises

1. $x > 67{,}000$

3. $x \leq 120$

5. $h \leq 1500$

7. $x \geq 93$

9. 3rd length: x

$87 + 64 + x \leq 291$

$151 + x \leq 291$

$x \leq 140$

140 cm or less

11. x = number of deliveries in week 6

$\frac{18 + 40 + 21 + 7 + 36 + x}{6} \geq 24$

$\frac{122 + x}{6} \geq 24$

$122 + x \geq 144$

$x \geq 22$

She must make 22 or more deliveries.

13. $A = LW, W = 8$

$8L \leq 60$

$L \leq 7.5$

Depth can be no more than 7.5 feet

15. x = number of miles

Cost for 2 days $= 2(37.50) = 75$

$0.195x + 75 \leq 150$

$0.195x \leq 75$

$x \leq 384.6$

They could drive no more than 384.6 miles.

17. $\frac{5}{9}(F - 32) < 110$

$5(F - 32) < 990$

$F - 32 < 198$

$F < 230$

Temperature less than 230^0F.

19. $A = 9.2x + 77.1, A = 307.1$

$9.2x + 77.1 > 307.1$

$9.2x > 230$

$x > 25$

After $1980 + 25 = 2005$

The budget will exceed \$307.1 billion after 2005.

21. x = hours of labor

$840 + 45x \leq 1200$

$45x \leq 360$

$x \leq 8$

It will take 8 hours or less.

23. $18n > 5000 + 7n$

$11n > 5000$

$n > 454.5$

Need to manufacture and sell more than 455 discs.

Cumulative Review

25. $10 - 3x > 14 - 2x$

$-x > 4$

$x < -4$

27. $30 - 2(x+1) \le 4x$

$30 - 2x - 2 \le 4x$

$-6x \le -28$

$x \ge \frac{14}{3}$

$x \ge 4\frac{2}{3}$

29. a. $8(365) = 2920$ gallons wasted

b. $\text{Cost} = \frac{2920}{1000}(8.50) = 24.82$

He will pay \$24.82 extra.

Putting Your Skills to Work

1. $a = \frac{22,766 - 15,303}{4} \approx 1865.75$

$b = 15,303$

$y = 1866x + 15,303$

2. 1995 - $x = 1995 - 1986 = 9$

$y = 1866(9) + 15,303$

$y = 32,097$ caribou

1984 - $x = 1984 - 1986 = -2$

$y = 1866(-2) + 15,303$

$y = 11,571$ caribou

3. 2003 - $x = 2003 - 1986 = 17$

$y = 1866(17) + 15,303$

$y = 47,025$ caribou

The measured value of 43,375 is $47,025 - 43,375 = 3650$ lower. The equation predicts a value that is too high for 2003.

4. $a = \frac{43,375 - 15,303}{17} \approx 1651$

$y = 1651 \times 15,303$

Chapter 3 Review Problems

1. $x+19$

2. $\frac{2}{3}x$

3. $\frac{1}{2}x$ or $\frac{x}{2}$

4. $x-18$

5. $3(x+4)$

6. $2x-3$

7. number of retired people: r
number of working people: $4r$
number of unemployed people: $\frac{1}{2}r$

8. $w=$ width of the rectangle
$3w+5=$ length of the redctangle

9. $b=$ the number of degrees in angle B
$2b=$ the number of degrees in angle A
$b-17=$ the number of degrees in angle C

10. number of algebra students: a
number of biology students: $a+29$
number of geology students: $\frac{1}{2}a$

11. $x=$ the number
$2x-18=-8$
$2x=10$
$x=5$
The number is 5.

12. $x=$ a number
$3x-6=15$
$3x=21$
$x=7$
The number is 7.

13. $x=$ cost of chair
$210+6x=450$
$6x=240$
$x=40$
1 chair costs $40.

14. $x=$ David's age
$2x=$ Jon's age
$2x=32$
$x=16$
David is 16 years old.

15. $t_1=$ time for first car
$t_2=$ time for other car
$330=50t_1$
$6.6=t_1$
$330=55t_2$
$6=t_2$
The first car took 6.6 hours.
The other car took 6 hours.

16. $x = \text{grade}$

$$\frac{85+78+65+92+70+x}{6} = 80$$
$$\frac{390+x}{6} = 80$$
$$390 + x = 480$$
$$x = 90$$

She needs a grade of 90.

17. x is the measure of the third sid

$$x + 2(31) = 85$$
$$x + 62 = 85$$
$$x = 23$$

The third side measures 23 feet.

18. 1st angle: x

2nd angle: $3x$

3rd angle: $2x - 12$

$$x + 3x + 2x - 12 = 180$$
$$6x - 12 = 180$$
$$6x = 192$$
$$x = 32$$
$$3x = 3(32) = 96$$
$$2x - 12 = 2(32) - 12 = 52$$

32^0, 96^0, and 52^0, respectively.

19. $x = \text{length of one piece}$

$\frac{3}{5}x = \text{length of other}$

$$x + \frac{3}{5}x = 50$$
$$5x + 3x = 250$$
$$8x = 250$$
$$x = 31.25$$
$$\frac{3}{5}x = 18.75$$

The lengths are 31.25 yd and 18.75 yd.

20. $x = \text{Jon's salary}$

$3000 + \frac{1}{2}x = \text{Lauren's salary}$

$$x + 3000 + \frac{1}{2}x = 48,000$$
$$\frac{3}{2}x + 3000 = 48,000$$
$$\frac{3}{2}x = 45,000$$
$$x = 30,000$$
$$3000 + \frac{1}{2}x = 3000 + \frac{1}{2}(30,000) = 18,000$$

Jon: \$30,000; Lauren: \$18,000

21. $x = \text{number of kilowatt-hours}$

$$25 + 0.15x = 71.50$$
$$0.15x = 46.50$$
$$x = 310$$

310 kilowatt-hours

22. x = number of miles driven

$$\begin{aligned}0.25x + 39(3) &= 187\\ 0.25x + 117 &= 187\\ 0.25x &= 70\\ x &= 280\end{aligned}$$

He drove 280 miles.

23. x = amount withdrawn

$$\begin{aligned}0.055(7400 - x) &= 242\\ 407 - 0.055x &= 242\\ -0.055x &= -165\\ x &= 3000\end{aligned}$$

They withdrew $3000.

24. x = original price

$$\begin{aligned}0.18x &= 36\\ x &= 200\end{aligned}$$

The original price was $200.

25. x = amount invested at 12%

$9000 - x$ = amount at 8%

$$\begin{aligned}0.12x + 0.08(9000 - x) &= 1000\\ 12x + 8(9000 - x) &= 100,000\\ 12x + 72,000 - 8x &= 100,000\\ 4x &= 28,000\\ x &= 7000\\ 9000 - x &= 2000\end{aligned}$$

They invested $7000 at 12% and $2000 at 8%.

26. x = amount at 4.5%

$5000 - x$ = amount at 6%

$$\begin{aligned}0.045x + 0.06(5000 - x) &= 270\\ 45x + 60(5000 - x) &= 270,000\\ 45x + 300,000 - 60x &= 270,000\\ -15x &= -30,000\\ x &= 2000\end{aligned}$$

$5000 - x = 5000 - 2000 = 3000$

$2000 at 4.5%; $3000 at 6%

27. x = number of dimes

$x + 3$ = number of quarters

$2(x + 3) = 2x + 6$ = number of nickels

$$\begin{aligned}0.05(2x + 6) + 0.10x + 0.25(x + 3) &= 3.75\\ 5(2x + 6) + 10x + 25(x + 3) &= 375\\ 10x + 30 + 10x + 25x + 75 &= 375\\ 45x &= 270\\ x &= 6\end{aligned}$$

$x + 3 = 6 + 3 = 9$

$2x + 6 = 2(6) + 6 = 18$

She has 18 nickels, 6 dimes, 9 quarters

28. $d-1 =$ number of nickels

$d =$ number of dimes

$d+2 =$ number of quarters

$$0.05(d-1)+0.10d+0.25(d+2)=3.65$$
$$5(d-1)+10d+25(d+2)=365$$
$$5d-5+10d+25d+50=365$$
$$40d+45=365$$
$$40d=320$$
$$d=8$$

$d-1=8-1=7$

$d+2=8+2=10$

7 nickels, 8 dimes, 10 quarters

29. $\text{S.A.}=4\pi r^2,\ r=\frac{d}{2}=\frac{32}{2}=16$

$\text{S.A.}=4(3.14)(16)^2$

$\text{S.A.}=3215.36$ sq. in.

30. $P=(\text{number of sides})(\text{length of side})$

$P=64$

$P=24$ feet

31. $x =$ measure of third angle

$$62+47+x=180$$
$$109+x=180$$
$$x=71$$

The angle is 71^0.

32. $A=\frac{1}{2}ab$

$=\frac{1}{2}(8)(10.5)$

$=42$

42 mi^2

33. $l=12,\ \ w=8,\ \ h=20$

$V=lwh=12(8)(20)=1920\text{ ft}^3$

34. $V=\frac{4}{3}\pi r^3$

$=\frac{4}{3}(3.14)(3)^3$

$=113.04$

113.04 cm^3

35. $A=\pi r^2$

$=3.14\left(\frac{18.00}{2}\right)^2$

$=3.14(9)^2$

$=254.34$

254.34 cm^2

36. $x =$ length of 3rd side

$P =$ sum of sides

$$17+17+x=46$$
$$34+x=46$$
$$x=12$$

The 3rd side is 12 in.

37. $C = 2.50\left[2(18)(22)+2(18)(19)-100\right]$
$= 2.50(1376)$
$= 3440$
\$3440

38. $r = 5,\ h = 14$
$S = 2\pi rh + 2\pi r^2$
$= 2(3.14)(5)(14)+2(3.14)(5)^2$
$= 439.60 + 157.00$
$= 596.60 \text{ m}^2$
$\text{Cost} = 40(596.60) = \$23,864.00$

39. $x =$ miles driven
$0.15x + 2(35) \le 100$
$15x + 7000 \le 10,000$
$15x \le 3000$
$x \le 200$
Distance must be no more than 200 miles.

40. $x =$ sales
$12,000 + 0.04x > 24,000$
$0.04x > 12,000$
$x > 300,000$
More than \$300,000 in sales

41. $d = rt$
$1590 = 8.75t + 20(t-24)$
$1590 = 28.75t - 480$
$1590 = 28.75t - 480$
$2070 = 28.75t$
$72 = t$
It took 72 months.
CP distance $= rt$
$= 8.75(72)$
$= 630$ miles
UP distance $= rt$
$= 20(72-24)$
$= 960$ miles

42. $x =$ amount spent on ties
$3(17.95) + x < 70$
$53.85 + x < 70$
$x < 16.15$
He must spend less than \$16.15 on ties.

43. $x =$ length of commercials
$4x + 4 =$ length of entertainment
$x + 4x + 4 = 30$
$5x = 26$
$x = 5.2$
$4x + 4 = 4(5.2) + 4 = 24.8$
Entertainment was 24.8 min.

44. x = weight of each ball

$3x = x + 8$

$2x = 8$

$x = 4$

4 oz

45. x = number of years to even the cost

$83x + 744 > 41x + 870$

$42x + 744 > 870$

$42x > 126$

$x > 3$

It will take more than 3 years.

46. x = number of months

$1100x + 42,000 < 1800x$

$42,000 < 700x$

$60 < x$

It will take more than 60 months.

47. x = height

$2.5x$ = width

$3(2.5x) + 2x = 38$

$7.5x + 2x = 38$

$9.5x = 38$

$x = 4$

$2.5x = 10$

Width = 10 feet, height = 4 feet

48.

Year	Amount	Total
1	1200	1200
2	2400	3600
3	3600	7200
4	4800	12,000
5	6000	18,000
6	7200	25,200
7	8400	33,600
8	9600	43,200
9	10,800	54,000
10	12,000	66,000

49. x = number of free throws

$3x$ = number of field goals

$1x + 2(3x) + 3(12) = 99$

$7x + 36 = 99$

$7x = 63$

$x = 9$

$3x = 27$

They made 9 free throws and 27 field goals.

50. x = width

$2x - 10$ = length

$P = 2L + 2W$

$76 = 2(2x - 10) + 2x$

$76 = 4x - 20 + 2x$

$76 = 6x - 20$

$96 = 6x$

$16 = x$

$2x - 10 = 22$

Length = 22 feet

Width = 16 feet

51. $x =$ hours Carl worked

$2x-8 =$ hours Ryan worked

$$x+2x-8=28$$
$$3x-8=28$$
$$3x=36$$
$$x=12$$
$$2x-8=2(12)-8=16$$

Carl worked 12 hours.

Ryan worked 16 hours.

52. $x =$ amount invested at 9.75%

$x+2000 =$ amount invested at 8.5%

$$0.0975x=0.085x+170$$
$$0.0125x=170$$
$$x=13,600$$
$$x+2000=13,600+2000=15,600$$

\$13,600 was invested at 9.75%.

\$15,600 was invested at 8.5%.

53. $x =$ measure of 2nd angle

$x+20 =$ measure of 1st angle

$2x =$ measure of 3rd angle

$$x+x+20+2x=180$$
$$4x+20=180$$
$$4x=160$$
$$x=40$$
$$x+20=60$$
$$2x=80$$

1st angle $= 60^0$

2nd angle $= 40^0$

3rd angle $= 80^0$

54. $A=\frac{1}{2}a(b_1+b_2)$, $a=12$, $b_1=15$, $b_2=9$

$$A=\frac{1}{2}(12)(15+9)=6(34)=204$$

Cost $= 6(204) = \$1224$

55. $x =$ score on the final exam

$$\frac{88+77+95+92+2x}{6}=90$$
$$\frac{352+2x}{6}=90$$
$$352+2x=540$$
$$2x=188$$
$$x=94$$

She must make 94 on the final exam.

56. $x =$ business miles driven

$2x =$ pleasure miles driven

$$x+2x=27,000$$
$$3x=27,000$$
$$x=9000$$

Deduction $= (0.33)(9000) = \$2970$

No, her correct deduction was \$2970.

57. $d=rt$, $r=\frac{32}{3}$, $t=60$

$$d=\frac{32}{3}(60)=640 \text{ miles}$$

Boston to Denver: $d=1800$

$$1800=\frac{32}{3}t$$
$$169 \approx t$$

It will take about 169 minutes or 2 hr 49 min.

58. $d = rt,\ d = 80,\ r = 0.8\left(\frac{50}{6}\right)$

$80 = 0.8\left(\frac{50}{6}\right)t$

$12 = t$

It will take him 12 seconds.

59. $x =$ number of Explorers

$2x =$ number of Caravans

$5x + 7(2x) = 95$

$5x + 14x = 95$

$19x = 95$

$x = 5$

$2x = 10$

They need 5 Explorers and 10 Caravans.

60. $x =$ smallest angle

$3x =$ largest angle

$2x - 12 =$ third angle

$x + 3x + 2x - 12 = 180$

$6x - 12 = 180$

$6x = 192$

$x = 32$

$3x = 3(32) = 96$

$2x - 12 = 2(32) - 12 = 52$

Smallest angle $= 32^\circ$

Largest angle $= 96^\circ$

Third angle $= 52^\circ$

61. $V = \frac{4}{3}\pi r^3,\ r = \frac{3}{2} = 1.5$

$V = \frac{4}{3}(3.14)(1.5)^3$

$V = 14.13$ cubic meters

$80 + 75 + 2(14.13) = 183.26$ kg

How Am I Doing? Chapter 3 Test

1. $x =$ number

$2x - 11 = 59$

$2x = 70$

$x = 35$

The number is 35.

2. $x =$ number

$\frac{1}{2}x + \frac{1}{9}x + \frac{1}{12}x = 25$

$36\left(\frac{1}{2}x + \frac{1}{9}x + \frac{1}{12}x\right) = 36(25)$

$18x + 4x + 3x = 900$

$25x = 900$

$x = 36$

The number is 36.

3. $x =$ number

$2(x + 5) = 3x + 14$

$2x + 10 = 3x + 14$

$10 = x + 14$

$-4 = x$

The number is -4.

4. $\frac{2}{3}x = \text{1st side}$

   $x = \text{2nd side}$

   $x - 14 = \text{3rd side}$

   $$\frac{2}{3}x + x + x - 14 = 66$$
   $$\frac{8}{3}x - 14 = 66$$
   $$\frac{8}{3}x = 80$$
   $$x = 30$$

   $\frac{2}{3}x = \frac{2}{3}(30) = 20$

   $x - 14 = 30 - 14 = 16$

   20 m, 30 m, and 16 m, respectively.

5. $w = \text{width}$

   $2w + 7 = \text{length}$

   $$2w + 2(2w + 7) = 134$$
   $$2w + 4w + 14 = 134$$
   $$6w + 14 = 134$$
   $$6w = 120$$
   $$w = 20$$

   $2w + 7 = 2(20) + 7 = 47$

   width = 20 m

   length = 47 m

6. $2x = \text{1st pollutant}$

   $x = \text{2nd pollutant}$

   $0.75x = \text{3rd pollutant}$

   $$2x + x + 0.75x = 15$$
   $$3.75x = 15$$
   $$x = 4$$

   $2x = 2(4) = 8$

   $0.75x = 0.75(4) = 3$

   1st pollutant: 8ppm;

   2nd pollutant: 4 ppm;

   3rd pollutant: 3 ppm

7. $x = \text{number of months}$

   $$116x + 200 = 1940$$
   $$116x = 1740$$
   $$x = 15$$

   15 months

8. $x = \text{last year's tuition}$

   $$x + 0.08x = 22,680$$
   $$1.08x = 22,680$$
   $$x = 21,000$$

   Last year's tuition was \$21,000

9. x = amount at 14%

$4000 - x$ = amount at 11%

$$0.14x + 0.11(4000 - x) = 482$$
$$100[0.14x + -.11(4000 - x)] = 100(482)$$
$$14x + 11(4000 - x) = 48,200$$
$$14x + 44,000 - 11x = 48,200$$
$$3x + 44,000 = 48,200$$
$$3x = 4200$$
$$x = 1400$$

$4000 - x = 4000 - 1400 = 2600$

\$1400 at 14%; \$2600 at 11%

10. $2x$ = number of nickels

$x - 1$ = number of dimes

x = number of quarters

$$0.05(2x) + 0.10(x - 1) + 0.25(x) = 3.50$$
$$5(2x) + 10(x - 1) + 25x = 350$$
$$10x + 10x - 10 + 25x = 350$$
$$45x - 10 = 350$$
$$45x = 360$$
$$x = 8$$

$2x = 2(8) = 16$

$x - 1 = 8 - 1 = 7$

16 nickels; 7 dimes; 8 quarters

11. $C = 2\pi r,\ r = 34$

$C \approx 2(3.14)(34) = 213.52$ inches

12. $A = \frac{1}{2}a(b_1 + b_2)$

$= \frac{1}{2}(16)(10 + 14)$

$= 192$

192 in.^{2}

13. $r = 10$

$V = \frac{4}{3}\pi r^3$

$= \frac{4}{3}(3.14)(10)^3$

$= 4187$ in.^{3}

14. $A = ab,\ a = 8,\ b = 12$

$A = 8(12)$

$= 96$ square centimeters

15. $C = 12(1.5) + 9(2) + \frac{1}{2}(2)(1.5)$

$= 18 + 18 + 1.5 = 37.5$ yd^2

Cost $= 12(37.5) = \$450$

16. x = score

$$\frac{76 + 84 + 78 + x}{4} \geq 80$$
$$\frac{238}{4} \geq 80$$
$$238 + x \geq 320$$
$$x \geq 82$$

At least an 82

17. $x = \text{sales}$

$$15{,}000 + 0.05(x - 10{,}000) > 20{,}000$$
$$15{,}000 + 0.05x - 500 > 20{,}000$$
$$0.05x + 14{,}500 > 20{,}000$$
$$0.05x > 5500$$
$$x > 110{,}000$$

She must have sales of more than \$110,000.

18. $A = LW$, $L = 150$, $W = 100$

$A = 150(100) = 15{,}000$ square feet

$\text{Time} = \dfrac{15{,}000}{25} = 600$ hours

Cumulative Test for Chapters 0-3

1. $2.4\overline{)8.856}$ = 3.69

$$\begin{array}{r} 72 \\ \hline 165 \\ 144 \\ \hline 216 \\ 216 \\ \hline \end{array}$$

2. $\dfrac{3}{8} + \dfrac{5}{12} + \dfrac{1}{2} = \dfrac{9}{24} + \dfrac{10}{24} + \dfrac{12}{24}$

$= \dfrac{9 + 10 + 12}{24}$

$= \dfrac{31}{24}$

3. $3(2x - 3y + 6) + 2(-3x - 7y)$

$= 6x - 9y + 18 - 6x - 14y$

$= -23y + 18$

4. $12 - 3(2 - 4) + 12 \div 4$

$= 12 - 3(-2) + 12 \div 4$

$= 12 + 6 + 3$

$= 21$

5.
$$H = \frac{1}{2}(3a + 5b)$$
$$2H = 2\left(\frac{1}{2}\right)(3a + 5b)$$
$$2H = 3a + 5b$$
$$2H + (-3a) = 3a + (-3a) + 5b$$
$$2H - 3a = 5b$$
$$\frac{2H - 3a}{5} = \frac{5b}{5}$$
$$\frac{2H - 3a}{5} = b$$

6. $5x - 3 \le 2(4x + 1) + 4$

$5x - 3 \le 8x + 2 + 4$

$5x - 3 \le 8x + 6$

$5x \le 8x + 9$

$-3x \le 9$

$x \ge -3$

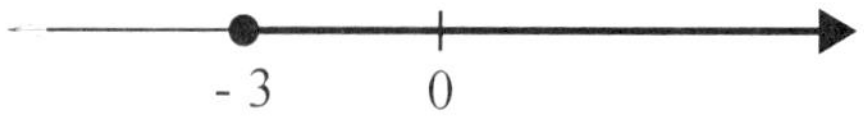

7. $x =$ the number

$$2x + 15 = 1$$
$$2x = -14$$
$$x = -7$$

The number is -7.

8. $x - 34 =$ Psychology students

$x =$ World History students

$$x + x - 34 = 134$$
$$2x - 34 = 134$$
$$2x = 168$$
$$x = 84$$
$$x - 34 = 84 - 34 = 50$$

50 Psychology students;

84 World History students

9. $w =$ width

$3w + 11 =$ length

$$P = 2L + 2W$$
$$2(3w + 11) + 2w = 78$$
$$6w + 22 + 2w = 78$$
$$8w = 56$$
$$w = 7$$
$$3w + 11 = 32$$

length = 32 cm, width = 7 cm

10. $x =$ last year's sales

$$x + 0.35x = 182,250$$
$$1.35x = 182,250$$
$$x = 135,000$$

\$135,000

11. $x =$ amount invested at 15%

$7000 - x =$ amount invested at 7%

$$0.15x + 0.07(7000 - x) = 730$$
$$15x + 7(7000 - x) = 73,000$$
$$15x + 49,000 - 7x = 73,000$$
$$8x = 24,000$$
$$x = 3000$$
$$7000 - x = 4000$$

He invested \$3000 at 15% and \$4000 at 7%.

12. $x + 3 =$ number of nickels

$3x =$ number of dimes

$x =$ number of quarters

$$0.05(x + 3) + 0.10(3x) + 0.25(x) = 2.55$$
$$5(x + 3) + 10(3x) + 25x = 255$$
$$5x + 15 + 30x + 25x = 255$$
$$60x + 15 = 255$$
$$60x = 240$$
$$x = 4$$

$$x + 3 = 4 + 3 = 7$$
$$3x = 3(4) = 12$$

7 nickels; 12 dimes; 4 quarters

13. $b = 25, \quad a = 13$

$$A = \frac{1}{2}ab = \frac{1}{2}(25)(13) = 162.5 \text{ m}^2$$

Cost = $4.50(162.5) = \$731.25$

14. $V = \frac{4}{3}\pi r^3$

$= \frac{4}{3}(3.14)(3.00)^3$

$= 113.04 \text{ in.}^3$

weight $= (113.04)(1.50) = 169.56$

169.56 lb

15. $d = rt,\ d = 1625,\ r = 65$

$1625 = 65t$

$25 = t$

It will take 25 days.

Chapter 4

4.1 Exercises

1. When you multiply exponential expressions with the same base, keep the base the same and add the exponents.

3. $\frac{2^2}{2^3} \stackrel{?}{=} \frac{1}{2^{3-2}}$
   $\frac{4}{8} \stackrel{?}{=} \frac{1}{2}$
   $\frac{1}{2} = \frac{1}{2}$

5. $6x^{11}y$: Coefficient is 6, bases are x, y and exponents are 11, 1.

7. $2 \cdot 2 \cdot a \cdot a \cdot a \cdot b = 2^2 a^3 b$

9. $(-3)(a)(a)(b)(c)(b)(c)(c) = -3a^2b^2c^3$

11. $(7^4)(7^6) = 7^{4+6} = 7^{10}$

13. $(5^{10})(5^{16}) = 5^{10+16} = 5^{26}$

15. $(x^4)(x^8) = x^{4+8} = x^{12}$

17. $(w^{12})(w^{20}) = w^{12+20} = w^{32}$

19. $-5x^4(4x^2) = -20x^{4+2}$
    $= -20x^6$

21. $(5x)(10x^2) = (5 \cdot 10)x^{1+2}$
    $= 50x^3$

23. $(3x^2y)(8x^3y^3) = 24x^{2+3}y^{1+3}$
    $= 24x^5y^4$

25. $\left(\frac{2}{5}xy^3\right)\left(\frac{1}{3}x^2y^2\right) = \frac{2}{15}x^{1+2}y^{3+2}$
    $= \frac{2}{15}x^3y^5$

27. $(1.1x^2z)(-2.5xy) = (1.1)(-2.5)x^{2+1}yz$
    $= -2.75x^3yz$

29. $(8a)(2a^3b)(0) = 0$

31. $(-16x^2y^4)(-5xy^3)$
    $= 80x^{2+1}y^{4+3}$
    $= 80x^3y^7$

33. $(-8x^3y^2)(3xy^5) = -24x^{3+1}y^{2+5}$
    $= -24x^4y^7$

35. $(-2x^3y^2)(0)(-3x^4y) = 0$

37. $\left(8a^4b^3\right)\left(-3x^2y^5\right) = -24a^4b^3x^2y^5$

39. $\left(2x^2y\right)\left(-3y^3z^2\right)\left(5xz^4\right) = -30x^{2+1}y^3z^{2+4}$
$= -30x^3y^3z^6$

41. $\dfrac{y^{12}}{y^5} = y^{12-5}$
$= y^7$

43. $\dfrac{y^5}{y^8} = \dfrac{1}{y^{3-5}}$
$= \dfrac{1}{y^3}$

45. $\dfrac{11^{18}}{11^{30}} = \dfrac{1}{11^{30-18}} = \dfrac{1}{11^{12}}$

47. $\dfrac{3^{18}}{3^{14}} = 3^{18-14} = 3^4$

49. $\dfrac{a^{13}}{4a^5} = \dfrac{a^{13-5}}{4}$
$= \dfrac{a^8}{4}$

51. $\dfrac{x^7}{y^9} = \dfrac{x^7}{y^9}$

51. $\dfrac{-12x^5y^3}{-24xy^3} = \dfrac{x^{5-1}y^{3-3}}{2} = \dfrac{x^4}{2}$

53. $\dfrac{48x^5y^3}{24xy^3} = 2x^{5-1}y^{3-3} = 2x^4$

55. $\dfrac{16x^5y}{-32x^2y^3} = \dfrac{x^{5-2}}{-2y^{3-1}}$
$= -\dfrac{x^3}{2y^2}$

57. $\dfrac{1.8f^4g^3}{54f^2g^8} = \dfrac{f^{4-2}}{30g^{8-3}} = \dfrac{f^2}{30g^5}$

59. $\dfrac{-51x^6y^8}{17x^3y^8} = -3x^{6-3}y^{8-8} = -3x^3$

61. $\dfrac{8^0x^2y^3}{16x^5y} = \dfrac{y^{3-1}}{16x^{5-2}} = \dfrac{y^2}{16x^3}$

63. $\dfrac{18a^6b^3c^0}{24a^5b^3} = \dfrac{3}{4}a^{6-5}b^{3-3}c^0 = \dfrac{3}{4}a$

65. $\dfrac{85a^2b}{45c^3} = \dfrac{17a^2b}{9c^3}$

67. $\dfrac{81x^8y^2z^4}{-3x^3yz} = -27x^5yz^3$
so, $\left(-3x^3yz\right)\left(-27x^5yz^3\right) = 81x^8y^2z^4$

69. $\left(x^2\right)^6 = x^{2\cdot 6} = x^{12}$

71. $\left(xy^2\right)^7 = x^7\left(y^2\right)^7 = x^7y^{14}$

73. $\left(rs^2\right)^6 = r^6 s^{2\cdot6} = r^6 s^{12}$

75. $\left(3a^3b^2c\right)^3 = 3^3 a^{3(3)} b^{2(3)} c^3$
$= 27a^9b^6c^3$

77. $\left(-3a^4\right)^2 = (-3)^2\left(a^4\right)^2 = 9a^8$

79. $\left(\frac{x}{2m^4}\right)^7 = \frac{x^7}{2^7 m^{4\cdot7}} = \frac{x^7}{128m^{28}}$

81. $\left(\frac{5x}{7y^2}\right)^2 = \frac{5^2x^2}{7^2y^{2(2)}} = \frac{25x^2}{49y^4}$

83. $\left(-3a^2b^3c^0\right)^4 = (-3)^4\left(a^2\right)^4\left(b^3\right)^4$
$= 81a^8b^{12}$

85. $\left(-2x^3y^0z\right)^3 = (-2)^3 x^{3\cdot3} y^{0\cdot3} z^3$
$= -8x^9z^3$

87. $\frac{(3x)^5}{\left(3x^2\right)^3} = \frac{3^5x^5}{3^3x^{2\cdot3}} = \frac{3^{5\cdot3}}{x^{6-5}} = \frac{9}{x}$

89. $\left(3ab^2\right)^3(ab) = \left(27a^3b^6\right)(ab) = 27a^4b^7$

91. $\left(\frac{8}{y^5}\right)^2 = \frac{8^2}{y^{5(2)}} = \frac{64}{y^{10}}$

93. $\left(\frac{2x}{y^3}\right)^4 = \frac{2^4x^4}{y^{3\cdot4}} = \frac{16x^4}{y^{12}}$

95. $\left(\frac{ab^2}{c^3d^4}\right)^4 = \frac{a^4\left(b^2\right)^4}{\left(c^3\right)^4\left(d^4\right)^4} = \frac{a^4b^8}{c^{12}d^{16}}$

97. $a^3 = -27$ $3c = 12$
$a = -3$ $c = 4$
$\left(x^b\right)^3 = x^9$ $\left(z^d\right)^3 = z^{21}$
$3b = 9$ $3d = 21$
$b = 3$ $d = 7$
$\left(y^c\right)^3 = y^{12}$ $-3x^3y^4z^7$

Cumulative Review

99. $-3-8 = -3+(-8) = -11$

101. $\left(\frac{2}{3}\right)\left(-\frac{21}{8}\right) = \left(\frac{2}{3}\right)\left(\frac{-3\cdot7}{4\cdot2}\right) = -\frac{7}{4}$

103. $\text{Percentage} = \frac{1{,}500{,}000}{8{,}511{,}960}$
$= 0.1762$
$= 17.6\%$

105. 2.3% of 1,500,000
$0.023(1{,}500{,}000) = 34{,}500$
34,500 sq km of rain forest lost

4.2 Exercises

1. $x^{-4} = \frac{1}{x^4}$

3. $3^{-4} = \frac{1}{3^4} = \frac{1}{81}$

5. $\frac{1}{y^{-8}} = y^8$

7. $\frac{x^{-4}y^{-5}}{z^{-6}} = \frac{z^6}{x^4y^5}$

9. $x^{-5}y^6 = \frac{y^6}{x^5}$

11. $(2x^{-3})^{-3} = 2^{-3}x^{-3(-3)}$
$= \frac{x^9}{2^3}$
$= \frac{x^9}{8}$

13. $3x^{-2} = \frac{3}{x^2}$

15. $(4x^2y)^{-2} = \frac{1}{(4x^2y)^2}$
$= \frac{1}{4^2x^{2(2)}y^2}$
$= \frac{1}{16x^4y^2}$

17. $\frac{3xy^{-2}}{z^{-3}} = \frac{3xz^3}{y^2}$

19. $\frac{(3x)^{-2}}{(3x)^{-3}} = \frac{(3x)^3}{(3x)^2}$
$= 3x$

21. $wx^{-5}y^3z^{-2} = \frac{wy^3}{x^5z^2}$

23. $(8^{-2})(2^3) = \frac{1}{8^2} \cdot 2^3$
$= \frac{1}{64} \cdot 8$
$= \frac{1}{8}$

25. $\left(\frac{3x^0y^2}{z^4}\right)^{-2} = \frac{3^{-2}x^{0(-2)}y^{2(-2)}}{z^{4(-2)}}$
$= \frac{x^0y^{-4}}{3^2z^{-8}}$
$= \frac{z^8}{9y^4}$

27. $\frac{x^{-2}y^{-3}}{x^4y^{-2}} = \frac{y^2}{x^4x^2y^3}$
$= \frac{1}{x^6y}$

29. $123,780 = 1.2378 \times 10^5$

31. $\left(-3x^{-4}y^{2}\right)^{-4} = \dfrac{1}{\left(-3x^{-4}y^{2}\right)^{4}}$
$= \dfrac{1}{81x^{-16}y^{8}}$
$= \dfrac{x^{16}}{81y^{8}}$

33. $889,610,000,000 = 8.8961\times10^{11}$

35. $0.00000001963 = 1.963\times10^{-8}$

37. $3.02\times10^{5} = 302,000$

39. $4.7\times10^{-4} = 0.00047$

41. $9.83\times10^{5} = 983,000$

43. $0.0000237 = 2.37\times10^{-5}$ miles per hour

45. $7\times10^{-6} = 0.000007$ m

47. $(56,000,000,000)(780,000,000)$
$= \left(5.6\times10^{10}\right)\left(7.8\times10^{8}\right)$
$= 43.68\times10^{18}$
$= 4.368\times10^{19}$

49. $\dfrac{(5,000,000)(16,000)}{8,000,000,000} = \dfrac{\left(5\times10^{6}\right)\left(1.6\times10^{4}\right)}{8\times10^{9}}$
$= \dfrac{8\times10^{10}}{8\times10^{9}}$
$= 1.0\times10^{1}$

51. $(0.003)^{4} = \left(3\times10^{-3}\right)^{4}$
$= 3^{4}\times10^{-3(4)}$
$= 81\times10^{-12}$
$= 8.1\times10^{-11}$

53. $(150,000,000)(0.00005)(0.002)(30,000)$
$= \left(1.5\times10^{8}\right)\left(5\times10^{-5}\right)\left(2\times10^{-3}\right)\left(3\times10^{4}\right)$
$= 45\times10^{4}$
$= 4.5\times10^{5}$

55. $\dfrac{6.816\times10^{12}}{2.92\times10^{8}} = 2.33\times10^{4}$ dollars

57. $d = (276)\left(3.09\times10^{13}\right)$
$= \left(2.76\times10^{2}\right)\left(3.09\times10^{13}\right) = 8.528\times10^{15}$
$r = 45,000 = 4.5\times10^{4}$
$d = rt$
$t = \dfrac{d}{r} = \dfrac{8.528\times10^{15}}{4.5\times10^{4}} = 1.90\times10^{11}$ hours

59. $d = rt$
$d = (0.00000275)(24)$
$= 2.75\times10^{-6}(24)$
$= 66\times10^{-6}$
$= 6.6\times10^{-5}$ miles

61. $\dfrac{3.2\times10^{9}}{20} = 0.16\times10^{9}$
$= \$1.6\times10^{8}$

63. $\text{Percent increase} = \frac{9.36\times10^{11} - 3.61\times10^{11}}{3.61\times10^{11}}$
$= 1.593$
$= 159.3\%$

Cumulative Review

65. $-2.7 - (-1.9) = -2.7 + 1.9 = -0.8$

67. $-\frac{3}{4} + \frac{5}{7} = \frac{-21}{28} + \frac{20}{28} = \frac{-21+20}{28} = -\frac{1}{28}$

69. x = Mario's salary
$x + 12,460$ = Gina's salary
$x - 8742$ = Alfonso's salary
$x + x + 12,460 + x - 8742 = 112,000$
$3x + 3718 = 112,000$
$3x = 108,282$
$x = 36,094$
$x + 12,460 = 48,554$
$x - 8742 = 27,352$
Mario earns $36,094
Gina earns $48,554
Alfonso earns $27,352

4.3 Exercises

1. A polynomial in x is the sum of a finite number of terms of the form ax^n, where a is any real number and n is a whole number. An example is $3x^2 - 5x - 9$.

3. The degree of a polynomial in x is the largest exponent of x in any of the terms of the polynomial.

5. $6x^3y$, degree 4: monomial

7. $20x^5 + 6x^3 - 7x$, degree 5: trinomial

9. $5xy^2 - 3x^2y^3$, degree 5: binomial

11. $(-3x + 15) + (8x - 43)$
$= (-3 + 8)x + 15 - 43$
$= 5x - 28$

13. $(6x^2 + 5x - 6) + (-8x^2 - 3x + 5)$
$= 6x^2 + 5x - 6 - 8x^2 - 3x + 5$
$= -2x^2 + 2x - 1$

15. $\left(\frac{1}{2}x^2 + \frac{1}{3}x - 4\right) + \left(\frac{1}{3}x^2 + \frac{1}{6}x - 5\right)$
$= \left(\frac{1}{2} + \frac{1}{3}\right)x^2 + \left(\frac{1}{3} + \frac{1}{6}\right)x - 4 - 5$
$= \frac{5}{6}x^2 + \frac{1}{2}x - 9$

17. $\left(3.4x^3 - 5.6x^2 - 7.1x + 3.4\right)$
$+\left(-1.7x^3 + 2.2x^2 - 6.1x - 8.8\right)$
$= (3.4 - 1.7)x^3 + (-5.6 + 2.2)x^2$
$+(-7.1 - 6.1)x + 3.4 - 8.8$
$= 1.7x^3 - 3.4x^2 - 13.2x - 5.4$

19. $(2x - 19) - (-3x + 5)$
$= (2x - 19) + (3x - 5)$
$= (2 + 3)x - 19 - 5$
$= 5x - 24$

21. $\left(\frac{2}{5}x^2 - \frac{1}{2}x + 5\right) - \left(\frac{1}{3}x^2 - \frac{3}{7}x - 6\right)$
$= \frac{2}{5}x^2 - \frac{1}{2}x + 5 - \frac{1}{3}x^2 + \frac{3}{7}x + 6$
$= \frac{1}{15}x^2 - \frac{1}{14}x + 11$

23. $\left(-3x^2 + 5x\right) - \left(2x^3 - 3x^2 + 10\right)$
$= \left(-3x^2 + 5x\right) + \left(-2x^3 + 3x^2 - 10\right)$
$= -2x^3 + (-3 + 3)x^2 + 5x - 10$
$= -2x^3 + 5x - 10$

25. $\left(0.5x^4 - 0.7x^2 + 8.3\right) - \left(5.2x^4 + 1.6x + 7.9\right)$
$= \left(0.5x^4 - 0.7x^2 + 8.3\right) + \left(-5.2x^4 - 1.6x - 7.9\right)$
$= (0.5 - 5.2)x^4 - 0.7x^2 - 1.6x + 8.3 - 7.9$
$= -4.7x^4 - 0.7x^2 - 1.6x + 0.4$

27. $(4x - 3) - (7x - 2) + (-5x + 8)$
$= 4x - 3 - 7x + 2 - 5x + 8$
$= -8x + 7$

29. $\left(5x^2y - 6xy^2 + 2\right) + \left(-8x^2y + 12xy^2 - 6\right)$
$= (5 - 8)x^2y + (-6 + 12)xy^2 + 2 - 6$
$= -3x^2y + 6xy^2 - 4$

31. $\left(3x^4 - 4x^2 - 18\right) - \left(2x^4 + 3x^3 + 6\right)$
$= \left(3x^4 - 4x^2 - 18\right) + \left(-2x^4 - 3x^3 - 6\right)$
$= (3 - 2)x^4 - 3x^3 - 4x^2 - 18 - 6$
$= x^4 - 3x^3 - 4x^2 - 24$

33. $0.04x + 5.2,\ x = 1975 - 1970 = 5$
$0.04(5) + 5.2 = 5.4$ mpg

35. $0.04x + 5.2 = 7.2$
$0.04x = 2$
$x = 50$
year $= 1970 + 50 = 2020$

37. $1.8x^2 + 22.2x + 325,\ x = 1990 - 1980 = 10$
$1.8(10)^2 \_ 22.2(10) + 325 = 727$
727,000 prisoners

39. $1.8x^2 + 22.2x + 325$

$2002 - 1980 = 22,\ 2006 - 1980 = 26$

$2002 : 1.8(22)^2 + 22.2(22) + 325$

$= 1{,}684{,}600$

$2006 : 1.8(26)^2 + 22.2(26) + 325$

$= 2{,}119{,}000$

Increase: $2{,}119{,}000 - 1{,}684{,}600$

$= 434{,}400$ prisoners

41. $x^2 + 12x + 2x^2 = 3x^2 + 12x$

Cumulative Review

43. $3y - 8x = 2$

$-8x = -3y + 2$

$x = \dfrac{-3y + 2}{-8}$

$x = \dfrac{3y - 2}{8}$

45. $A = \dfrac{1}{2}h(b + c)$

$2A = h(b + c)$

$\dfrac{2A}{h} = b + c$

$\dfrac{2A}{h} - c = b$ or $b = \dfrac{2A - hc}{h}$

47. x = amount spent in 1990

$x + 0.9x = 1.324$

$1.9x = 1.324$

$x = 0.6968$

\$696.8 billion was spent

How Am I Doing? Sections 4.1-4.3

1. $(-5xy)(3x^2y^5) = -15x^3y^6$

2. $-\dfrac{35xy^6}{25x^8y^3} = -\dfrac{7y^3}{5x^7}$

3. $\dfrac{60x^7y^0}{15x^2y^9} = \dfrac{4x^5}{y^9}$

4. $(-3x^5y)^4 = (-3)^4x^{20}y^4 = 81x^{20}y^4$

5. $(-4x^{-3}y^4)^{-2} = 4^{-2}x^6y^{-8}$

$= \dfrac{x^6}{4^2y^8}$

$= \dfrac{x^6}{16y^8}$

6. $\dfrac{3x^{-2}y^6}{9xy^{-9}} = \dfrac{y^{15}}{3x^3}$

7. $58{,}740 = 5.874 \times 10^4$

8. $0.00009362 = 9.362 \times 10^{-5}$

9. $(2x^2 + 0.5x - 2) + (0.3x^2 - 0.9x - 3.4)$

$= 2.3x^2 - 0.4x - 5.4$

10. $\left(5x^2+8x-14\right)-\left(-2x^2-9x+12\right)$
$=5x^2+8x-14+2x^2+9x-12$
$=7x^2+17x-26$

11. $\left(\frac{1}{2}x^3+\frac{1}{4}x^2-2x\right)-\left(\frac{1}{3}x^3-\frac{1}{8}x^2-5x\right)$
$=\frac{1}{2}x^3+\frac{1}{4}x^2-2x-\frac{1}{3}x^3+\frac{1}{8}x^2+5x$
$=\frac{1}{6}x^3+\frac{3}{8}x^2+3x$

12. $\left(\frac{1}{16}x^2+\frac{1}{8}\right)+\left(\frac{1}{4}x^2-\frac{3}{10}x-\frac{1}{2}\right)$
$=\frac{1}{16}x^2+\frac{1}{8}+\frac{1}{4}x^2-\frac{3}{10}x-\frac{1}{2}$
$=\frac{5}{16}x^2-\frac{3}{10}x-\frac{3}{8}$

4.4 Exercises

1. $-2x\left(6x^3-x\right)=-12x^4+2x^2$

3. $3x^2(7x-3)=21x^3-9x^2$

5. $2x^3\left(-2x^3+5x-1\right)$
$=-4x^6+10x^4-2x^3$

7. $\frac{1}{2}\left(2x+3x^2+5x^3\right)$
$=x+\frac{3}{2}x^2+\frac{5}{2}x^3$

9. $\left(5x^3-2x^2+6x\right)\left(-3xy^2\right)$
$=-15x^4y^2+6x^3y^2-18x^2y^2$

11. $\left(3x^3+x^2-8x\right)(3xy)$
$=9x^4y+3x^3y-24x^2y$

13. $\left(x^3-3x^2+5x-2\right)(3x)$
$=3x^4-9x^3+15x^2-6x$

15. $\left(x^2y^2-6xy+8\right)(-2xy)$
$=-2x^3y^3+12x^2y^2-16xy$

17. $\left(-7x^3+3x^2+2x-1\right)\left(4x^2y\right)$
$=-28x^5y+12x^4y+8x^3y-4x^2y$

19. $\left(3d^4-4d^2+6\right)\left(-2c^2d\right)$
$=-6c^2d^5+8c^2d^3-12c^2d$

21. $6x^3\left(2x^4-x^2+3x+9\right)$
$=12x^7-6x^5+18x^4+54x^3$

23. $-7x^4\left(5x^3-8x^2+3x\right)$
$=-35x^7+56x^6-21x^5$

25. $(x+10)(x+3)=x^2+13x+30$

27. $(x+6)(x+2)=x^2+8x+12$

29. $(x-8)(x+2)=x^2-6x-16$

31. $(x-5)(x-4)=x^2-9x+20$

33. $(5x-2)(-4x-3)=-20x^2-7x+6$

35. $(7yx-4)(x+2y)=7x^2+14xy-4x-8y$

37. $(5x+2)(3x-y)=15x^2-5xy+6x-2y$

39. $(4y+1)(5y-3)=20y^2-7y-3$

41. The signs are incorrect.
$(x-2)(-3)=-3x+6$

43. $(5x+2)(5x+2)=25x^2+20x+4$

45. $(4x-3y)(5x-2y)=20x^2-23xy+6y^2$

47. $(7x-2)^2=(7x-2)(7x-2)$
$=49x^2-28x+4$

49. $(4a+2b)^2=(4a+2b)(4a+2b)$
$=16a^2+16ab+4b^2$

51. $(0.2x+3)(4x-0.3)$
$=0.8x^2-0.06x+12x-0.9$
$=0.8x^2+11.94x-0.9$

53. $\left(\frac{1}{2}x+\frac{1}{3}\right)\left(\frac{1}{2}x-\frac{1}{4}\right)=\frac{1}{4}x^2-\frac{1}{8}x+\frac{1}{6}x-\frac{1}{12}$
$=\frac{1}{4}x^2+\frac{1}{24}x-\frac{1}{12}$

55. $(2a-3b)(x-5b)=2ax-10ab-3bx+15b^2$

57. $(7x+3y)(5x+9y)=35x^2+78xy+27y^2$

59. $(5y-3z)(4y-2z)=20y^2-22yz+6z^2$

61. $(4x-6)(7x+3)=28x^2+12x-42x-18$
$=28x^2-30x-18$

Cumulative Review

63. $3(w-7)-(4-w)=11w$
$3w-21-4+w=11w$
$4w-25=11w$
$-25=7w$
$-\frac{25}{7}=w$

65. x = number of tiger's eye marbles
$2x$ = number of cat's eye marbles
$0.11x+0.07(2x)=6.00$
$0.11x+0.14x=6.00$
$0.25x=6.00$
$x=24$
$2x=2(24)=48$
24 tiger's eye marbles
48 cat's eye marbles

67. $1.2x+16,\ x=1994-1992=2$
$1.2(2)+16=2.4+16$
$=18.4$
They spent 18.4 hours.

69. $1.2x+16,\ x=2008-1992=16$
$1.2(16)+16=19.2+16$
$=35.2$
They will spent 32.5 hours.

4.5 Exercises

1. In the special case of $(a+b)(a-b)$, a binomial times a binomial is a <u>binomial</u>.

3. $(4x-7)^2=16x^2-56x+49$
The student left out the middle term which comes from the product of the two outer terms and the product of the two inner terms.

5. $(y-7)(y+7)=y^2-7^2=y^2-49$

7. $(x-9)(x+9)=x^2-(9)^2=x^2-81$

9. $(7x-4)(7x+4)=(7x)^2-(4)^2=49x^2-16$

11. $(2x-7)(2x+7)=(2x)^2-(7)^2=4x^2-49$

13. $(5x-3y)(5x+3y)=(5x)^2-(3y)^2$
$=25x^2-9y^2$

15. $(0.6x+3)(0.6x-3)=(0.6x)^2-3^2$
$=0.36x^2-9$

17. $(3y+1)^2=(3y)^2+2(3y)(1)+1^2$
$=9y^2+6y+1$

19. $(5x-4)^2=(5x)^2-2(5x)(4)+4^2$
$=25x^2-40x+16$

21. $(7x+3)^2=(7x)^2+2(7x)(3)+3^2$
$=49x^2+42x+9$

23. $(3x-7)^2=(3x)^2-2(3x)(7)+7^2$
$=9x^2-42x+49$

25. $\left(\frac{2}{3}x+\frac{1}{4}\right)^2$
$=\left(\frac{2}{3}x\right)^2+2\left(\frac{2}{3}x\right)\left(\frac{1}{4}\right)+\left(\frac{1}{4}\right)^2$
$=\frac{4}{9}x^2+\frac{1}{3}x+\frac{1}{16}$

27. $(9xy+4z)^2=(9xy)^2+2(9xy)(4z)+(4z)^2$
$=81x^2y^{22}+72xyz+16z^2$

29. $(7x+3y)(7x-3y)=(7x)^2-(3y)^2$
$=49x^2-9y^2$

31. $(7c-6d)^2=(7c)^2-2(7c)(6d)$
$=49c^2-84cd+36d^2$

33. $(9a-10b)(9a+10b)=(9a)^2-(10b)^2$
$=81a^2-100b^2$

35. $(5x+9y)^2=(5x)^2+2(5x)(9y)+(9y)^2$
$=25x^2+90xy+81y^2$

37. $(x^2+5x-3)(x-2)$
$=x^3+5x^2-3x-2x^2-10x+6$
$=x^3+3x^2-13x+6$

39. $(4x+1)(x^3-2x^2+x-1)$
$=4x(x^3-2x^2+x-1)+1(x^3-2x^2+x-1)$
$=4x^4-8x^3+4x^2-4x+x^3-2x^2+x-1$
$=4x^4-7x^3+2x^2-3x-1$

41. $(a^2-3a+2)(a^2+4a-3)$
$=a^2(a^2+4a-3)-3a(a^2+4a-3)$
$+2(a^2+4a-3)$
$=a^4+4a^3-3a^2-3a^3-12a^2+9a+2a^2$
$+8a-6$
$=a^4+a^3-13a^2+17a-6$

43. $(x+3)(x-1)(3x-8)$
$=(x^2+2x-3)(3x-8)$
$=3x^3+6x^2-9x-8x^2-16x+24$
$=3x^3-2x^2-25x+24$

45. $(3x+5)(x-2)(x-4)$
$=(3x+5)(x^2-6x+8)$
$=3x(x^2-6x+8)+5(x^2-6x+8)$
$=3x^3-18x^2+24x+5x^2-30x+40$
$=3x^3-13x^2-6x+40$

47. $(a-5)(2a+3)(a+5)$
$=(2a^2-7a-15)(a+5)$
$=2a^3-7a^2-15a+10a^2-35a-75$
$=2a^3+3a^2-50a-75$

49. $V=(2x+1)(3x-2)(4x+3)$
$=(6x^2-x-2)(4x+3)$
$=24x^3-4x^2-8x+18x^2-3x-6$
$=24x^3+14x^2-11x-6$

Cumulative Review

51. Let x = amount invested at 7%
$18,000-x$ = amount invested at 11%
$0.07x+0.11(18,000-x)=1540$
$7x+11(18,000-x)=154,000$
$7x+198,000-11x=154,000$
$-4x=-44,000$
$x=11,000$
$18,000-x=7000$
She invested \$11,000 at 7% and \$7000 at 11%.

53. $F = 1.8C + 32$
$F = 1.8(-33.4) + 32$
$F = -60.12 + 32$
$F = -28.12^0 \text{F}$

4.6 Exercises

1. $\dfrac{25x^4 - 15x^2 + 20x}{5x} = \dfrac{25x^4}{5x} - \dfrac{15x^2}{5x} + \dfrac{20x}{5x}$
$= 5x^3 - 3x + 4$

3. $\dfrac{8y^4 - 12y^3 - 4y^2}{4y^2} = \dfrac{8y^4}{4y^2} - \dfrac{12y^3}{4y^2} - \dfrac{4y^2}{4y^2}$
$= 2y^2 - 3y - 1$

5. $\dfrac{81x^7 - 236x^5 - 63x^3}{9x^3} = \dfrac{81x^7}{9x^3} - \dfrac{36x^5}{9x^3} - \dfrac{63x^3}{9x^3}$
$= 79x^4 - 4x^2 - 7$

7. $(48x^7 - 54x^4 + 36x^3) \div 6x^3$
$= \dfrac{48x^7}{6x^3} - \dfrac{54x^4}{6x^3} + \dfrac{36x^3}{6x^3}$
$= 8x^4 - 9x + 6$

9. $$\begin{array}{r@{}l} & 3x + 5 \\ 2x+1 \big) & 6x^2 + 13x + 5 \\ & \underline{6x^2 + \ \ 3x} \\ & \qquad 10x + 5 \\ & \qquad \underline{10x + 5} \\ & \qquad\qquad 0 \end{array}$$

$$\frac{6x^2 + 13x + 5}{2x + 1} = 3x + 5$$

Check: $(3x+5)(2x+1)$
$= 6x^2 + 3x + 10x + 5$
$= 6x^2 + 13x + 5$

11. $$\begin{array}{r@{}l} & x - 3 \\ x-5 \big) & x^2 - 8x - 17 \\ & \underline{x^2 - 5x} \\ & \quad -3x - 17 \\ & \quad \underline{-3x + 15} \\ & \qquad\quad -32 \end{array}$$

$$\frac{x^2 - 8x - 17}{x - 5} = x - 3 - \frac{32}{x - 5}$$

Check: $(x-5)\left(x - 3 - \dfrac{32}{x-5}\right)$
$= x^2 - 3x - 5x + 15 - 32$
$= x^2 - 8x - 17$

13. $$\begin{array}{r@{}l} & 3x^2 - 4x + 8 \\ x+1 \big) & 3x^3 - \ x^2 + 4x - 2 \\ & \underline{3x^3 + 3x^2} \\ & \quad -4x^2 + 4x - 2 \\ & \quad \underline{-4x^2 - 4x} \\ & \qquad\qquad 8x - 2 \\ & \qquad\qquad \underline{8x + 8} \\ & \qquad\qquad\quad -10 \end{array}$$

$$\frac{3x^3 - x^2 + 4x - 2}{x + 1} = 3x^2 - 4x + 8 - \frac{10}{x + 1}$$

Check: $(x+1)\left(3x^2-4x+8-\dfrac{10}{x+1}\right)$

$= 3x^3-x^2+4x+8-10$

$= 3x^3-x^2+4x-2$

15.
$$\begin{array}{r} 2x^2-3x-2 \\ 2x+5\overline{\big)4x^3+4x^2-19x-15} \\ \underline{4x^3+10x^2} \\ -6x^2-19x \\ \underline{-6x^2-15x} \\ -4x-15 \\ \underline{-4x-10} \\ -5 \end{array}$$

$$\frac{4x^3+4x^2-19x-15}{2x+5}=2x^2-3x-2-\frac{5}{2x+5}$$

Check:

$(2x+5)\left(2x^2-3x-2-\dfrac{5}{2x+5}\right)$

$= 4x^3+4x^2-19x-10-5$

$= 4x^3+4x^2-19x-15$

17.
$$\begin{array}{r} 2x^2+3x-1 \\ 5x-2\overline{\big)10x^3+11x^2-11x+2} \\ \underline{10x^3-4x^2} \\ 15x^2-11x \\ \underline{15x^2-6x} \\ -5x+2 \\ \underline{-5x+2} \\ 0 \end{array}$$

$$\frac{10x^3+11x^2-11x+2}{5x-2}=2x^2+3x-1$$

19.
$$\begin{array}{r} 6y^2+3y-8 \\ 2y-3\overline{\big)12y^3-12y^2-25y+31} \\ \underline{12y^3-18y^2} \\ 6y^2-25y \\ \underline{6y^2-9y} \\ -16y+31 \\ \underline{-16y+24} \\ 7 \end{array}$$

$$\frac{12y^3-12y^2-25y+31}{2y-3}$$
$$=6y^2+3y-8+\frac{7}{2y-3}$$

21.
$$\begin{array}{r} y^2-4y-1 \\ y+3\overline{\big)y^3-y^2-13y-12} \\ \underline{y^3+3y^2} \\ -4y^2-13y-12 \\ \underline{-4y^2-12y} \\ -y-12 \\ \underline{-y-3} \\ -9 \end{array}$$

$$\left(y^3-y^2-13y-12\right)\div(y+3)$$
$$=y^2-4y-1-\frac{9}{y+3}$$

23.
$$
\begin{array}{r}
y^3 + 2y^2 - 5y - 10 \\
y-2 \overline{) y^4 + 0y^3 - 9y^2 + 0y - 5} \\
\underline{y^4 - 2y^3 \qquad\qquad\qquad} \\
2y^3 - 9y^2 + 0y - 5 \\
\underline{2y^3 - 4y^2 \qquad\qquad} \\
-5y^2 + 0y - 5 \\
\underline{-5y^2 + 10y \qquad} \\
-10y - 5 \\
\underline{-10y + 20} \\
-25
\end{array}
$$

$$(y^4 - 9y^2 - 5) \div (y - 2)$$
$$= y^3 + 2y^2 - 5y - 10 - \frac{25}{y-2}$$

25.
$$
\begin{array}{r}
2y^2 + \frac{1}{2}y + \frac{7}{8} \\
4y-1 \overline{) 8y^3 + 0y^2 + 3y - 7} \\
\underline{8y^3 - 2y^2 \qquad\qquad\quad} \\
2y^2 + 3y \qquad \\
\underline{2y^2 - \frac{1}{2}y \qquad} \\
\frac{7}{2}y - 7 \\
\underline{\frac{7}{2}y - \frac{7}{8}} \\
-\frac{49}{8}
\end{array}
$$

$$\frac{8y^3 + 3y - 7}{4y - 1} = 2y^2 + \frac{1}{2}y + \frac{7}{8} - \frac{49/8}{4y-1}$$

Cumulative Review

27. $110,000 - 0.3(110,000) = 77,000$

77,000 gallons

29. x = number of cats in the study

$0.92x = 184$

$x = 200$

200 cats were in the study

31. a. $\frac{4+1+9+6+1}{5} = \frac{21}{5}$
$= 4.2$

4.2 hurricanes per year

b. $\frac{9+7+8+7+6}{5} = \frac{37}{5} = 7.4$

7.4 hurricanes per year

c. Percent increase $= \frac{7.4 - 4.2}{4.2}$
$= 0.762$
$= 76.2\%$

d. $100\% + 76.2\% = 176.2\%$

176.2% of $7.4 = 13.0$

13.0 hurricanes

Putting Your Skills to Work

1. $45,678 - 32,152 = \$13,526$

2. $42(13,526) = \$568,082$

3. $\text{Percentage growth} = \dfrac{100,987 - 74,560}{74,560}$
$= \dfrac{26,427}{74,560}$
$= 0.3544$
$= 35.4\%$

4. $100\% + 35.4\% = 135.4\%$
135.4% of 100,987
$1.354(100,987) = \$136,777$ in 2006

5. $I = 35x^3 - 620x^2 + 2820x + 43,112$
$x = 2004 - 1998 = 6$
$I = 35(6)^3 - 620(6)^2 + 2820(6) + 43,112$
$I = \$45,272$ in 2004

6. $x = 2006 - 1998 = 8$
$I = 35(8)^3 - 620(8)^2 + 2820(8) + 43,112$
$I = \$43,912$ in 2006

Chapter 4 Review Problems

1. $(-6a^2)(3a^5) = -18a^7$

2. $(5^{10})(5^{13}) = 5^{10+13} = 5^{23}$

3. $(3xy^2)(2x^3y^4) = 6x^4y^6$

4. $(2x^3y^4)(-7xy^5) = -14x^4y^9$

5. $\dfrac{7^{15}}{7^{27}} = \dfrac{1}{7^{27-15}} = \dfrac{1}{7^{12}}$

6. $\dfrac{x^{12}}{x^{17}} = \dfrac{1}{x^{17-12}} = \dfrac{1}{x^5}$

7. $\dfrac{y^{30}}{y^{16}} = y^{30-16} = y^{14}$

8. $\dfrac{9^{13}}{9^{24}} = \dfrac{1}{9^{11}}$

9. $\dfrac{-15xy^2}{25x^6y^6} = -\dfrac{3}{5x^{6-1}y^{6-2}} = -\dfrac{3}{5x^5y^4}$

10. $\dfrac{-12a^3b^6}{18a^2b^{12}} = -\dfrac{2a}{3b^6}$

11. $(x^3)^8 = x^{3(8)} = x^{24}$

12. $(b^5) = b^{30}$

13. $(-3a^3b^2)^2 = (-3)^2 a^{3(2)}b^{2(2)} = 9a^6b^4$

14. $(3x^3y)^4 = 3^4(x^3)^4 y^4$
$= 81x^{12}y^4$

15. $\left(\frac{5ab^2}{c^3}\right)^2 = \frac{5^2a^2\left(b^2\right)^2}{\left(c^3\right)^2} = \frac{25a^2b^4}{c^6}$

16. $\left(\frac{x^0y^3}{4w^5z^2}\right)^3 = \frac{\left(y^3\right)^3}{4^3\left(w^5\right)^3\left(z^2\right)^3} = \frac{y^9}{64w^{15}z^6}$

17. $a^{-3}b^5 = \frac{b^5}{a^3}$

18. $m^8p^{-5} = \frac{m^8}{p^5}$

19. $\frac{2x^{-6}}{y^{-3}} = \frac{2y^3}{x^6}$

20. $\left(3x^{-4}y^3\right)^{-2} = 3^{-2}\left(x^{-4}\right)^{-2}\left(y^3\right)^{-2}$
$= \frac{x^8y^{-6}}{3^2}$
$= \frac{x^8}{9y^6}$

21. $\left(5x^2y^{-4}\right)^{-2} = 5^{-2}\left(x^2\right)^{-2}\left(y^{-4}\right)^{-2}$
$= \frac{x^{-4}y^8}{5^2}$
$= \frac{y^8}{25x^4}$

22. $\frac{3x^{-3}}{y^{-2}} = \frac{3y^2}{x^3}$

23. $\frac{4x^{-5}y^{-6}}{w^{-2}z^8} = \frac{4w^2}{x^5y^6z^8}$

24. $\frac{3^{-3}a^{-2}b^5}{c^{-3}d^{-4}} = \frac{b^5c^3d^4}{3^3a^2} = \frac{b^5c^3d^4}{27a^2}$

25. $156,340,200,000 = 1.563402\times10^{11}$

26. $179,632 = 1.79632\times10^5$

27. $0.0078 = 7.8\times10^{-3}$

28. $0.00006173 = 6.173\times10^{-5}$

29. $1.2\times10^5 = 120,000$

30. $6.034\times10^6 = 6,034,000$

31. $3\times10^6 = 3,000,000$

32. $2.5\times10^{-1} = 0.25$

33. $4.32\times10^{-5} = 0.0000432$

34. $6\times10^{-9} = 000000006\times10^{-9}$
$= 0.000000006$

35. $\dfrac{(28,000,000)(5,000,000,000)}{7,000}$

$= \dfrac{(2.8\times10^7)(5\times10^9)}{7\times10^3}$

$= \dfrac{14\times10^{16}}{7\times10^3} = 2\times10^{13}$

36. $(3.12\times10^5)(2.0\times10^6)(1.5\times10^8)$

$= 9.36\times10^{5+6+8} = 9.36\times10^{19}$

37. $(1.6\times10^{-3})(3.0\times10^{-5})(2.0\times10^{-2})$

$= 9.6\times10^{-10}$

38. $\dfrac{(0.00078)(0.000005)(0.00004)}{0.002}$

$= \dfrac{(7.8\times10^{-4})(5.0\times10^{-6})(4.0\times10^{-5})}{2.0\times10^{-3}}$

$= \dfrac{156\times10^{-15}}{2.0\times10^{-3}} = 78\times10^{-12} = 7.8\times10^{-11}$

39. $r = 40,000 = 4\times10^4$

$t = 365(24) = 8760 = 8.76\times10^3$

$d = rt = (4\times10^4)(8.76\times10^3)$

$= 35.04\times10^7$

$= 3.504\times10^8$ km

40. Seconds in one day $= 60\times60\times24 = 86,400$

$= 8.64\times10^4$

Cycles of radiation in one second

$= 9,192,631,770 = 9.19\times10^9$

Cycles of radiation in one day

$= (8.64\times10^4)(9.19\times10^9)$

$= 79.4016\times10^{13} = 7.94\times10^{14}$

41. $\dfrac{60}{1\times10^{-8}} = 60\times10^8 = 6\times10^9$

In one minute the computer can perform 6×10^9 operations.

42. $(4.3x^2 - 2.6x - 3.1) + (7.1x^2 - 4.7x + 8.4)$

$= 11.4x^2 - 7.3x + 5.3$

43. $(1.2x^2 - 3.4x + 6) + (5.5x^2 - 7.6x - 3)$

$= (1.2 + 5.5)x^2 + (-3.4 - 7.6)x + 6 - 3$

$= 6.7x^2 - 11x + 3$

44. $(x^3 + x^2 - 6x + 2) - (2x^3 - x^2 - 5x - 6)$

$= (x^3 + x^2 - 6x + 2) + (-2x^3x^2 + 5x + 6)$

$= (x^3 - 2x^3) + (x^2 + x^2)$

$+ (-6x + 5x) + (2 + 6)$

$= -x^3 + 2x^2 - x + 8$

45. $(4x^3 - x^2 - x + 3) - (-3x^3 + 2x^2 + 5x - 1)$

$= (4x^3 - x^2 - x + 3) + (3x^3 - 2x^2 - 5x + 1)$

$= (4 + 3)x^3 + (-1 - 2)x^2 + (-1 - 5)x + 3 + 1$

$= 7x^3 - 3x^2 - 6x + 4$

46. $\left(\frac{3}{5}x^2y-\frac{1}{3}x+\frac{3}{4}\right)-\left(\frac{1}{2}x^2y+\frac{2}{7}x+\frac{1}{3}\right)$
$=\frac{3}{5}x^2y-\frac{1}{3}x+\frac{3}{4}-\frac{1}{2}x^2y-\frac{2}{7}x-\frac{1}{3}$
$=\frac{1}{10}x^2y-\frac{13}{21}x+\frac{5}{12}$

47. $\frac{1}{2}x^2-\frac{3}{4}x+\frac{1}{5}-\left(\frac{1}{4}x^2-\frac{1}{2}x+\frac{1}{10}\right)$
$=\frac{1}{2}x^2-\frac{3}{4}x+\frac{1}{5}-\frac{1}{4}x^2+\frac{1}{2}x-\frac{1}{10}$
$=\left(\frac{1}{2}-\frac{1}{4}\right)x^2+\left(-\frac{3}{4}+\frac{1}{2}\right)x+\frac{1}{5}-\frac{1}{10}$
$=\frac{1}{4}x^2-\frac{1}{4}x+\frac{1}{10}$

48. $(5x^2+3x)+(-6x^2+2)-(5x-8)$
$=(5x^2+3x)+(-6x^2+2)+(-5x+8)$
$=(5x^2-6x^2)+(3x-5x)+(2+8)$
$=-x^2-2x+10$

49. $(2x^2-7)-(3x^2-4)+(-5x^2-6x)$
$=(2x^2-7)+(-3x^2+4)+(-5x^2-6x)$
$=(2-3-5)x^2-6x-7+4$
$=-6x^2-6x-3$

50. $(3x+1)(5x-1)=15x^2-3x+5x-1$
$=15x^2+2x-1$

51. $(7x-2)(4x-3)=28x^2-21x-8x+6$
$=28x^2-29x+16$

52. $(2x+3)(10x+9)$
$=20x^2+18x+30x+27$
$=20x^2+48x+27$

53. $5x(2x^2-6x+3)=10x^3-30x^2+15x$

54. $(6x^3-2x^2+x-4)(-3x)$
$=-18x^4+6x^3-3x^2+12x$

55. $(2xy^2-3xy-4y)(-3x^2y)$
$=-6x^3y^3+9x^3y^2+12x^2y^2$

56. $(5a+7b)(a-3b)=5a^2-15ab+7ab-21b^2$
$=5a^2-8ab-21b^2$

57. $(2x^2-3)(4x^2-5y)$
$=8x^4-10x^2y-12x^2+15y$

58. $-3x^2y(5x^4y+3x^2-2)$
$=-3x^2y(5x^4y)-3x^2y(3x^2)-3x^2y(-2)$
$=-15x^6y^2-9x^4y+6x^2y$

59. $(3x-2)^2$
$=9x^2+2(-2)(3)x+4$
$=9x^2-12x+4$

60. $(5x+3)(5x-3)=(5x)^2-3^2=25x^2-9$

61. $(7x+6y)(7x-6y)$
$=(7x)^2-(6y)^2$
$=49x^2-36y^2$

62. $(5a-2b)^2=(5a)^2-2(5a)(2b)+(2b)^2$
$=25a^2-20ab+4b^2$

63. $(8x+9y)^2=64x^2+2(8x)(9y)+81y^2$
$=64x^2+144xy+81y^2$

64. $(x^2+7x+3)(4x-1)$
$=4x^3+28x^2+12x-x^2-7x-3$
$=4x^3+27x^2+5x-3$

65. $(x-6)(2x-3)(x+4)$
$=(2x^2-15x+18)(x+4)$
$=2x^3-15x^2+18x+8x^2-60x+72$
$=2x^3-7x^2-42x+72$

66. $(12y^3+18x^2+24y)\div(6y)$
$=\frac{12y^3}{6y}+\frac{18y^2}{6y}+\frac{24y}{6y}$
$=2y^2+3y+4$

67. $(30x^5+35x^4-90x^3)\div(5x^2)$
$=\frac{30x^5}{5x^2}+\frac{35x^4}{5x^2}-\frac{90x^3}{5x^2}$
$=6x^3+7x^2-18x$

68. $(16x^3y^2-24x^2y+32xy^2)\div(4xy)$
$=\frac{16x^3y^2}{4xy}-\frac{24x^2y}{4xy}+\frac{32xy^2}{4xy}$
$=4x^2y-6x+8y$

69. $(106x^6-24x^5+38x^4+26x^3)\div(2x^3)$
$=\frac{106x^6}{2x^3}-\frac{24x^5}{2x^3}+\frac{38x^4}{2x^3}+\frac{26x^3}{2x^3}$
$=53x^3-12x^2+19x+13$

70.
$$\begin{array}{r|l} & 3x-2 \\ \hline 5x+7 & 15x^2+11x-14 \\ & \underline{15x^2+21x} \\ & \quad -10x-14 \\ & \quad \underline{-10x-14} \\ & \qquad 0 \end{array}$$

$$\frac{15x^2+11x-14}{5x+7}=3x-2$$

71.
$$\begin{array}{r|l} & 3x-7 \\ \hline 4x+9 & 12x^2-x-62 \\ & \underline{12x^2+27x} \\ & \quad -28x-63 \\ & \quad \underline{-28x-63} \\ & \qquad 0 \end{array}$$

$$\frac{12x^2-x-63}{4x+9}=3x-7$$

72.
$$\begin{array}{r} 3x^2+2x+4 \\ 2x-1\overline{)6x^3+x^2+6x+5} \\ \underline{6x^3-3x^2} \\ 4x^2+6x \\ \underline{4x^2-2x} \\ 8x+5 \\ \underline{8x-4} \\ 9 \end{array}$$

$$(6x^3+x^2+6x+5)\div(2x-1)$$
$$=3x^2+2x+4+\frac{9}{2x-1}$$

73.
$$\begin{array}{r} 2x^2-5x+13-\frac{27}{x+2} \\ x+2\overline{)2x^3-x^2+3x-1} \\ \underline{2x^3+4x^2} \\ -5x^2+3x \\ \underline{-5x^2-10x} \\ 13x-1 \\ \underline{13x+26} \\ -27 \end{array}$$

74.
$$\begin{array}{r} 4x+1 \\ 3x+2\overline{)12x^2+11x+2} \\ \underline{12x^2+8x} \\ 3x+2 \\ \underline{3x+2} \end{array}$$

$$(12x^2+11x+2)\div(3x+2)=4x+1$$

75.
$$\begin{array}{r} 4x-5+\frac{11}{2x+1} \\ 2x+1\overline{)8x^2-6x+6} \\ \underline{8x^2+4x} \\ -10x+6 \\ \underline{-10x-5} \\ 11 \end{array}$$

76.
$$\begin{array}{r} x^2+3x+8 \\ x-3\overline{)x^3+0x^2-x-24} \\ \underline{x^3-3x^2} \\ 3x^2-x \\ \underline{3x^2-9x} \\ 8x-24 \\ \underline{8x-24} \end{array}$$

77.
$$\begin{array}{r} 2x^2+4x+5 \\ x-2\overline{)2x^3+0x^2-3x+1} \\ \underline{2x^3-4x^2} \\ 4x^2-3x \\ \underline{4x^2-8x} \\ 5x+1 \\ \underline{5x-10} \\ 11 \end{array}$$

$$(2x^3-3x+1)\div(x-2)$$
$$=2x^2+4x+5+\frac{11}{x-2}$$

78. $(4^6)(4^{23})=4^{29}$

79. $\frac{6^{50}}{6^{35}} = 6^{15}$

80. $(-5x^3y^4)(-25x^5y^{10}) = 125x^8y^{14}$

81. $\frac{-36x^6y^5}{18xy^8} = -\frac{2x^5}{y^3}$

82. $(-4x^5y^6)^3 = (-4)^3(x^5)^3(y^6)^3$
$= -64x^{15}y^{18}$

83. $0.0000786 = 7.86 \times 10^{-5}$

84. $(-6x^3 - 7x + 8) - (-2x^3 + 5x - 9)$
$= -6x^3 - 7x + 8 + 2x^3 - 5x + 9$
$= -4x^3 - 12x + 17$

85. $-6x^2(2x^3 + 5x^2 - 2)$
$= -12x^5 - 30x^4 + 12x^2$

86. $(7x-3)(2x-1)(x+2)$
$= (14x^2 - 13x + 3)(x+2)$
$= 14x^3 - 13x^2 + 3x + 28x^2 - 26x + 6$

87. $$\begin{array}{r} 2x^2 + 5x + 3 \\ 3x-2\,\overline{)\,6x^3 + 11x^2 - x - 2} \\ \underline{6x^3 - 4x^2} \\ 15x^2 - x \\ \underline{15x^2 - 10x} \\ 9x - 2 \\ \underline{9x - 6} \\ 4 \end{array}$$

$\frac{6x^3 + 11x^2 - x - 2}{3x-2} = 2x^2 + 5x + 3 + \frac{4}{3x-2}$

88. $\frac{\$12.9 \times 10^9}{2.8 \times 10^8 \text{ people}} = \$4.607 \times 10^1/\text{person}$
$= \$46.07/\text{person}$

89. $1.132 \times 10^9 + 1.31 \times 10^8$
$= 1.132 \times 10^9 + 0.131 \times 10^9$
$= 1.263 \times 10^9$ people

90. $(9.11 \times 10^{-28})(30,000)$
$= (9.11 \times 10^{-28})(3 \times 10^4)$
$= 27.33 \times 10^{-24}$
$= 2.733 \times 10^{-23}$ grams

91. $(3.9\times10^{26})(60)(60)(24)$
$= 3.9\times10^{26}(8.64\times10^{4})$
$= 33.696\times10^{30}$
$= 3.3696\times10^{31}$ joules

92. $A = 2x(2y+1) - xy$
$= 4xy + 2x - xy$
$= 3xy + 2x$

93. $A = 2x(x) - 4y(y) = 2x^2 - 4y^2$

How Am I Doing? Chapter 4 Test

1. $(3^{10})(3^{24}) = 3^{10+24} = 3^{34}$

2. $\dfrac{25^{18}}{25^{34}} = \dfrac{1}{25^{16}}$

3. $(8^4)^6 = 8^{4\cdot 6} = 8^{24}$

4. $(-3xy^4)(-4x^3y^6) = 12x^{1+3}y^{4+6} = 12x^4y^{10}$

5. $\dfrac{-35x^8y^{10}}{25x^5y^{10}} = -\dfrac{7x^3}{5}$

6. $(-5xy^6)^3 = (-5)^3x^3(y^6)^3 = -125x^3y^{18}$

7. $\left(\dfrac{7a^7b^2}{3c^0}\right)^2 = \dfrac{7^2a^{7(2)}b^{2(2)}}{3^2c^{0(2)}} = \dfrac{49a^{14}b^4}{9}$

8. $\dfrac{(3x^2)^3}{(6x)^2} = \dfrac{3^3(x^2)^3}{6^2x^2} = \dfrac{27x^6}{36x^2} = \dfrac{3x^4}{4}$

9. $4^{-3} = \dfrac{1}{4^3} = \dfrac{1}{64}$

10. $8(x+d) = 3(y-d)$
$8x + 8d = 3y - 3d$
$8x + 11d = 3y$
$11d = 3y - 8x$
$d = \dfrac{3y-8x}{11}$

11. $\dfrac{3x^{-3}y^2}{x^{-4}y^{-5}} = 3xy^7$

12. $0.0005482 = 5.482\times10^{-4}$

13. $(5x+6)(4x-3) = 20x^2 + 9x - 18$

14. $(4.0\times10^{-3})(3.0\times10^{-8})(2.0\times10^{4})$
$= 24.0\times10^{-7} = 2.4\times10^{-6}$

15. $(2x^2 - 3x - 6) + (-4x^2 + 8x + 6)$
$= (2-4)x^2 + (-3+8)x + 6 - 6$
$= -2x^2 + 5x$

16. $\left(3x^3-4x^2+3\right)-\left(14x^3+-7x+11\right)$
$=3x^2-4x^2+3-14x^3-7x-11$
$=-11x^3-4x^2+7x-8$

17. $-7x^2\left(3x^3-4x^2+6x-2\right)$
$=-21x^5+28x^4-42x^3+14x^2$

18. $\left(5x^2y^2-6xy+2\right)\left(3x^2y\right)$
$=5x^2y^2\left(3x^2y\right)-6xy\left(3x^2y\right)+2\left(3x^2y\right)$
$=15x^4y^3-18x^3y^2+6x^2y$

19. $(5a-4b)(2a+3b)$
$=10a^2+15ab-8ab-12b^2$
$=10a^2+7ab-12b^2$

20. $(3x+2)(2x+1)(x-3)$
$=\left(6x^2+7x+2\right)(x-3)$
$=6x^3+7x^2+2x-18x^2-21x-6$
$=6x^3-11x^2-19x-6$

21. $\left(7x^2+2y^2\right)^2$
$=49x^4+2\left(7x^2\right)\left(2y^2\right)+4y^4$
$=49x^4+28x^2y^2+4y^4$

22. $(5s-11t)(5s+11t)=(5s)^2-(11t)^2$
$=25s^2-121t^2$

23. $(3x-2)\left(4x^3-2x^2+7x-5\right)$
$=3x\left(4x^3-2x^2+7x-5\right)$
$-2\left(4x^3-2x^2+7x-5\right)$
$=12x^4-6x^3+21x^2-15x-8x^3+4x^2$
$-14x+10$
$=12x^4-14x^3+25x^2-29x+10$

24. $\left(3x^2-5xy\right)\left(x^2+3xy\right)$
$=3x^4+9x^3y-5x^3y-15x^2y^2$
$=3x^4+4x^3y-15x^2y^2$

25. $\dfrac{15x^6-5x^4+25x^3}{5x^3}=\dfrac{15x^6}{5x^3}-\dfrac{5x^4}{5x^3}+\dfrac{25x^3}{5x^3}$
$=3x^3-x+5$

26.
$$\begin{array}{r} 2x^2-7x+4 \\ 4x+3\overline{)8x^3-22x^2-5x+12} \\ \underline{8x^3+\ 6x^2}\qquad\qquad \\ -28x^2-\ 5x\qquad \\ \underline{-28x^2-21x}\qquad \\ 16x+12 \\ \underline{16x+12} \end{array}$$

$\dfrac{8x^3-22x^2-5x+12}{4x+3}=2x^2-7x+4$

27.
$$
\begin{array}{r}
2x^2+6x+12 \\
x-3\overline{)2x^3+0x^2-6x-36} \\
\underline{2x^3-6x^2} \qquad\qquad \\
6x^2-6x \qquad \\
\underline{6x^2-18x} \qquad \\
12x-36 \\
\underline{12x-36}
\end{array}
$$

$$\frac{2x^3-6x-36}{x-3}=2x^2+6x+12$$

28. $\dfrac{2.618\times10^{11}}{86}=0.03044\times10^{11}$

$=3.044\times10^9$ barrels per year

29. $d=rt$

$=2.49\times10^4(7)(24)$

$=418\times10^4$

$=4.18\times10^6$ miles

Cumulative Test for Chapters 0-4

1. $\dfrac{5}{12}-\dfrac{7}{8}=\dfrac{10}{24}-\dfrac{21}{24}=-\dfrac{11}{24}$

2. $(-3.7)\times(0.2)=-0.74$

3. $\left(-4\dfrac{1}{2}\right)\div\left(5\dfrac{1}{4}\right)=\left(-\dfrac{9}{2}\right)\div\left(\dfrac{21}{4}\right)$

$=\left(-\dfrac{3\cdot3}{2}\right)\left(\dfrac{2\cdot2}{3\cdot7}\right)$

$=-\dfrac{6}{7}$

4. 6% of $1842.5=0.06(1842.5)$

$=110.55$

5. $7x(3x-4)-5x(2x-3)-(3x)^2$

$=21x^2-28x-10x^2+15x-9x^2$

$=2x^2-13x$

6. $2x^2-3xy+y^2=2(-2)^2-3(-2)(3)+3^2$

$=2(4)+18+9$

$=8+18+9=35$

7. $7x-3(4-2x)=14x-(3-x)$

$7x-12+6x=14x-3+x$

$13x-12=15x-3$

$-9=2x$

$-\dfrac{9}{2}=x$

8. $\dfrac{2}{3}x+6=4(x-11)$

$3\left(\dfrac{2}{3}x\right)+3(6)=3(4)(x-11)$

$2x+18=12x-132$

$18=10x-132$

$150=10x$

$15=x$

9. $4 - 7x < 11$
   $-7x < 7$
   $x > -1$

10. $8(x+d) = 3(y-d)$
    $8x + 8d = 3y - 3d$
    $8x + 11d = 3y$
    $11d = 3y - 8x$
    $d = \dfrac{3y - 8x}{11}$

11. $\dfrac{11{,}904}{x} = 0.96$
    $11{,}904 = 0.96x$
    $12{,}400 = x$
    VBM has 12,400 employees.

12. $(0.06)(3320) = 199.20$
    \$199.20

13. $(5x+6)(4x-3) = 20x^2 + 9x - 18$

14. $(3x-5)^2 = (3x)^2 - 2(3x)(5) + 5^2$
    $= 9x^2 - 30x + 25$

15. $(3x+2)(2x+1)(x-4)$
    $= (3x+2)(2x^2 - 7x - 4)$
    $= 6x^3 - 21x^2 - 12x + 4x^2 - 14x - 8$
    $= 6x^3 - 17x^2 - 26x - 8$

16. $(-4x^2y^5)(5xy^3) = (-4)(5)x^{2+1}y^{5+3}$
    $= -20x^3y^8$

17. $\dfrac{14x^8y^3}{-21x^5y^{12}} = -\dfrac{2x^3}{3y^9}$

18. $(-2x^3y^2z^0)^4 = (-2)^4(x^3)^4(y^2)^4(z^0)^4$
    $= 16x^{12}y^8$

19. $\dfrac{9x^{-3}y^{-4}}{w^2z^{-8}} = \dfrac{9z^8}{x^3y^4w^2}$

20. $1{,}360{,}000{,}000{,}000{,}000 = 1.36 \times 10^{15}$

21. $0.00056 = 5.6 \times 10^{-4}$

22. $\dfrac{(2.0 \times 10^{-12})(8.0 \times 10^{-20})}{4.0 \times 10^3} = \dfrac{16.0 \times 10^{-32}}{4.0 \times 10^3}$
    4.0×10^{-35}

23. $(x^3 - 3x^2 - 5x + 20) - (-4x^3 - 10x^2 + x - 30)$
    $= (x^3 - 3x^2 - 5x + 20) + (4x^3 + 10x^2 - x + 30)$
    $= (x^3 + 4x^3) + (-3x^2 + 10x^2) + (-5x - x)$
    $+ (20 + 30)$
    $= 5x^3 + 7x^2 - 6x + 50$

24. $-6xy^2(6x^2 - 3xy + 8y^2)$
    $= -36x^3y^2 + 18x^2y^3 - 48xy^4$

25. $(2x^2 - 3x + 1)(3x - 5)$
    $= 6x^3 - 9x^2 + 3x - 10x^2 + 15x - 5$
    $= 6x^3 - 19x^2 + 18x - 5$

26.
$$
\begin{array}{r}
x+5 \\
x-3\overline{\big)\,x^2+2x-12} \\
\underline{x^2-3x} \\
5x-12 \\
\underline{5x-12} \\
3
\end{array}
$$

$$\frac{2x^3-6x-36}{x-3}=2x^2+6x+12$$

Chapter 5

5.1 Exercises

1. $3x^2$ and $5x^3$ are called factors.

3. The factoring is not complete because $6a^3 + 3a^2 - 9a$ contains a common factor of $3a$.

5. $3a^2 + 3a = 3a(a+1)$
   Check:
   $3a(a+1) = 3a^2 + 3a$

7. $21abc - 14ab^2 = 7ab(3c - 2b)$
   Check:
   $7ab(3c - 2b) = 21abc - 14ab^2$

9. $2\pi rh + 2\pi r^2 = 2\pi r(h + r)$

11. $5x^3 + 25x^2 - 15x = 5x(x^2 + 5x - 3)$
    Check:
    $5x(x^2 + 5x - 3) = 5x^3 + 25x^2 - 15x$

13. $12ab - 28bc + 20ac = 4(3ab - 7bc + 5ac)$
    Check:
    $4(3ab - 7bc + 5ac) = 12ab - 28bc + 20ac$

15. $16x^5 + 24x^3 - 32x^2 = 8x^2(2x^3 + 3x - 4)$

17. $14x^2y - 35xy - 63x = 7x(2xy - 5y - 9)$

19. $54x^2 - 45xy + 18x = 9x(6x - 5y + 2)$

21. $3xy^2 - 2ay + 5xy - 2y$
    $= y(3xy - 2a + 5x - 2)$
    Check:
    $y(3xy - 2a + 5x - 2)$
    $= 3xy^2 - 2ay + 5xy - 2y$

23. $24x^2y - 40xy^2 = 8xy(3x - 5y)$

25. $7x^3y^2 + 21x^2y^2 = 7x^2y^2(x + 3)$

27. $9x^2y - 18xy^2 - 27xy = 9xy(x - 2y - 3)$

29. $7a(x + 2y) - b(x + 2y) = (7a - b)(x + 2y)$

31. $3x(x - 4) - 2(x - 4) = (x - 4)(3x - 2)$

33. $6b(2a - 3c) - 5d(2a - 3c)$
    $= (2a - 3c)(6b - 5d)$

35. $7c(b - a^2) - 5d(b - a^2) + 2f(b - a^2)$
    $= (b - a^2)(7c - 5d + 2f)$

37. $2a(xy-3)-4(xy-3)-z(xy-3)$
$=(xy-3)(2a-4-z)$

39. $4a^3(a-3b)+(a-3b)=(a-3b)(4a^3+1)$

41. $(2a+3)-7x(2a+3)=(2a+3)(1-7x)$

43. $A=2.786a+2.786b+2.786c+2.786d$
$=2.786(a+b+c+d)$ in^2

Cumulative Review

45. $x=$ 1st odd integer
$x+2=$ 2nd odd integer
$x+4=$ 3rd odd integer
$x+x+2+x+4=x+40$
$3x+6=x+40$
$2x=34$
$x=17$
$x+2=19,\ x+4=21$
The 3 integers are 17, 19, 21

47. $l=rt,\ l_1=l_2,\ r_1=6.72,\ t_1=2,\ t_2=6$
$6.72(2)=r_2(6)$
$r_2=2.24$ feet per minute

49. 15% of 258,000,000
$=-0.15(258{,}000{,}000)$
$=38{,}700{,}000$ tons

51. $\frac{38{,}700{,}000}{174{,}000{,}000}=0.2224$ tons per person
$\frac{0.2224 \text{ tons}}{1 \text{ person}}\times\frac{2000 \text{ pounds}}{1 \text{ ton}}$
$=445$ pounds/person

5.2 Exercises

1. We must remove a common factor of 5 from the last two terms.
$3x^2-6xy+5x-10y$
$=3x(x-2y)+5(x-2y)$
$=(x-2y)(3x+5)$

3. $ab-3a+4b-12=a(b-3)+4(b-3)$
$=(a+4)(b-3)$
Check:
$(a+4)(b-3)=ab-3a+4b-12$

5. $x^3-4x^2+3x-12$
$=x^2(x-4)+3(x-4)$
$=(x-4)(x^2+3)$
Check:
$(x-4)(x^2+3)=x^3+3x-4x^2-12$
$=x^3-4x^2+3x-12$

7. $2ax + 6bx - ay - 3by$
$= 2x(a+3b) - y(a+3b)$
$= (a+3b)(2x-y)$
Check:
$(a+3b)(2x-y) = 2ax - ay + 6bx - 3by$
$= 2ax + 6bx - ay - 3by$

9. $3ax + bx - 6a - 2b$
$= x(3a+b) - 2(3a+b)$
$= (3a+b)(x-2)$
Check:
$(3a+b)(x-2)$
$= 3ax - 6a + bx - 2b$
$= 3ax + bx - 6a - 2b$

11. $5a + 12bc + 10b + 6ac$
$= 5a + 10b + 6ac + 12bc$
$= 5(a+2b) + 6c(a+2b)$
$= (a+2b)(5+6c)$
Check:
$(a+2b)(5+6c)$
$= 5a + 6ac + 10b + 12bc$
$= 5a + 12bc + 10b + 6ac$

13. $5a - 5b - 2ax + 2xb$
$= 5(a-b) - 2x(a-b)$
$= (a-b)(5-2x)$
Check:
$(a-b)(5-2x)$
$= 5a - 2ax - 5b + 2bx$
$= 5a - 5b - 2ax + 2xb$

15. $y^2 - 2y - 3y + 6$
$= y(y-2) - 3(y-2)$
$= (y-2)(y-3)$

17. $14 - 7y + 2y - y^2$
$= 7(2-y) + y(2-y)$
$= (7+y)(2-y)$

19. $6ax - y + 2ay - 3x$
$= 6ax - 3x + 2ay - y$
$= 3x(2a-1) + y(2a-1)$
$= (2a-1)(3x+y)$

21. $2x^2 + 8x - 3x - 12$
$= 2x(x+4) - 3(x+4)$
$= (2x-3)(x+4)$

23. $t^3 - t^2 + t - 1 = t^2(t-1) + (t-1)$
$= (t-1)(t^2+1)$

25. $28x^2 + 8xy^2 + 21xw + 6y^2w$
$= 4x(7x+2y^2) + 3w(7x+2y^2)$
$= (7x+2y^2)(4x+3w)$

27. Rearrange the terms so that factoring gives the same expression in parentheses.
$6a^2 - 12bd - 8ad + 9ab$
$= 6a^2 - 8ad + 9ab - 12bd$
$= 2a(3a-4d) + 3b(3a-4d)$
$= (3a-4d)(2a+3b)$

Cumulative Review

29. $x =$ number of seconds

$73 - 4x = 41$

$-4x = -32$

$x = 8$

The brakes were applied for 8 seconds

31. $x =$ dollar value in 1997

$0.119x = 15.1$

$x = 126.9$

The dollar value in 1997 was $126.9 million

33. $100\% + 33\% = 133\%$

$2000: 133\%$ of 5.5 million

$= 1.33(5{,}500{,}000)$

$= 7{,}315{,}000$ tons

$100\% + 28\% = 128\%$

$2005: 128\%$ of $7{,}315{,}000$

$= 1.28(7{,}315{,}000)$

$= 9{,}363{,}000$ tons

5.3 Exercises

1. Find two numbers whose <u>product</u> is 6 and whose <u>sum</u> is 5.

3. $x^2 + 2x + 1$; Product: 1, Sum: 2, + signs

$x^2 + 2x + 1 = (x+1)(x+1)$

5. $x^2 + 12x + 20$; Product: 20, Sum: 12, + signs

$x^2 + 12x + 20 = (x+10)(x+2)$

7. $x^2 - 4x + 3$; Product: 3, Sum: -4, $-$ signs

$x^2 - 4x + 3 = (x-1)(x-3)$

9. $x^2 - 11x + 28$; Product: 28, Sum: -11, $-$ signs

$x^2 - 11x + 28 = (x-4)(x-7)$

11. $x^2 + x - 20$; Product: -20, Sum: 1, Opposite signs with larger absolute value +.

$x^2 + x - 20 = (x+5)(x-4)$

13. $x^2 - 13x - 14$; Product: -14, Sum: -13, Opposite signs with larger absolute value $-$.

$x^2 - 13x - 14 = (x-14)(x+1)$

15. $x^2 + 2x - 35$; Product: -35, Sum: 2 Opposite signs with larger absolute value +.

$x^2 + 2x - 35 = (x+7)(x-5)$

17. $x^2 - 2x - 24$; Product: -24, Sum: -2 Opposite signs with larger absolute value $-$.

$x^2 - 2x - 24 = (x-6)(x+4)$

19. $x^2 + 12x + 32$; Product: 32, Sum: 12, + signs

$x^2 + 12x + 32 = (x+4)(x+8)$

21. $x^2 - 10x + 24 = (x-6)(x-4)$
Check: $(x-6)(x-4) = x^2 - 4x - 6x + 24$
$= x^2 - 10x + 24$

23. $x^2 + 13x + 30 = (x+3)(x+10)$
Check: $(x+3)(x+10) = x^2 + 10x + 3x + 30$
$= x^2 + 13x + 30$

25. $x^2 - 6x + 5$; Product: 5, Sum: -6,
$-$ signs
$x^2 - 6x + 5 = (x-1)(x-5)$

27. $a^2 + 6a - 16 = (a+8)(a-2)$
Check: $(a+8)(a-2) = a^2 - 2a + 8a - 16$
$= a^2 + 6a - 16$

29. $x^2 - 12x + 32 = (x-4)(x-8)$
Check: $(x-4)(x-8) = x^2 - 8x - 4x + 32$
$= x^2 - 12x + 32$

31. $x^2 + 4x - 21 = (x+7)(x-3)$
Check: $(x+7)(x-3) = x^2 - 3x + 7x - 21$
$= x^2 + 4x - 21$

33. $x^2 + 15x + 56 = (x+7)(x+8)$
Check: $(x+7)(x+8) = x^2 + 8x + 7x + 56$
$= x^2 + 15x + 56$

35. $x^2 - 21x - 46 = (x-23)(x+2)$
Check: $(x-23)(x+2) = x^2 + 2x - 23x - 46$
$= x^2 - 21x - 46$

37. $x^2 + 9x - 36 = (x+12)(x-3)$
Check: $(x+12)(x-3) = x^2 - 3x + 12x - 36$
$= x^2 + 9x - 36$

39. $x^2 - 2xy - 15y^2 = (x-5y)(x+3y)$
Check:
$(x-5y)(x+3y) = x^2 + 3xy - 5xy - 15y^2$
$= x^2 - 2xy - 15y^2$

41. $x^2 - 16xy + 63y^2 = (x-9y)(x-7y)$
Check:
$(x-9y)(x-7y) = x^2 - 7xy - 9xy + 63y^2$
$= x^2 - 16xy + 63y^2$

43. $4x^2 + 24x + 20 = 4(x^2 + 6x + 5)$
$= 4(x+1)(x+5)$

45. $6x^2 + 18x + 12 = 6(x^2 + 3x + 2)$
$= 6(x+2)(x+1)$

47. $2x^2 - 20x + 42 = 2(x^2 - 10x + 21)$
$= 2(x-7)(x-3)$

49. $3x^2 - 6x - 72 = 3(x^2 - 2x - 24)$
$3(x-6)(x+4)$

51. $7x^2 + 21x - 70 = 7(x^2 + 3x - 10)$
$= 7(x+5)(x-2)$

53. $3x^2 - 18x + 15 = 3(x^2 - 6x + 5)$
$= 3(x-5)(x-1)$

55. $A = x^2 - \pi r^2,\ r = \dfrac{x}{2}$
$A = x^2 - \pi\left(\dfrac{x}{2}\right)^2$
$A = x^2\left(1 - \dfrac{3.14}{4}\right) = 64(0.215)$
$= 13.76$ sq. in.

57. $A = lw$
$A = 10(12) - x(x+2)$
$= 120 - x^2 - 2x$
$= 120 - 2x - x^2$
$= (10-x)(12+x)$

59. $A = P + rPt$
$A - P = rPt$
$\dfrac{A-P}{rP} = t$

61. $c =$ car's speed
$c = \dfrac{d}{t} = \dfrac{d}{2}$
$c + 20 =$ train's speed
$\dfrac{\text{d}}{2} + 20 = \dfrac{d}{1.5}$
$1.5d + 60 = 2d$
$60 = 0.5d$
$120 = d$
120 miles

63. $x =$ number of 30-min periods after midnight
$4(65) + 85x = 515$
$260 + 85x = 515$
$85x = 255$
$x = 3$
The reception ended at 1:30 a.m.

65. $M = 3$
$T = 19 + 2M$
$T = 19 + 2(3)$
$T = 25$
25^0 C

5.4 Exercises

1. $4x^2 + 13x + 3$; Grouping number: 12
$4x^2 + 13x + 3 = 4x^2 + 12x + x + 3$
$= 4x(x+3) + 1(x+3)$
$= (4x+1)(x+1)$
Check:
$(4x+1)(x+3) = 4x^2 + 12x + x + 3$
$= 4x^2 + 13x + 3$

3. $5x^2 + 7x + 2$; Grouping number: 10
$5x^2 + 7x + 2 = 5x^2 + 5x + 2x + 2$
$= 5x(x+1) + 2(x+1)$
$= (5x+2)(x+1)$
Check:
$(5x+2)(x+1) = 5x^2 + 5x + 2x + 2$
$= 5x^2 + 7x + 2$

5. $3x^2 - 4x - 7$; Grouping number: -21

$$\begin{aligned} 3x^2 - 4x - 7 &= 3x^2 - 7x + 3x - 7 \\ &= x(3x-7) + (3x-7) \\ &= (3x-7)(x+1) \end{aligned}$$

Check:

$$\begin{aligned} (3x-7)(x+1) &= 3x^2 + 3x - 7x - 7 \\ &= 3x^2 - 4x - 7 \end{aligned}$$

7. $2x^2 - 5x - 3$; Grouping number: -6

$$\begin{aligned} 2x^2 - 5x - 3 &= 2x^2 - 6x + x - 3 \\ &= 2x(x-3) + (x-3) \\ &= (x-3)(2x+1) \end{aligned}$$

Check:

$$\begin{aligned} (x-3)(2x+1) &= 2x^2 + x - 6x - 3 \\ &= 2x^2 - 5x - 3 \end{aligned}$$

9. $9x^2 + 9x + 2$; Grouping number: 18

$$\begin{aligned} 9x^2 + 9x + 2 &= 9x^2 + 3x + 6x + 2 \\ &= 3x(3x+1) + 2(3x+1) \\ &= (3x+1)(3x+2) \end{aligned}$$

Check:

$$\begin{aligned} (3x+1)(3x+2) &= 9x^2 + 6x + 3x + 2 \\ &= 9x^2 + 9x + 2 \end{aligned}$$

11. $15x^2 - 34x + 15$; Grouping number: 225

$$\begin{aligned} 15x^2 - 34x + 15 &= 15x^2 - 25x - 9x + 15 \\ &= 5x(3x-5) - 3(3x-5) \\ &= (5x-3)(3x-5) \end{aligned}$$

Check:

$$\begin{aligned} (5x-3)(3x-5) &= 15x^2 - 25x - 9x + 15 \\ &= 15x^2 - 34x + 15 \end{aligned}$$

13. $2x^2 + 3x - 20$; Grouping number: -40

$$\begin{aligned} 2x^2 + 3x - 20 &= 2x^2 + 8x - 5x - 20 \\ &= 2x(x+4) - 5(x+4) \\ &= (x+4)(2x-5) \end{aligned}$$

Check:

$$\begin{aligned} (x+4)(2x-5) &= 2x^2 - 5x + 8x - 20 \\ &= 2x^2 + 3x - 20 \end{aligned}$$

15. $8x^2 + 10x - 3$; Grouping number: -24

$$\begin{aligned} 8x^2 + 10x - 3 &= 8x^2 + 12x - 2x - 3 \\ &= 4x(2x+3) - (2x+3) \\ &= (2x+3)(4x-1) \end{aligned}$$

Check:

$$\begin{aligned} (2x+3)(4x-1) &= 8x^2 - 2x + 12x - 3 \\ &= 9x^2 + 10x - 3 \end{aligned}$$

17. $6x^2 - 5x - 6$; Grouping number: -36

$$6x^2 - 5x - 6 = 6x^2 - 9x + 4x - 6$$
$$= 3x(2x - 3) + 2(2x - 3)$$
$$= (2x - 3)(3x + 2)$$

Check:

$$(2x - 3)(3x + 2) = 6x^2 + 4x - 9x - 6$$
$$= 6x^2 - 5x - 6$$

19. $6x^2 + x - 15 = 6x^2 + 10x - 9x - 15$
$$= 2x(3x + 5) - 3(3x + 5)$$
$$= (3x + 5)(2x - 3)$$

21. $7x^2 - 5x - 18 = 7x^2 - 14x + 9x - 18$
$$= 7x(x - 2) + 9(x - 2)$$
$$= (7x + 9)(x - 2)$$

23. $9y^2 - 13y + 4 = 9y^2 - 9y - 4y + 4$
$$= 9y(y - 1) - 4(y - 1)$$
$$= (y - 1)(9y - 4)$$

25. $5a^2 - 13a - 6 = 5a^2 - 15a + 2a - 6$
$$= 5a(a - 3) + 2(a - 3)$$
$$= (a - 3)(5a + 2)$$

27. $12x^2 - 20x + 3 = 12x^2 - 18x - 2x + 3$
$$= 6x(2x - 3) - (2x - 3)$$
$$= (2x - 3)(6x - 1)$$

29. $15x^2 + 4x - 4 = 15x^2 + 10x - 6x - 4$
$$= 5x(3x + 2) - 2(3x + 2)$$
$$= (3x + 2)(5x - 2)$$

31. $12x^2 + 28x + 15 = 12x^2 + 18x + 10x + 15$
$$= 6x(2x + 3) + 5(2x + 3)$$
$$= (6x + 5)(2x + 3)$$

33. $12x^2 - 16x - 3 = 12x^2 + 2x - 18x - 3$
$$= 2x(6x + 1) - 3(6x + 1)$$
$$= (6x + 1)(2x - 3)$$

35. $2x^4 + 15x^2 - 8 = 2x^4 + 16x^2 - x^2 - 8$
$$= 2x^2(x^2 + 8) - (x^2 + 8)$$
$$= (x^2 + 8)(2x^2 - 1)$$

37. $2x^2 + 11xy + 15y^2$
$$= 2x^2 + 6xy + 5xy + 15y^2$$
$$= 2x(x + 3y) + 5y(x + 3y)$$
$$= (x + 3y)(2x + 5y)$$

39. $5x^2 + 16xy - 16y^2 = 5x^2 + 20xy - 4xy - 16y^2$
$$= 5x(x + 4y) - 4y(x + 4y)$$
$$= (x + 4y)(5x - 4y)$$

41. $4x^2 + 34x + 42$
$$= 2(2x^2 + 17x + 21)$$
$$= 2(x + 7)(2x + 3)$$

43. $8x^2 - 26x + 6$
$= 2(4x^2 - 13x + 3)$
$= 2(x - 3)(4x - 1)$

45. $10x^2 - 25x - 15 = 5(2x^2 - 5x - 3)$
$= 5(2x^2 - 6x + x - 3)$
$= 5[2x(x - 3) + 1(x - 3)]$
$= 5(2x + 1)(x - 3)$

47. $6x^3 + 9x^2 - 60x = 3x(2x^2 + 3x - 20)$
$= 3x(2x - 5)(x + 4)$

49. $5x^2 + 3x - 2 = 5x^2 + 5x - 2x - 2$
$= 5x(x + 1) - 2(x + 1)$
$= (x + 1)(5x - 2)$

51. $12x^2 - 38x + 20 = 2(6x^2 - 19x + 10)$
$= 2(6x^2 - 15x - 4x + 10)$
$= 2[3x(2x - 5) - 2(2x - 5)]$
$= 2(2x - 5)(3x - 2)$

53. $12x^2 - 20x + 3 = 12x^2 - 18x - 2x + 3$
$= 6x(2x - 3) - (2x - 3)$
$= (2x - 3)(6x - 1)$

55. $8x^2 + 16x - 10 = 2(4x^2 + 8x - 5)$
$= 2(4x^2 + 10x - 2x - 5)$
$= 2[2x(2x + 5) - (2x + 5)]$
$= 2(2x + 5)(2x - 1)$

Cumulative Review

57. $7x - 3(4 - 2x) = 2(x - 3) - (5 - x)$
$7x - 12 + 6x = 2x - 6 - 5 + x$
$13x - 12 = 3x - 11$
$10x = 1$
$x = \frac{1}{10}$

59. Percent increase $= \frac{1809 - 1430}{1430}$
$= 0.265$
$= 26.5\%$

61. China:

Percent increase $= \frac{1131 - 778}{778}$
$= 0.454$
$= 45.5\%$

China had the greatest percent.

How Am I Doing? Sections 5.1-5.4

1. $6xy-15z+21=3(2xy-5z+7)$

2. $20x^2-32xy+12x=4x(5x-8y+3)$

3. $7(4x-5)-b(4x-5)=(4x-5)(7-b)$

4. $2x(8y+3z)-5y(8y+3z)$
$=(8y+3z)(2x-5y)$

5. $18+3x-6y-xy=3(6+x)-y(6+x)$
$=(6+x)(3-y)$

6. $15x-9xb+20w-12bw$
$=3x(5-3b)+4w(5-3b)$
$=(5-3b)(3x+4w)$

7. $x^3-5x^2-3x+15=x^2(x-5)-3(x-5)$
$=(x-5)(x^2-3)$

8. $7a+21b+2ab+6b^2=7(a+3b)+2b(a+3b)$
$=(a+3b)(7+2b)$

9. $x^2-15x+56=x^2-7x-8x+56$
$=x(x-7)-8(x-7)$
$=(x-7)(x-8)$

10. $x^2+12x-64=x^2+16x-4x-64$
$=x(x+16)-4(x+16)$
$=(x+16)(x-4)$

11. $x^2+13xy+40y^2=x^2+8xy+5xy+40y^2$
$=x(x+8y)5y(x+8y)$
$=(x+8y)(x+5y)$

12. $7x^2-14x-245=7(x^2-2x-35)$
$=7(x^2-7x+5x-35)$
$=7[x(x-7)+5(x-7)]$
$=7(x-7)(x+5)$

13. $10x^2+x-2=10x^2+5x-4x-2$
$=5x(2x+1)-2(2x+1)$
$=(2x+1)(5x-2)$

14. $3x^2-23x+14=3x^2-21x-2x+14$
$=3x(x-7)-2(x-7)$
$=(x-7)(3x-2)$

15. $6x^2+17xy+12y^2=6x^2+9xy+8xy+12y^2$
$=3x(2x+3y)+4y(2x+3y)$
$=(2x+3y)(3x+4y)$

16. $14x^3 - 20x^2 - 16x = 2x(7x^2 - 10x - 8)$
$= 2x(7x^2 - 14x \_ 4x - 8)$
$= 2x[7x(x-2) + 4(x-2)]$
$= 2x(x-2)(7x+4)$

5.5 Exercises

1. $9x^2 - 1 = (3x)^2 - (1)^2 = (3x+1)(3x-1)$

3. $81x^2 - 16 = (9x)^2 - (4)^2 = (9x+4)(9x-4)$

5. $x^2 - 49 = (x)^2 - (7)^2 = (x+7)(x-7)$

7. $4x^2 - 25 = (2x)^2 - (5)^2 = (2x+5)(2x-5)$

9. $x^2 - 25 = (x)^2 - (5)^2 = (x+5)(x-5)$

11. $1 - 16x^2 = (1)^2 - (4x)^2 = (1+4x)(1-4x)$

13. $16x^2 - 49y^2 = (4x)^2 - (7y)^2$
$= (4x+7y)(4x-7y)$

15. $36x^2 - 169y^2 = (6x)^2 - (13y)^2$
$= (6x+13y)(6x-13y)$

17. $81x^2 - 100 = (9x)^2 - (10)^2$
$= (9x+10)(9x-10)$

19. $25a^2 - 81b^2 = (5a)^2 - (9b)^2$
$= (5a+9b)(5a-9b)$

21. $9x^2 + 6x + 1 = (3x)^2 + 2(3x)(1) + (1)^2$
$= (3x+1)^2$

23. $y^2 - 10y + 25 = y^2 - 2(y)(5) + (5)^2$
$= (y-5)^2$

25. $36x^2 - 60x + 25 = (6x)^2 - 2(6x)(5) + (5)^2$
$= (6x-5)^2$

27. $49x^2 + 28x + 4 = (7x)^2 + 2(7x)(2) + 2^2$
$= (7x+2)^2$

29. $x^2 + 14x + 49 = (x)^2 + 2(x)(7) + (7)^2$
$= (x+7)^2$

31. $25x^2 - 40x + 16 = (5)^2 - 2(5x)(4) + 4^2$
$= (5x-4)^2$

33. $81x^2 + 36xy + 4y^2$
$= (9x)^2 + 2(9x)(2y) + (2y)^2$
$= (9x+2y)^2$

35. $25x^2 - 30xy + 9y^2$
$= (5x)^2 - 2(5x)(3y) + (3y)^2$
$= (5x-3y)^2$

37. $16a^2+72ab+81b^2$
$=(4a)^2+2(4a)(9b)+(9b)^2$
$=(4a+9b)^2$

39. $49x^2-42xy+9y^2$
$=(7x)^2-2(7x)(3y)+(3y)^2$
$=(7x-3y)^2$

41. $49x^2+42x+9=(7x)^2+2(7x)(3)+(3)^2$
$=(7x+3)^2$

43. $49x^2-9=(7x)^2-(3)^2$
$=(7x+3)(7x-3)$

45. $x^4-36=(x^2)^2-(6)^2=(x^2+6)(x^2-6)$

47. $9x^4-12x^2+4=(3x^2)^2-2(3x)(2)+(2)^2$
$=(3x^2-2)^2$

49. You cannot factor $9x^2+1$ because there are no combinations of the product of two binomials that give $9x^2+1$.

51. $16=4^2$
$56=(2)(4)(7)$
$c=7^2$
$c=49$
There is only one answer.

53. $16x^2-36=4(4x^2-9)$
$=4(2x+3)(2x-3)$

55. $147x^2-3y^2=3(49x^2-y^2)$
$=3\left[(7x)^2-y^2\right]$
$=3(7x-y)(7x+y)$

57. $12x^2-36x+27=3(4x^2-12x+9)$
$=3(2x-3)^2$

59. $98x^2+84x+18=2(49x^2+42x+9)$
$=2(7x+3)^2$

61. $x^2-15x+44=x^2-11x-4x+44$
$=x(x-11)-4(x-11)$
$=(x-11)(x-4)$

63. $2x^2+5x-3=(2x-1)(x+3)$

65. $16x^2-121=(4x+11)(4x-11)$

67. $9x^2+42x+49=(3x+7)^2$

69. $3x^2+6x-45=3(x^2+2x-15)$
$=3(x+5)(x-3)$

71. $5x^2-80=5(x^2-16)$
$=5(x+4)(x-4)$

73. $7x^2+42x+63=7(x^2+6x+9)$
$=7\left[(x^2)+2(x)(3)+(3)^2\right]$
$=7(x+3)^2$

75. $2x^2-32x+126=2(x^2-16x+63)$
$=2(x-9)(x-7)$

Cumulative Review

77.
$$\begin{array}{r} x^2+3x+4 \\ x-2\overline{)x^3+x^2-2x-11} \\ \underline{x^3-2x^2} \\ 3x^2-2x \\ \underline{3x^2-6x} \\ 4x-11 \\ \underline{4x-8} \\ -3 \end{array}$$

$(x^3+x^2-2x-11)\div(x-2)$
$=x^2+3x+4-\dfrac{3}{x-2}$

79. Daily diet $=0.2(150)=3$ ounces
$.4(3)=1.2$ ounces of greens
$.35(3)=1.05$ ounces of bulk vegtable
$.25(3)=0.75$ ounces of fruit

81. $A=6288-700M$
$A=6288-700(3.5)$
$A=6288-2450$
$A=3838$
3838 feet above sea level

5.6 Exercises

1. $6a^2+2ab-3a=a(6a+2b-3)$
Check: $a(6a+2b)=6a^2+2ab-3a$

3. $16x^2-25y^2=(4x)^2-(5y)^2$
$=(4x+5y)(4x-5y)$
Check:
$(4x+5y)(4x-5y)=(4x)^2-(5y)^2$
$=16x^2-25y^2$

5. $9x^2-12xy+4y^2=(3x-2y)^2$
Check:
$(3x-2y)^2=9x^2+2(3x)(-2y)+(-2y)^2$
$=9x^2-12xy+4y^2$

7. $x^2+8x+15=(x+5)(x+3)$
Check:
$(x+5)(x+3)=x^2+3x+5x+15$
$=x^2+8x+15$

9. $15x^2+7x-2=(5x-1)(3x+2)$

Check:

$(5x-1)(3x+2)=15x^2+10x-3x-2$
$=15x^2+7x-2$

11. $ax-3cx+3ay-9cy$
$=x(a-3c)+3y(a-3c)$
$=(a-3c)(x+3y)$

Check:

$(x+3y)(a-3c)=ax-3cx+3ay-9cy$

13. $y^2+14y+49=(y)^2+2(y)(7)+(7)^2$
$=(y+7)^2$

15. $4x^2-12x+9=(2x-3)(2x-3)$
$=(2x-3)^2$

17. $2x^2-11x+12=(2x-3)(x-4)$

19. $x^2-3xy-70y^2=(x-10y)(x+7y)$

21. $ax-5a+3x-15=a(x-5)+3(x-5)$
$=(a+3)(x-5)$

23. $45x-5x^3=5x(9-x^2)$
$=5x(3+x)(3-x)$

25. $5x^3y^3-10x^2y^3+5xy^3=5xy^3(x^2-2x+1)$
$=5xy^3(x-1)^2$

27. $27xyz^2-12xy=3xy(9z^2-4)$
$=3xy(3z+2)(3z-2)$

29. $3x^2+6x-105=3(x^2+2x-35)$
$=3(x+7)(x-5)$

31. $3x^3+3x^2-36x=3x(x^2+x-12)$
$=3x(x^2+4x-3x-12)$
$=3x[x(x+4)-3(x+4)]$
$=3x(x+4)(x-3)$

33. $7x^2-2x^4+4=-1(2x^4-7x^2-4)$
$=-1(2x^2+1)(x^2-4)$
$=-1(2x^2+1)(x+2)(x-2)$

35. $6x^2-3x+2$ is prime.

37. $5x^2+10xy-30y=5(x^2+2xy-6y)$

39. $30x^3+3x^2y-6xy^2=3x(10x^2+xy-2y^2)$
$=3x(2x+y)(5x-2)$

41. $24x^2-58x+30=2(12x^2-29x+15)$
$=2(12x^2-20x-9x+15)$
$=2[4x(3x-5)-3(3x-5)]$
$=2(3x-5)(4x-3)$

43. A polynomial that cannot be factored by the methods of this chapter is called prime.

Cumulative Review

45. 20% of what = 19.56

$0.2x = 19.56$

$x = 97.8$

Sale price = 97.80 − 19.56 = \$78.24

47. Percentage increase $= \dfrac{65-57}{57}$

$= 0.140$

$= 14\%$

2010 : 100% + 14% = 114%

114% of 65 = 1.14(65) = \$74

49. Average rate of change $= \dfrac{65-57}{10} = 0.8$

$y = 0.8x + 57$

5.7 Exercises

1. $x^2 - 4x - 21 = 0$

$(x+3)(x-7) = 0$

$x + 3 = 0 \qquad x - 7 = 0$

$x = -3 \qquad x = 7$

Check:

$(-3)^2 - 4(-3) - 21 \stackrel{?}{=} 0 \qquad (7)^2 - 4(7) - 21 \stackrel{?}{=} 0$

$9 + 12 - 21 \stackrel{?}{=} 0 \qquad 49 - 28 - 21 \stackrel{?}{=} 0$

$0 = 0 \qquad 0 = 0$

3. $x^2 + 16x + 39 = 0$

$(x+13)(x+3) = 0$

$x + 13 = 0 \qquad x + 3 = 0$

$x = -13 \qquad x = -3$

Check:

$(-13)^2 + 16(-13) + 39 \stackrel{?}{=} 0$

$169 - 208 + 39 \stackrel{?}{=} 0$

$0 = 0$

$(-3)^2 + 16(-3) + 39 \stackrel{?}{=} 0$

$9 - 48 + 39 \stackrel{?}{=} 0$

$0 = 0$

5. $2x^2 - 7x + 6 = 0$

$2x^2 - 3x - 4x + 6 = 0$

$x(2x-3) - 2(2x-3) = 0$

$(2x-3)(x-2) = 0$

$2x - 3 = 0 \qquad x - 2 = 0$

$x = \dfrac{3}{2} \qquad x = 2$

Check:

$2\left(\frac{3}{2}\right)^2 - 7\left(\frac{3}{2}\right) + 6 \stackrel{?}{=} 0$ $\quad$ $2(2)^2 - 7(2) + 6 \stackrel{?}{=} 0$

$\frac{9}{2} - \frac{21}{2} + \frac{12}{2} \stackrel{?}{=} 0$ $\quad$ $8 - 14 + 6 \stackrel{?}{=} 0$

$0 = 0$ $\quad$ $0 = 0$

7. $6x^2 - 13x = -6$

$6x^2 - 13x + 6 = 0$

$(3x-2)(2x-3) = 0$

$3x - 2 = 0$ $\quad$ $2x - 3 = 0$

$3x = 2$ $\quad$ $2x = 3$

$x = \frac{2}{3}$ $\quad$ $x = \frac{3}{2}$

Check:

$6\left(\frac{2}{3}\right)^2 - 13\left(\frac{2}{3}\right) \stackrel{?}{=} -6$ $\quad$ $6\left(\frac{3}{2}\right)^2 - 13\left(\frac{3}{2}\right) \stackrel{?}{=} -6$

$\frac{8}{3} - \frac{26}{3} \stackrel{?}{=} -6$ $\quad$ $\frac{27}{2} - \frac{39}{2} \stackrel{?}{=} -6$

$-6 = -6$ $\quad$ $-6 = -6$

9. $x^2 + 13x = 0$

$x(x+13) = 0$

$x = 0$ $\quad$ $x + 13 = 0$

$x = -13$

Check:

$(-13)^2 + (13)(-13) \stackrel{?}{=} 0$ $\quad$ $0^2 + 13(0) \stackrel{?}{=} 0$

$169 - 169 \stackrel{?}{=} 0$ $\quad$ $0 \stackrel{?}{=} 0$

$0 = 0$

11. $8x^2 = 72$

$8x^2 - 72 = 0$

$8(x^2 - 9) = 0$

$8(x-3)(x+3) = 0$

$x - 3 = 0$ $\quad$ $x + 3 = 0$

$x = 3$ $\quad$ $x = -3$

Check:

$8(-3)^2 \stackrel{?}{=} 72$ $\quad$ $8(3)^2 \stackrel{?}{=} 72$

$72 = 72$ $\quad$ $72 = 72$

13. $5x^2 + 3x = 8x$

$5x^2 - 5x = 0$

$5x(x-1) = 0$

$5x = 0$ $\quad$ $x - 1 = 0$

$x = 0$ $\quad$ $x = 1$

Check:

$5(0)^2 + 3(0) \stackrel{?}{=} 8(0)$ $\quad$ $5(1)^2 + 3(1) \stackrel{?}{=} 8(1)$

$0 + 0 \stackrel{?}{=} 0$ $\quad$ $5 + 3 \stackrel{?}{=} 8$

$0 = 0$ $\quad$ $8 = 8$

15. $6x^2 = 16x - 8$

$6x^2 - 16x + 8 = 0$

$2(3x^2 - 8x + 4) = 0$

$2(2x-2)(x-2) = 0$

$3x - 2 = 0$ $\quad$ $x - 2 = 0$

$3x = 2$ $\quad$ $x = 2$

$x = \frac{2}{3}$

Check:

$$6\left(\frac{2}{3}\right)^2 \stackrel{?}{=} 16\left(\frac{2}{3}\right) - 8$$

$$\frac{24}{9} \stackrel{?}{=} \frac{32}{3} - \frac{24}{3}$$

$$\frac{8}{3} = \frac{8}{3}$$

$$6(2)^2 \stackrel{?}{=} 16(2) - 8$$

$$24 \stackrel{?}{=} 32 - 8$$

$$24 = 24$$

19. $4x^2 - 3x + 1 = -7x$

$$4x^2 + 4x + 1 = 0$$

$$(2x+1)(2x+1) = 0$$

$$2x + 1 = 0$$

$$2x = -1$$

$$x = -\frac{1}{2}$$

Check:

$$4\left(-\frac{1}{2}\right)^2 - 3\left(-\frac{1}{2}\right) + 1 \stackrel{?}{=} -7\left(-\frac{1}{2}\right)$$

$$1 + \frac{3}{2} + 1 \stackrel{?}{=} \frac{7}{2}$$

$$\frac{7}{2} = \frac{7}{2}$$

21. $\frac{x^2}{2} - 8 + x = -8$

$$x^2 - 16 + 2x = -16$$

$$x^2 + 2x = 0$$

$$x(x+2) = 0$$

$x = 0$ $\quad$ $x + 2 = 0$

$x = -2$

Check:

$$\frac{(0)^2}{2} - 8 + (0) \stackrel{?}{=} -8 \qquad \frac{(-2)^2}{2} - 8 + (-2) \stackrel{?}{=} -8$$

$$0 - 8 + 0 \stackrel{?}{=} -8 \qquad 2 - 8 - 2 \stackrel{?}{=} -8$$

$$-8 = -8 \qquad -8 = -8$$

23. $\frac{x^2 + 7x}{4} = -3$

$$x^2 + 7x = -12$$

$$x^2 + 7x + 12 = 0$$

$$(x+4)(x+3) = 0$$

$x + 4 = 0$ $\quad$ $x + 3 = 0$

$x = -4$ $\quad$ $x = -3$

Check:

$$\frac{(-4)^2 + 7(-4)}{4} \stackrel{?}{=} -3 \qquad \frac{(-3)^2 + 7(-3)}{4} \stackrel{?}{=} -3$$

$$\frac{16 - 28}{4} \stackrel{?}{=} -3 \qquad \frac{9 - 21}{4} \stackrel{?}{=} -3$$

$$\frac{-12}{4} \stackrel{?}{=} -3 \qquad \frac{-12}{4} \stackrel{?}{=} -3$$

$$-3 = -3 \qquad -3 = -3$$

25. $\frac{10x^2 - 25x}{12} = 5$

$$10x^2 - 25x = 60$$

$$10x^2 - 25x - 60 = 0$$

$$5(2x^2 - 5x - 12) = 0$$

$$5(2x+3)(x-4) = 0$$

$2x+3=0 \qquad x-4=0$

$x=-\frac{3}{2} \qquad x=4$

Check:

$$\frac{9\left(-\frac{5}{3}\right)^2-12\left(-\frac{5}{3}\right)}{3}\stackrel{?}{=}15 \qquad \frac{9(3)^2\ 12(3)}{3}\stackrel{?}{=}15$$

$$\frac{25+20}{3}\stackrel{?}{=}15 \qquad \frac{81-36}{3}\stackrel{?}{=}15$$

$$15=15 \qquad 15=15$$

27. An equation in the form $ax^2+bx=0$ can always be solved by factoring out x.

29. $x =$ length

$\frac{1}{2}x+3 =$ width

$$x\left(\frac{1}{2}x+3\right)=140$$

$$\frac{x^2}{2}+3x=140$$

$$x^2+6x=280$$

$$x^2+6x-280=0$$

$$(x+20)(x-14)=0$$

$x+20=0 \qquad x-14=0$

$x=-20 \qquad x=14$

Width not negative $\quad \frac{1}{2}x+3=10$

The length is 14 m and the width is 10 m.

31. $G=\frac{x^2-3x+2}{2},\ x=13$

$$G=\frac{13^2-3(13)+2}{2}$$

$$G=\frac{169-39+2}{2}$$

$G=66$ groups

33. $G=\frac{x^2-3x+2}{2}$

$$72=x^2-3x+2$$

$$0=x^2-3x-70$$

$$0=(x-10)(x+7)$$

$x-10=0 \qquad x+7=0$

$x=10 \qquad x=-7$

Not possible

There are 10 students.

35. $s=-5t^2+vt+h$

$v=13 \qquad h=6$

$$s=-5t^2+13t+6$$

When the ball hits the ground $s=0$.

$$0=-5t^2+13t+6$$

$$0=5t^2-13t-6$$

$$0=(5t+2)(t-3)$$

$5t+2=0 \qquad t-3=0$

$t=-\frac{2}{5} \qquad t=3$

Time can't be negative.

The ball will hit the ground at $t=3$ sec.

$$s = -5(2)^2 + 13(2) + 6$$
$$= -20 + 26 + 6$$
$$= 12$$

After 2 seconds the ball is 12 m from the ground.

37. $C = 2x^2 - 7x$
$15 = 2x^2 - 7x$
$2x^2 - 7x - 15 = 0$
$(2x+3)(x-5) = 0$

$2x + 3 = 0$ $\qquad$ $x - 5 = 0$

$x = -\frac{3}{2}$ $\qquad$ $x = 5$

The number cannot be negative, so 5 additional helicopters were produced.

Use $T = 0.5(x^2 - x)$ in Exercise 37.

39. $x = 70$
$T = 0.5\left[(70)^2 - 70\right]$
$= 0.5(4900 - 70)$
$= 2415$ telephone calls

41. $T = 153$
$0.5(x^2 - x) = 306$
$x^2 - x - 306 = 0$
$(x+17)(x-18) = 0$

$x + 17 = 0$ $\qquad$ $x - 18 = 0$

$x = -17$ $\qquad$ $x = 18$

41. The negative number of people is not allowed. 18 people worked that day.

Cumulative Review

43. $(2x^2y^3)(-5x^3y) = -10x^5y^4$

45. $\dfrac{21a^5b^{10}}{-14ab^{12}} = -\dfrac{3a^4}{2b^2}$

Putting Your Skills to Work

1. Increase from 2001 to 2002 $= 294 - 258 = 36$ / yr
Reviews in 2005 $= 294 + 3(36) = 402$

2. Increase from 2001 to 2002 $= 94 - 89 = 5$
Reviews in 2006 $= 94 + 4(5) = 114$

3. 1999: Percentage reversals $= \dfrac{57}{195} = 0.29 = 29\%$
2000: Percentage reversals $= \dfrac{84}{247} = 0.34 = 34\%$
The percentage increased by 5.

4. 2001: Percentage reversals $= \dfrac{89}{258} = 0.34 = 34\%$
2002: Percentage reversals $= \dfrac{94}{294} = 0.32 = 32\%$
The percentage decreased by 2.

5. $y = 30x + 294,\ x = 2$
$y = 30(2) + 294$
$y = 354$

6. $y = 11x + 99,\ x = 3$
   $y = 11(3) + 99$
   $y = 132$

7. $y = 30x + 294$
   $y = 6(5x + 49)$

8. $y = 11x + 99$
   $y = 11(x + 9)$

Chapter 5 Review Problems

1. $12x^3 - 20x^2y = 4x^2(3x - 5y)$

2. $10x^3 - 35x^3y = 5x^3(2 - 7y)$

3. $7x^2y - 14xy^2 - 21x^3y^3$
   $= 7xy(x - 2y - 3x^2y^2)$

4. $50a^4b^5 - 25a^4b^4 + 75a^5b^5$
   $= 25a^4b^4(2b - 1 + 3ab)$

5. $3a^3 + 6a^2 - 9ab + 12a = 3a(a^2 + 2a - 3b + 4)$

6. $2x - 4y + 6z + 12 = 2(x - 2y + 3z + 6)$

7. $2a(a + 3b) - 5(a + 3b) = (a + 3b)(2a - 5)$

8. $15x^3y + 6xy^2 + 3xy = 3xy(5x^2 + 2y + 1)$

9. $3ax - 7a - 6x + 14 = a(3x - 7) - 2(3x - 7)$
   $= (3x - 7)(a - 2)$

10. $a^2 + 5ab - 4a - 20b = a(a + 5b) - (4a + 5b)$
    $= (a + 5b)(a - 4)$

11. $x^2y + 3y - 2x^2 - 6 = y(x^2 + 3) - 2(x^2 + 3)$
    $= (x^2 + 3)(y - 2)$

12. $30ax - 15ay + 42x - 21y$
    $= 3(10ax - 5ay + 14x - 7y)$
    $= 3[5a(2x - y) + 7(2x - y)]$
    $= 3(2x - y)(5a + 7)$

13. $15x^2 - 3x + 10x - 2 = 3x(5x - 1) + 2(5x - 1)$
    $= (5x - 1)(3x + 2)$

14. $30w^2 - 18w + 5wz - 3z$
    $= 6w(5w - 3) + z(5w - 3)$
    $= (5w - 3)(6w + z)$

15. $x^2 - 2x - 35 = (x - 7)(x + 5)$

16. $x^2 - 10x + 24 = (x - 6)(x - 4)$

17. $x^2 + 14x + 48 = (x + 6)(x + 8)$

18. $x^2 + 8xy + 15y^2 = (x + 3y)(x + 5y)$

19. $x^4 + 13x^2 + 42 = (x^2 + 6)(x^2 + 7)$

20. $x^4 - 2x^2 - 35 = (x^2 - 7)(x^2 + 5)$

21. $6x^2 + 30x + 36 = 6(x^2 + 5x + 6)$
$= 6(x + 2)(x + 3)$

22. $3x^2 + 39x + 36 = 3(x^2 + 13x + 12)$
$= 3(x + 12)(x + 1)$

23. $2x^2 - 28x + 96 = 2(x^2 - 14x + 48)$
$= 2(x - 8)(x - 6)$

24. $4x^2 - 44x + 120 = 4(x^2 - 11x + 30)$
$= 4(x - 6)(x - 5)$

25. $4x^2 + 7x - 15 = (4x - 5)(x + 3)$

26. $12x^2 + 11x - 5 = 12x^2 - 4x + 15x - 5$
$= 4x(3x - 1) + 5(3x - 1)$
$= (3x - 2)(2x - 3)$

27. $15x^2 + 7x - 4 = (3x - 1)(5x + 4)$

28. $6x^2 - 13x + 6 = 6x^2 - 4x - 9x + 6$
$= 2x(3x - 2) - 3(3x - 2)$
$= (3x - 2)(2x - 3)$

29. $2x^2 - x - 3 = (2x - 3)(x + 1)$

30. $3x^2 + 2x - 8 = (3x - 4)(x + 2)$

31. $20x^2 + 48x - 5 = (10x - 1)(2x + 5)$

32. $20x^2 + 21x - 5 = 20x^2 - 4x + 25x - 5$
$= 4x(5x - 1) + 5(5x - 1)$
$= (5x - 1)(4x + 5)$

33. $6a^2 + 11a - 10 = (3a - 2)(2a + 5)$

34. $6a^2 - 19a + 10 = (3a - 2)(2a - 5)$

35. $6x^2 + 4x - 10 = 2(3x^2 + 2x - 5)$
$= 2(3x + 5)(x - 1)$

36. $6x^2 - 4x - 10 = 2(3x^2 - 2x - 5)$
$= 2(3x^2 - 5x + 3x - 5)$
$= 2[x(3x - 5) + 1(3x - 5)]$
$= 2(3x - 5)(x + 1)$

37. $4x^2 - 26x + 30 = 2(2x^2 - 13x + 15)$
$= 2(2x - 3)(x - 5)$

38. $4x^2 - 20x - 144 = 4(x^2 - 5x - 36)$
$= 4(x - 9)(x + 4)$

39. $12x^2 - 22x - 20 = 2(6x^2 - 11x - 10)$
$= 2(3x + 2)(2x - 5)$

40. $18x^2 + 51x - 42 = 3(6x^2 + 17x - 14)$
$= 3(3x - 2)(2x + 7)$

41. $6x^2 - 19xy + 10y^2 = (2x - 5y)(3x - 2y)$

42. $6x^2 - 32xy + 10y^2 = 2(3x^2 - 16xy + 5y^2)$
$= 2(3x^2 - 15xy - xy + 5y^2)$
$= 2[3x(x - 5y) - y(x - 5y)]$
$= 2(x - 5y)(3x - y)$

43. $49x^2 - y^2 = (7x)^2 - (y)^2$
$= (7x + y)(7x - y)$

44. $16x^2 - 36y^2 = 4(4x^2 - 9y^2)$
$= 4[(2x)^2 - (3y)^2]$
$= 4(2x - 3y)(2x + 3y)$

45. $25x^2 + 30x + 9 = (5x + 3)^2$

46. $49x^2 - 28x + 4 = (7x - 2)^2$

47. $25x^2 - 36 = (5x)^2 - (6)^2$
$= (5x + 6)(5x - 6)$

48. $100x^2 - 9 = (10x)^2 - 3^2$
$= (10x - 3)(10x + 3)$

49. $y^2 - 36x^2 = (y + 6x)(y - 6x)$

50. $9y^2 - 25x^2 = (3y + 5x)(3y - 5x)$

51. $36x^2 + 12x + 1 = (6x + 1)^2$

52. $25x^2 - 20x + 4 = (5x)^2 - 2(5x)(2) + 2^2$
$= (5x - 2)^2$

53. $16x^2 - 24xy + 9y^2 = (4x - 3y)^2$

54. $49x^2 - 28xy + 4y^2$
$= (7x)^2 - 2(7x)(2y) + (2y)^2$
$= (7x - 2y)^2$

55. $2x^2 - 18 = 2(x^2 - 9) = 2(x + 3)(x - 3)$

56. $3x^2 - 75 = 3(x^2 - 25)$
$= 3(x^2 - 5^2)$
$= 3(x - 5)(x + 5)$

57. $72x^2 - 192x + 128 = 8(9x^2 - 24x + 16)$
$= 8(3x - 4)^2$

58. $50x^2 - 120x + 72 = 2(25x^2 - 60x + 36)$
$= 2(5x - 6)^2$

59. $4x^2 - 9y^2 = (2x)^2 - (3y)^2$
$= (2x + 3y)(2x - 3y)$

60. $x^2 + 6x + 9 = x^2 + 2(x)(3) + 3^2$
$= (x + 3)^2$

61. $x^2 - 9x + 18 = (x - 6)(x - 3)$

62. $x^2 + 13x - 30 = (x + 15)(x - 2)$

63. $6x^2 + x - 7 = (6x + 7)(x - 1)$

64. $10x^2 + x - 2 = 10x^2 - 4x + 5x - 2$
$= 2x(5x - 2) + 1(5x - 2)$
$= (5x - 2)(2x + 1)$

65. $24x - 60 = 12(2x - 5)$

66. $8x^2y^2 - 4xy = 4xy(2xy - 1)$

67. $50x^3y^2 + 30x^2y^2 - 10x^2y^2$
$= 50x^3y^2 + 20x^2y^2$
$= 10x^2y^2(5x + 2)$

68. $26a^3b - 13ab^3 + 52a^2b^4$
$= 13ab(2a^2 - b^2 + 4ab^3)$

69. $x^3 - 16x^2 + 64x = x(x^2 - 16x + 64)$
$= x(x - 8)^2$

70. $2x^2 + 40x + 200 = 2(x^2 + 20x + 100)$
$= 2[x^2 + 2(x)(10) + 10^2]$
$= 2(x + 10)^2$

71. $3x^2 - 18x + 27 = 3(x^2 - 6x + 9)$
$= 3(x - 3)^2$

72. $25x^3 - 60x^2 + 36x = x(25x^2 - 60x + 36)$
$= x[(5x)^2 - 2(5x)(6) + 6^2]$
$= x(5x - 6)^2$

73. $7x^2 - 9x - 10 = (7x + 5)(x - 2)$

74. $4x^2 - 13x - 12 = (4x + 3)(x - 4)$

75. $9x^3y - 4xy^3 = xy(9x^2 - 4y^2)$
$= xy(3x + 2y)(3x - 2y)$

76. $3x^3a^3 - 11x^4a^2 - 20x^5a$
$= x^3a(3a^2 - 11xa - 20x^2)$
$= x^3a(3a^2 + 4xa - 15xa - 20x^2)$
$= x^3a[a(3a + 4x) - 5x(3a + 4x)]$
$= x^3a(3a + 4x)(a - 5x)$

77. $12a^2 + 14ab - 10b^2 = 2(6a^2 + 7ab - 5b^2)$
$= 2(3a + 5b)(2a - b)$

78. $121a^2 + 66ab + 9b^2 = (11a + 3b)^2$

79. $7a - 7 - ab + b = 7(a-1) - b(a-1)$
$= (a-1)(7-b)$

80. $8b - 10 + 28bc - 35c$
$= 2(4b-5) + 7c(4b-5)$
$= (4b-5)(2+7c)$

81. $18b - 21 + 14x - 12bx$
$= 18b - 21 - 12bx + 14x$
$= 3(6b-7) - 2x(6b-7)$
$= (6b-7)(3-2x)$

82. $5xb - 35x + 4by - 28y$
$= 5x(b-7) + 4y(b-7)$
$= (b-7)(5x+4y)$

83. $2a^2x - 15ax + 7x = x(2a^2 - 15a + 7)$
$= x(2a-1)(a-7)$

84. $x^5 - 17x^3 + 16x = x(x^4 - 17x^2 + 16)$
$= x(x^2 - 16)(x^2 - 1)$
$= x(x^2 - 4^2)(x^2 - 1^2)$
$= x(x-4)(x+4)(x-1)(x+1)$

85. $x^4 - 81y^{12} = (x^2)^2 - (9y^6)^2$
$= (x^2 + 9y^6)(x^2 - 9y^6)$
$= (x^2 + 9y^6)(x + 3y^3)(x - 3y^3)$

86. $6x^4 - x^2 - 15 = 6x^4 - 10x^2 + 9x^2 - 15$
$= 2x^2(3x^2 - 5) + 3(3x^2 - 5)$
$= (3x^2 - 5)(2x^2 + 3)$

87. $28yz - 16xyz + x^2yz = yz(28 - 16x + x^2)$
$= yz(14 - x)(2 - x)$

88. $12x^3 + 17x^2 + 6x = x(12x^2 + 17x + 6)$
$= x(12x^2 + 8x + 9x + 6)$
$= x[4x(3x+2) + 3(3x+2)]$
$= x(3x+2)(4x+3)$

89. $16w^2 - 2w - 5 = (2w+1)(8w-5)$

90. $12w^2 - 12w + 3 = 3(4w^2 - 4w + 1)$
$= 3\left[(2w)^2 - 2(2w)(1) + 1^2\right]$
$= 3(2w-1)^2$

91. $4y^3 + 10y^2 - 6y = 2y(2y^2 + 5y - 3)$
$= 2y(2y-1)(y+3)$

92. $10y^2 + 33y - 7 = 10y^2 - 2y + 35y - 7$
$= 2y(5y-1) + 7(5y-1)$
$= (5y-1)(2y+7)$

93. $8y^{10} - 16y^8 = 8y^8(y^2 - 2)$

94. $9x^4 - 144 = 9(x^2 - 16)$
$= 9(x+4)(x-4)$

95. $x^2 - 6x + 12$ is prime.

96. $8x^2 - 19x - 6$ is prime.

97. $8y^5 + 4y^3 - 60y = 4y(2y^4 + y^2 - 15)$
$= 4y(2y^2 - 5)(y^2 + 3)$

98. $9xy^2 + 3xy - 42x = 3x(3y^2 + y - 14)$
$= 3x(3y^2 + 7y - 6y - 14)$
$= 3x[y(3y+7) - 2(3y+7)]$
$= 3x(3y+7)(y-2)$

99. $16x^4y^2 - 56x^2y + 49 = (4x^2y - 7)^2$

100. $128x^3y - 2xy = 2xy(64x^2 - 1)$
$= 2xy[(8x)^2 - 1^2]$
$= 2xy(8x-1)(8x+1)$

101. $2ax + 5a - 10b - 4bx$
$= a(2x+5) - 2b(2x+5)$
$= (2x+5)(a-2b)$

102. $2x^3 - 9 + x^2 - 18x = 2x^3 + x^2 - 18x - 9$
$= x^2(2x+1) - 9(2x+1)$
$= (2x+1)(x^2 - 9)$
$= (2x+1)(x^2 - 3^2)$
$= (2x+1)(x-3)(x+3)$

103. $x^2 - 3x - 18 = 0$
$(x-6)(x+3) = 0$
$x - 6 = 0$ $x + 3 = 0$
$x = 6$ $x = -3$

104. $2x^2 + 5x - 12 = 0$
$(2x-3)(x+4) = 0$
$2x - 3 = 0$ $x + 4 = 0$
$2x = 3$ $x = -4$
$x = \frac{3}{2}$

105. $5x^2 = 2x - 7x^2$
$12x^2 - 2x = 0$
$2x(6x-1) = 0$
$2x = 0$ $6x - 1 = 0$
$x = 0$ $x = \frac{1}{6}$

106.
$$5x^2 - x = 4x^2 + 12$$
$$x^2 - x = 12$$
$$x^2 - x - 12 = 0$$
$$(x-4)(x+3) = 0$$
$$x - 4 = 0 \qquad x + 3 = 0$$
$$x = 4 \qquad x = -3$$

107.
$$2x^2 + 9x - 5 = 0$$
$$(2x-1)(x+5) = 0$$
$$2x - 1 = 0 \qquad x + 5 = 0$$
$$x = \frac{1}{2} \qquad x = -5$$

108.
$$x^2 + 11x + 24 = 0$$
$$(x+8)(x+3) = 0$$
$$x + 8 = 0 \qquad x + 3 = 0$$
$$x = -8 \qquad x = -3$$

109.
$$x^2 + 14x + 45 = 0$$
$$(x+9)(x+5) = 0$$
$$x + 9 = 0 \qquad x + 5 = 0$$
$$x = -9 \qquad x = -5$$

110.
$$5x^2 = 7x + 6$$
$$5x^2 - 7x - 6 = 0$$
$$(5x+3)(x-2) = 0$$
$$5x + 3 = 0 \qquad x - 2 = 0$$
$$5x = -3 \qquad x = 2$$
$$x = -\frac{3}{5} \qquad x = 2$$

111.
$$3x^2 + 6x = 2x^2 - 9$$
$$x^2 + 6x + 9 = 0$$
$$(x+3)^2 = 0$$
$$x + 3 = 0$$
$$x = -3$$

112.
$$4x^2 + 9x - 9 = 0$$
$$(4x-3)(x+3) = 0$$
$$4x - 3 = 0 \qquad x + 3 = 0$$
$$4x = 3 \qquad x = -3$$
$$x = \frac{3}{4} \qquad x = -3$$

113.
$$5x^2 - 11x + 2 = 0$$
$$(5x-1)(x-2) = 0$$
$$5x - 1 = 0 \qquad x - 2 = 0$$
$$x = \frac{1}{5} \qquad x = 2$$

114.
$$x = \text{altitude}$$
$$x + 3 = \text{base}$$
$$A = \frac{1}{2}ab$$
$$35 = \frac{1}{2}x(x+3)$$
$$70 = x^2 + 3x$$
$$0 = x^2 + 3x - 70$$
$$0 = (x+10)(x-7)$$

$x+10=0 \qquad x-7=0$

$x=-10 \qquad x=-7$

Since length cannot be negative, $x=7$ and $x+3=7+3=10$.

altitude: 7 cm; base 10 cm

115. $x = \text{width}$

$2x+1 = \text{length}$

$A = \text{length} \times \text{width}$

$105 = x(2x+1)$

$105 = 2x^2 + x$

$0 = 2x^2 + x - 105$

$0 = (2x+15)(x-7)$

$2x+15=0 \quad \text{or} \quad x-7=0$

$x = -\frac{15}{2} \qquad x=7$

$2x+1=15$

Length cannot be negative so width = 7 feet and length = 15 feet.

116. $h = -16t^2 + 80t + 96$

$0 = -16t^2 + 80t + 96$

$0 = -16(t^2 - 5t - 6)$

$0 = -16(t-6)(t+1)$

$t-6=0 \qquad t+1=0$

$t=6 \qquad t=-1$

Since the time must be positive, $t = 6$ seconds.

117. $480 = -5x^2 + 100x$

$5x^2 - 100x + 480 = 0$

$5(x^2 - 20x + 96) = 0$

$x-12=0 \quad \text{or} \quad x-8=0$

$x=12 \qquad x=8$

The current is 12 amperes or 8 amperes.

How Am I Doing? Chapter 5 Test

1. $x^2 + 12x - 28 = (x+14)(x-2)$

2. $16x^2 - 81 = (4x+9)(4x-9)$

3. $10x^2 + 27x + 5 = 10x^2 + 2x + 25x + 5$
$= 2x(5x+1) + 5(5x+1)$
$= (5x+1)(2x+5)$

4. $9a^2 - 30ab + 25b^2$
$= (3a)^2 - 2(3a)(5b) + (5b)^2$
$= (3a-5b)^2$

5. $7x - 9x^2 + 14xy = x(7 - 9x + 14y)$

6. $10xy + 15by - 8x - 12b$
$= 5y(2x+3b) - 4(2x+3b)$
$= (2x+3b)(5y-4)$

7. $6x^3 - 20x^2 + 16x = 2x(3x^2 - 10x + 8)$
$= 2x(3x^2 - 6x - 4x + 8)$
$= 2x[3x(x-2) - 4(x-2)]$
$= 2x(x-2)(3x-4)$

8. $5a^2c - 11ac + 2c = c(5a^2 - 11a + 2)$
$= c(5a-1)(a-2)$

9. $81x^2 - 100 = (9x+10)(9x-10)$

10. $9x^2 - 15x + 4 = (3x-1)(3x-4)$

11. $20x^2 - 45 = 5(4x^2 - 9)$
$= 5(2x+3)(2x-3)$

12. $36x^2 + 1 = (6x)^2 + (1)^2$
It is prime.

13. $3x^3 + 11x^2 + 10x = x(3x^2 + 11x + 10)$
$= x(3x+5)(x+2)$

14. $60xy^2 - 20x^2y - 45y^3$
$= -5y(-12xy + 4x^2 + 9y^2)$
$= -5y(4x^2 - 12xy + 9y^2)$
$= -5y[(2x)^2 - 2(2x)(3y) + (3y)^2]$
$= -5y(2x-3y)^2$

15. $81x^2 - 1 = (9x)^2 - (1)^2$
$= (9x+1)(9x-1)$

16. $81y^4 - 1 = (9y^2+1)(9y^2-1)$
$= (9y^2+1)(3y+1)(3y-1)$

17. $2ax + 6a - 5x - 15 = 2a(x+3) - 5(x+3)$
$= (x+3)(2a-5)$

18. $aw^2 - 8b + 2bw^2 - 4a$
$= aw^2 - 4a + 2bw^2 - 8b$
$= a(w^2-4) + 2b(w^2-4)$
$= (w^2-4)(a+2b)$
$= (w-2)(w+2)(a+2b)$

19. $3x^2 - 3x - 90 = 3(x^2 - x - 30)$
$= 3(x-6)(x+5)$

20. $2x^3 - x^2 - 15x = x(2x^2 - x - 15)$
$= x(2x^2 - 6x + 5x - 15)$
$= x[2x(x-3) + 5(x-3)]$
$= x(x-3)(2x+5)$

21. $x^2 + 14x + 45 = 0$
$(x+9)(x+5) = 0$
$x + 9 = 0 \qquad x + 5 = 0$
$x = -9 \qquad x = -5$

22. $14+3x(x+2)=-7x$

$$14+3x^2+6x=-7x$$
$$3x^2+13x+14=0$$
$$(3x+7)(x+2)=0$$
$$3x+7=0 \qquad x+2=0$$
$$3x=-7 \qquad x=-2$$
$$x=-\frac{7}{3} \qquad x=-2$$

23. $2x^2+x-10=0$

$$(2x+5)(x-2)=0$$
$$2x+5=0 \qquad x-2=0$$
$$x=-\frac{5}{2} \qquad x=2$$

24. $x^2-3x-28=0$

$$(x-7)(x+4)=0$$
$$x-7=0 \qquad x+4=0$$
$$x=7 \qquad x=-4$$

25. $x = \text{width}$

$$2x-1=\text{length}$$
$$x(2x-1)=91$$
$$2x^2-x=91$$
$$2x^2-x-91=0$$
$$(2x+13)(x-7)=0$$
$$2x+13=0 \qquad x-7=0$$
$$2x=-13 \qquad x=7$$
$$x=-\frac{13}{2}$$

Since the width cannot be negative, $x = 7$ and $2x-1=2(7)-1=13$.

width = 7 miles, length = 13 miles.

Cumulative Test Chapters 0 - 5

1. $\frac{72}{480}=0.15=15\%$

2. $(0.13)(3.8)=0.494$

3. $4.3-2(3.5-7.2)-6.1$
$=4.3-2(-3.8)-6.1$
$=4.3+7.6-6.1$
$=5.8$

4. $(-2x^3y^4)(-4xy^6)=(-2)(-4)x^{3+1}y^{4+6}$
$=8x^4y^{10}$

5. $(-3)^4=81$

6. $(9x-4)(3x+2)=27x^2+18x-12x-8$
$=27x^2+6x-8$

7. $(2x^2-6x+1)(x-3)$
$=2x^3-6x^2+x-6x^2+18x-3$
$=2x^3-12x^2+19x-3$

8. $3x-4 \geq 6x+5$
$$-4 \geq 3x+5$$
$$-9 \geq 3x$$
$$-3 \geq x$$

9. $3x-(7-5x)=3(4x-5)$
$$3x-7+5x=12x-15$$
$$-7+8x=12x-15$$
$$8=4x$$
$$2=x$$

10. $\frac{1}{2}x-3=\frac{1}{4}(3x+3)$
$$4\left(\frac{1}{2}x-3\right)=4\left[\frac{1}{4}(3x+3)\right]$$
$$2x-12=3x+3$$
$$-12=x+3$$
$$-15=x$$

11. $3(q+1)=8q-p$
$$3q+3=5q-p$$
$$p+3=5q$$
$$\frac{p+3}{5}=q$$

12. $6x^2-5x+1=6x^2-3x-2x+1$
$$=3x(2x-1)-1(2x-1)$$
$$=(3x+2)(3x-1)$$

13. $6x^2+5x-4=(3x+4)(2x-1)$

14. $9x^2+3x-2=9x^2+6x-3x-2$
$$=3x(3x+2)-1(3x+2)$$
$$=(3x+2)(3x-1)$$

15. $121x^2-64y^2=(11x)^2-(8y)^2$
$$=(11x+8y)(11x-8y)$$

16. $4x+120-80x^2=-80x^2+4x+120$
$$=-4(20x^2-x-30)$$
$$=-4(20x^2+24x-25x-30)$$
$$=-4[4x(5x+6)-5(5x+6)]$$
$$=-4(5x+6)(4x-5)$$

17. x^2+5x+9 is prime.

18. $16x^3+40x^2+25x=x(16x^2+40x+25)$
$$=x\left[(4x)^2+2(4x)(5)+5^2\right]$$
$$=x(4x+5)^2$$

19. $81x^4-16b^4=(9x^2)^2-(4b^2)^2$
$$=(9x^2+4b^2)(9x^2-4b^2)$$
$$=(9x^2+4b^2)(3x+2b)(3x-2b)$$

20. $2ax-4bx+3a-6b$
$$=2x(a-2b)+3(a-2b)$$
$$=(a-2b)(2x+3)$$

21. $15x^2+6x-9=3\left(5x^2+2x-3\right)$
$$=3\left(x^2+5\right)\left(x^2+3\right)$$

22. $x^2+5x-24=0$
$$(x+8)(x-3)=0$$
$$x+8=0 \qquad x-3=0$$
$$x=-8 \qquad x=3$$

23. $3x^2-11x+10=0$
$$(3x-5)(x-2)=0$$
$$3x-5=0 \qquad x-2=0$$
$$x=\frac{5}{3} \qquad x=2$$

24. $x=\text{length}$
$$x-5=\text{width}$$
$$A=LW$$
$$300=x(x-5)$$
$$300=x^2-5x$$
$$0=x^2-5x-300$$
$$0=(x-20)(x+15)$$

$$x-20=0 \qquad x+15=0$$
$$x=20 \qquad x=-15$$
Not possible
$$x-5=15$$
$$\text{length}=20\text{ ft., width}=15\text{ ft.}$$

Chapter 6

6.1 Exercises

1. $\frac{3a-9b}{a-3b}=\frac{3(a-3b)}{a-3b}=3$

3. $\frac{6x+18}{x^2+3x}=\frac{6(x+3)}{x(x+3)}=\frac{6}{x}$

5. $\frac{9x^2+6x+1}{1-9x^2}=\frac{(3x+1)(3x+1)}{(1+3x)(1-3x)}=\frac{3x+1}{1-3x}$

7. $\frac{3a^2b(a-2b)}{6ab^2}=\frac{a(a-2b)}{2b}$

9. $\frac{x^2+x-2}{x^2-x}=\frac{(x+2)(x-1)}{x(x-1)}=\frac{x+2}{x}$

11. $\frac{x^2-3x-10}{3x^2+5x-2}=\frac{(x-5)(x+2)}{(3x-1)(x+2)}$
$=\frac{x-5}{3x-1}$

13. $\frac{x^2+4x-21}{x^3-49x}=\frac{(x+7)(x-3)}{x(x^2-49)}$
$=\frac{(x+7)(x-3)}{x(x+7)(x-7)}$
$=\frac{x-3}{x(x-7)}$

15. $\frac{3x^2+7x-6}{x^2+7x+12}=\frac{(3x-2)(x+3)}{(x+4)(x+3)}=\frac{3x-2}{x+4}$

17. $\frac{3x^2-8x+5}{4x^2-5x+1}=\frac{(3x-5)(x-1)}{(4x-1)(x-1)}=\frac{3x-5}{4x-1}$

19. $\frac{5x^2-27x+10}{5x^2+3x-2}=\frac{(5x-2)(x-5)}{(5x-2)(x+1)}$
$=\frac{x-5}{x+1}$

21. $\frac{10-2x}{4x^2-20x}=\frac{2(5-x)}{4x(x-5)}$
$=\frac{-2(x-5)}{4x(x-5)}$
$=\frac{-1}{2x}$

23. $\frac{2x^2-7x-15}{25-x^2}=\frac{(2x+3)(x-5)}{(5-x)(5+x)}$
$=\frac{-2x-3}{x+5}=\frac{-2x-3}{5+x}$

25. $\frac{(4x+5)^2}{8x^2+6x-5}=\frac{(4x+5)^2}{(4x+5)(2x-1)}$
$=\frac{(4x+5)(4x+5)}{(4x+5)(2x-1)}=\frac{4x+5}{2x-1}$

27. $\frac{2x^2+9x-18}{30-x-x^2}=\frac{2x^2+9x-18}{-(x^2+x-30)}$

$$= \frac{(2x-3)(x+6)}{-(x+6)(x-5)}$$
$$= \frac{2x-3}{-x+5}$$

29. $\frac{a^2+2ab-3b^2}{2a^2+5ab-3b^2} = \frac{(a+3b)(a-b)}{(2a-b)(a+3b)} = \frac{a-b}{2a-b}$

31. $\frac{9x^2-4y^2}{9x^2+12xy+4y^2} = \frac{(3x-2y)(3x+2y)}{(3x+2y)(3x+2y)}$
$$= \frac{3x-2y}{3x+2y}$$

33. $\frac{x^3-bx^2-abx+ax^2}{bx^2-b^2x-b^2c+bcx} = \frac{x(x^2-bx-ab+ax)}{b(x^2-bx-bc+cx)}$
$$= \frac{x[x(x-b)+a(-b+x)]}{b[x(-b)+c(-b+x)]}$$
$$= \frac{x(x-b)(x+a)}{b(x-b)(x+c)}$$
$$= \frac{x(x+a)}{b(x+c)}$$

Cumulative Review

35. $(3x-7)^2 = (3x)^2 - 2(3x)(7) + (7)^2$
$$= 9x^2 - 42x + 49$$

37. $(2x+3)(x-4)(x-2)$
$$= (2x+3)(x^2-6x+8)$$
$$= 2x^3 - 12x^2 + 16x + 3x^2 - 18x + 24$$
$$= 2x^3 - 9x^2 - 2x + 24$$

39. $\frac{4\frac{7}{8}}{3} = \frac{\frac{39}{8}}{\frac{3}{1}} = \frac{39}{8}\cdot\frac{1}{3} = \frac{13}{8} = 1\frac{5}{8}$ acre

41. Spanish $= 341 + 17 = 358$ million
Chinese Mandarin $= 2(358) + 158 = 874$
874,000,000 people

6.2 Exercises

1. Factor the numerator and denominator completely and divide out any common factors.

3. $\frac{2x-10}{x-4}\cdot\frac{x^2+5x+4}{x^2-4x-5} = \frac{2(x-5)}{x-4}\cdot\frac{(x+4)(x+1)}{(x-5)(x+1)}$
$$= \frac{2(x+4)}{x-4}$$

5. $\frac{24x^3}{4x^2-36}\cdot\frac{2x^2+6x}{16x^2} = \frac{3(8x^3)}{4(x^2-9)}\cdot\frac{2x(x+3)}{2\cdot 8x^2}$
$$= \frac{3\cdot 8x^3}{4(x+3)(x-3)}\cdot\frac{2x(x+3)}{2\cdot 8x^2}$$
$$= \frac{3x^2}{4(x-3)}$$

7. $\frac{x^2+3x-10}{x^2+x-20}\cdot\frac{x^2-3x-4}{x^2+4x+3}$
$$= \frac{(x-2)(x+5)}{(x-4)(x+5)}\cdot\frac{(x-4)(x+1)}{(x+3)(x+1)}$$
$$= \frac{x-2}{x+3}$$

9. $\frac{x+6}{x-8} \div \frac{x+5}{x^2-6x-16} = \frac{x+6}{x-8} \cdot \frac{x^2-6x-16}{x+5}$

$= \frac{x+6}{x-8} \cdot \frac{(x-8)(x+2)}{x+5}$

$= \frac{(x+6)(x+2)}{x+5}$

11. $(6x-5) \div \frac{36x^2-25}{6x^2+17x+10}$

$= \frac{6x-5}{1} \cdot \frac{6x^2+17x+10}{36x^2-25}$

$= \frac{6x-5}{1} \cdot \frac{(6x+5)(x+2)}{(6x+5)(6x-5)}$

$= x+2$

13. $\frac{3x^2+12xy+12y^2}{x^2+4xy+3y^2} \div \frac{4x+8y}{x+y}$

$= \frac{3(x+2y)(x+2y)}{(x+3y)(x+y)} \div \frac{4(x+2y)}{x+y}$

$= \frac{3(x+2y)(x+2y)}{(x+3y)(x+y)} \cdot \frac{x+y}{4(x+2y)} = \frac{3(x+2y)}{4(x+3y)}$

15. $\frac{(x+5)^2}{3x^2-7x+2} \cdot \frac{x^2-4x+4}{(x+5)}$

$= \frac{(x+5)(x+5)}{(3x-1)(x-2)} \cdot \frac{(x-2)(x-2)}{(x+5)}$

$= \frac{(x+5)(x-2)}{3x-1}$

17. $\frac{x^2+x-30}{10-2x} \div \frac{x^2+4x-12}{5x+15}$

$= \frac{x^2+x-30}{10-2x} \cdot \frac{5x+15}{x^2+4x-12}$

$= \frac{(x+6)(x-5)}{2(5-x)} \cdot \frac{5(x+3)}{(x+6)(x-2)}$

$= -\frac{5(x+3)}{2(x-2)}$

19. $\frac{y^2+4y-12}{y^2+2y-24} \cdot \frac{y^2-16}{y^2+2y-8}$

$= \frac{(y+6)(y-2)}{(y+6)(y-4)} \cdot \frac{(y+4)(y-4)}{(y+4)(y-2)}$

$= 1$

21. $\frac{x^2+7x+12}{2x^2+9x+4} \div \frac{x^2+6x+9}{2x^2-x-1}$

$= \frac{(x+4)(x+3)}{(2x+1)(x+4)} \cdot \frac{(2x+1)(x-1)}{(x+3)(x+3)}$

$= \frac{x-1}{x+3}$

23. By definition, the denominator of a rational expression cannot have the value zero. So the original expression cannot have a replacement value of 2 (or the first denominator would be zero). If we multiply the first fraction by the reciprocal of the second fraction, then the denominator $x+7$ cannot be zero. Thus x also cannot have a replacement value of -7.

Cumulative Review

25. $6x^2 + 3x - 18 = 5x - 2 + 6x^2$

$$3x - 18 = 5x - 2$$
$$-18 = 2x - 2$$
$$-16 = 2x$$
$$-8 = x$$

27. $\frac{1}{16}(80) = 5$

5 milligrams of medication

29. Harold: length $= x$, width $= x$

George: length $= 3x + x$, width $= x - 2$

$A = LW$

$36 = (3 + x)(x - 2)$

$36 = x^2 + x - 6$

$0 = x^2 + x - 6$

$0 = (x + 7)(x - 6)$

$x + 7 = 0 \qquad x - 6 = 0$

$x = -7 \qquad x = 6$

Not possible

Harold: 6 ft by 6 ft

George: 9 ft by 4 ft

6.3 Exercises

1. The LCD would be a product that contains each factor. However, any repeated factor in any one denominator must be repeated the greatest number of times it occurs in any one denominator. So the LCD would be $(x+5)(x+3)^2$.

3. $\frac{x}{x+5} + \frac{2x+1}{5+x} = \frac{x + 2x + 1}{x+5}$

$$= \frac{3x+1}{x+5}$$

5. $\frac{3x}{x+3} - \frac{x+5}{x+3} = \frac{3x - (x+5)}{x+3}$

$$= \frac{3x - x - 5}{x+3}$$
$$= \frac{2x-5}{x+3}$$

7. $\frac{8x+3}{5x+7} - \frac{6x+10}{5x+7} = \frac{8x + 3 - (6x+10)}{5x+7}$

$$= \frac{8x + 3 - 6x - 10}{5x+7}$$
$$= \frac{2x-7}{5x+7}$$

9. $3ab$

a^2, b^2

LCD $= 3a^2b^2$

11. $18x^2y^5 = 2\cdot 3^2x^2y^5$
$30x^3y^3 = 2\cdot 3\cdot 5x^3y^3$
$\text{LCD} = 2\cdot 3^2\cdot 5x^3y^5 = 90x^3y^5$

13. $5x-15 = 5(x-3)$
$3x-9 = 3(x-3)$
$\text{LCD} = 3\cdot 5(x-3)$

15. $x+3 = x+3$
$x^2-9 = (x+3)(x-3)$
$\text{LCD} = (x+3)(x-3) = x^2-9$

17. $3x^2+14x-5 = (3x-1)(x+5)$
$9x^2-6x+1 = (3x-1)(3x-1)$
$\text{LCD} = (x+5)(3x-1)^2$

19. $\frac{7}{ab}+\frac{3}{b} = \frac{7}{ab}+\frac{3}{b}\cdot\frac{a}{a}$
$= \frac{7}{ab}+\frac{3a}{ab}$
$= \frac{7+3a}{ab}$

21. $\frac{3}{x+7}+\frac{8}{x^2-49} = \frac{3}{x+7}\cdot\frac{x-7}{x-7}+\frac{8}{(x+7)(-7)}$
$= \frac{3x-21+8}{(x+7)(x-7)} = \frac{3x-13}{(x+7)(x-7)}$

23. $\frac{3y}{y+2}+\frac{y}{y-2}$
$= \frac{3y}{y+2}\cdot\frac{y-2}{y-2}+\frac{y}{y-2}\cdot\frac{y+2}{y+2}$
$= \frac{3y^2-6y}{(y+2)(y-2)}+\frac{y^2+2y}{(y-2)(y+2)}$
$= \frac{3y^2-6y+y^2+2y}{(y+2)(y-2)} = \frac{4y^2-4y}{(y+2)(y-2)}$

25. $\frac{6}{5a}+\frac{5}{3a+2} = \frac{6}{5a}\cdot\frac{3a+2}{3a+2}+\frac{5}{3a+2}\cdot\frac{5a}{5a}$
$= \frac{6(3a+2)+5\cdot 5a}{5a(3a+2)}$
$= \frac{18a+12+25a}{5a(3a+2)}$
$= \frac{43a+12}{5a(3a+2)}$

27. $\frac{2}{3xy}+\frac{1}{6yz} = \frac{2}{3xy}\cdot\frac{2z}{2z}+\frac{1}{6yz}\cdot\frac{x}{x}$
$= \frac{4z}{6xyz}+\frac{x}{6xyz} = \frac{4z+x}{6xyz}$

29. $\frac{5x+6}{x-3}-\frac{x-2}{2x-6} = \frac{5x+6}{x-3}\cdot\frac{2}{2}-\frac{x-2}{2(x-3)}$
$= \frac{2(5x+6)-(x-2)}{2(x-3)}$
$= \frac{10x+12-x+2}{2(x-3)}$
$= \frac{9x+14}{2(x-3)}$

31. $\frac{3x}{x^2-25}-\frac{2}{x+5}=\frac{3x}{(x+5)(x-5)}-\frac{2}{x+5}\cdot\frac{x-5}{x-5}$
$=\frac{3x-2(x-5)}{(x+5)(x-5)}$
$=\frac{3x-2x+10}{(x+5)(x-5)}$
$=\frac{x+10}{(x+5)(x-5)}$

33. $\frac{a+3b}{2}-\frac{a-b}{5}=\frac{a+3b}{2}\cdot\frac{5}{5}-\frac{a-b}{5}\cdot\frac{2}{2}$
$=\frac{5(a+3b)-2(a-b)}{2\cdot 5}$
$=\frac{5a+15b-2a+2b}{10}$
$=\frac{3a+17b}{10}$

35. $\frac{8}{2x-3}-\frac{6}{x+2}$
$=\frac{8}{2x-3}\cdot\frac{x+2}{x+2}-\frac{6}{x+2}\cdot\frac{2x-3}{2x-3}$
$=\frac{8x+16}{(2x-3)(x+2)}-\frac{12x-18}{(2x-3)(x+2)}$
$=\frac{8x+16-(12x-18)}{(2x-3)(x+2)}=\frac{8x+16-12x+18}{(2x-3)(x+2)}$
$=\frac{-4x+34}{(2x-3)(x+2)}$

37. $\frac{x}{x^2+2x-3}-\frac{x}{x^2-5x+4}$
$=\frac{x}{(x+3)(x-1)}-\frac{x}{(x-4)(x-1)}$
$=\frac{x}{(x+3)(x-1)}\cdot\frac{x-4}{x-4}-\frac{x}{(x-4)(x-1)}\cdot\frac{x+3}{x+3}$
$=\frac{x^2-4x}{(x+3)(x-1)(x-4)}-\frac{x^2+3x}{(x+3)(x-1)(x-4)}$
$=\frac{x^2-4x-(x^2+3x)}{(x+3)(x-1)(x-4)}$
$=\frac{x^2-4x-x^2-3x}{(x+3)(x-1)(x-4)}$
$=\frac{-7x}{(x+3)(x-1)(x-4)}$

39. $\frac{2}{x^2+5x+6}+\frac{3}{x^2+7x+10}$
$=\frac{2}{(x+2)(x+3)}\cdot\frac{x+5}{x+5}+\frac{3}{(x+2)(x+5)}\cdot\frac{x+3}{x+3}$
$=\frac{2x+10+3x+9}{(x+2)(x+3)(x+5)}=\frac{5x+19}{(x+2)(x+3)(x+5)}$

41. $\frac{3x-8}{x^2-5x+6}+\frac{x+2}{x^2-6x+8}$
$=\frac{3x-8}{(x-2)(x-3)}+\frac{x+2}{(x-4)(x-2)}$
$=\frac{3x-8}{(x-2)(x-3)}\cdot\frac{x-4}{x-4}+\frac{x+2}{(x-4)(x-2)}\cdot\frac{x-3}{x-3}$
$=\frac{3x^2-20x+32}{(x-2)(x-3)(x-4)}+\frac{x^2-x-6}{(x-2)(x-3)(x-4)}$
$=\frac{3x^2-20x+32+x^2-x-6}{(x-2)(x-3)(x-4)}$
$=\frac{4x^2-21x+26}{(x-2)(x-3)(x-4)}$
$=\frac{(x-2)(4x-13)}{(x-2)(x-3)(x-4)}=\frac{4x-13}{(x-3)(x-4)}$

43. $\frac{6x}{y-2x}-\frac{5x}{2x-y}=\frac{6x}{-(2x-y)}-\frac{5x}{2x-y}$

$=\frac{-6x-5x}{2x-y}$

$=\frac{-11x}{2x-y}$

$=\frac{11x}{y-2x}$

45. $\frac{3y}{8y^2+2y-1}-\frac{5y}{2y^2-9y-5}$

$=\frac{3y}{(4y-1)(2y+1)}-\frac{5y}{(2y+1)(y-5)}$

$=\frac{3y}{(4y-1)(2y+1)}\cdot\frac{y-5}{y-5}-\frac{5y}{(2y+1)(y-5)}\cdot\frac{4y-1}{4y-1}$

$\frac{3y^2-15y}{(4y-1)(2y+1)(y-5)}-\frac{20y^2-5y}{(4y-1)(2y+1)(y-5)}$

$=\frac{3y^2-15y-(20y^2-5y)}{(4y-1)(2y+1)(y-5)}$

$=\frac{3y^2-15y-20y^2+5y}{(4y-1)(2y+1)(y-5)}$

$=\frac{-17y^2-10y}{(4y-1)(2y+1)(y-5)}$

47. $\frac{4y}{y^2+4y+3}+\frac{2}{y+1}=\frac{4y}{(y+3)(y+1)}+\frac{2}{y+1}$

$=\frac{4y}{(y+3)(y+1)}+\frac{2}{y+1}\cdot\frac{y+3}{y+3}$

$=\frac{4y}{(y+3)(y+1)}+\frac{2y+6}{(y+3)(y+1)}$

$=\frac{4y+2y+6}{(y+3)(y+1)}=\frac{6y+6}{(y+3)(y+1)}$

$=\frac{6(y+1)}{(y+3)(y+1)}=\frac{6}{y+3}$

49. $\frac{2x}{x-3}+\frac{3x}{x+2}+\frac{7}{x^2-x-6}$

$=\frac{2x}{x-3}\cdot\frac{x+2}{x+2}+\frac{3x}{x+2}\cdot\frac{x-3}{x-3}+\frac{7}{(x+2)(x-3)}$

$=\frac{2x^2+4x+3x^2-9x+7}{(x+2)(x-3)}$

$=\frac{5x^2-5x+7}{(x+2)(x-3)}$

Cumulative Review

51. $\frac{1}{3}(x-2)+\frac{1}{2}(x+3)=\frac{1}{4}(3x+1)$

$12\left[\frac{1}{3}(x-2)+\frac{1}{2}(x+3)\right]=12\left(\frac{1}{4}\right)(3x+1)$

$4(x-2)+6(x+3)=3(3x+1)$

$4x-8+6x+18=9x+3$

$10x+10=9x+3$

$x=-7$

53. $\frac{1}{2}x < \frac{1}{3}x + \frac{1}{4}$

$12\left(\frac{1}{2}x\right) < 12\left(\frac{1}{3}x + \frac{1}{4}\right)$

$6x < 4x + 3$

$2x < 3$

$x < \frac{3}{2}$

55. Let x = number of days.

$1.50(2x) > 50$

$3x > 50$

$x > \frac{50}{3}$

$x > 16\frac{2}{3}$

At least 17 days per month

57. (a) Drop $20 - 18 = 18$ million

Rate $= \frac{18}{20} = 0.9$ million

$= 900,000$ fish/month

(b) x = number of months

$0.9x = 2$

$x \approx 2.2$

It would be destroyed after 3 months.

How Am I Doing? Sections 6.1-6.3

1. $\frac{8x-48}{x^2-6x} = \frac{8(x-6)}{x(x-6)} = \frac{8}{x}$

2. $\frac{2x^2-7x-15}{x^2-12x+35} = \frac{(2x+3)(x-5)}{(x-5)(x-7)}$

$= \frac{2x+3}{x-7}$

3. $\frac{y^2+6y+9}{27x^2-3x^2y^2} = \frac{(y+3)(y+3)}{3x^2(9-y^2)}$

$= \frac{(y+3)(y+3)}{3x^2(3+y)(3-y)}$

$= \frac{y+3}{3x^2(3-y)}$

4. $\frac{5x^2-23x+12}{5x^2+7x-6} = \frac{(5x-3)(x-4)}{(5x-3)(x+2)}$

$= \frac{x-4}{x+2}$

5. $\frac{8x^3}{3x+12} \cdot \frac{9x+36}{16x^2} = \frac{8x^3}{3(x+4)} \cdot \frac{9(x+4)}{2 \cdot 8x^2}$

$= \frac{3x}{2}$

6. $\dfrac{x-5}{x^2+5x-14}\cdot\dfrac{x^2+12x+35}{15-3x}$
$=\dfrac{x-5}{(x+7)(x-2)}\cdot\dfrac{(x+7)(x+5)}{3(5-x)}$
$=-\dfrac{x+5}{3(x-2)}$

7. $\dfrac{x^2-9}{2x+6}\div\dfrac{2x^2-5x-3}{4x^2-1}$
$=\dfrac{x^2-9}{2x+6}\cdot\dfrac{4x^2-1}{2x^2-5x-3}$
$=\dfrac{(x+3)(x-3)}{2(x+3)}\cdot\dfrac{(2x+1)(2x-1)}{(2x+1)(x-3)}$
$=\dfrac{2x-1}{2}$

8. $\dfrac{3a^2+7a+2}{4a^2+11a+6}\div\dfrac{16a^2-9}{6a^2-9}$
$=\dfrac{3a^2+7a+2}{4a^2+11a+6}\cdot\dfrac{16a^2-9}{6a^2-13a-5}$
$=\dfrac{(3a+1)(a+2)}{(4a+3)(a+2)}\cdot\dfrac{(4a+3)(4a-3)}{(2a-5)(3a+1)}$
$=\dfrac{4a-3}{2a-5}$

9. $\dfrac{x-3y}{xy}-\dfrac{4a-y}{ay}=\dfrac{x-3y}{xy}\cdot\dfrac{a}{a}-\dfrac{4a-y}{ay}\cdot\dfrac{x}{x}$
$=\dfrac{a(x-3y)-x(4a-y)}{axy}$
$=\dfrac{ax-3ay-4ax+xy}{axy}$
$=\dfrac{xy-3ax-3ay}{axy}$

10. $\dfrac{7}{2x-4}+\dfrac{-14}{x^2-4}=\dfrac{7}{2(x-2)}\cdot\dfrac{x+2}{x+2}$
$+\dfrac{-14}{(x+2)(x-2)}\cdot\dfrac{2}{2}$
$=\dfrac{7(x+2)=14\cdot 2}{2(x+2)(x-2)}$
$=\dfrac{7x+14-28}{2(x+2)(x-2)}$
$=\dfrac{7x-14}{2(x+2)(x-2)}$
$=\dfrac{7(x-2)}{2(x+2)(x-2)}$
$=\dfrac{7}{2(x+2)}$

11. $\dfrac{2x}{x^2+10x+21}+\dfrac{x-3}{x+7}$
$=\dfrac{2x}{(x+3)(x+7)}+\dfrac{x-3}{x+7}\cdot\dfrac{x+3}{x+3}$
$=\dfrac{2x+(x-3)(x+3)}{(x+3)(x+7)}$
$=\dfrac{2x+x^2-9}{(x+3)(x+7)}$
$=\dfrac{x^2+2x-9}{(x+3)(x+7)}$

12. $$\frac{2}{x^2+3x-10}-\frac{5x}{x^2-6x+8}$$
$$=\frac{2}{(x+5)(x-2)}\cdot\frac{x-4}{x-4}-\frac{5x}{(x-2)(x-4)}\cdot\frac{x+5}{x+5}$$
$$=\frac{2(x-4)-5(x+5)}{(x-2)(x-4)(x+5)}$$
$$=\frac{2x-8-5x^2\quad 25x}{(x-2)(x-4)(x+5)}$$
$$=\frac{-5x^2-23x-8}{(x-2)(x-4)(x+5)}$$

6.4 Exercises

1. $$\frac{\frac{3}{x}}{\frac{2}{x^2}+\frac{5}{x}}=\frac{\frac{3}{x}}{\frac{2}{x^2}+\frac{5}{x}}\cdot\frac{x^2}{x^2}$$
$$=\frac{\frac{3x^2}{x}}{\frac{2x^2}{x^2}+\frac{5x^2}{x}}$$
$$=\frac{3x}{2+5x}$$

3. $$\frac{\frac{1}{x}+\frac{1}{y}}{\frac{1}{xy}}=\frac{\frac{1}{x}+\frac{1}{y}}{\frac{1}{xy}}\cdot\frac{(xy)}{(xy)}$$
$$=\frac{\frac{xy}{x}+\frac{xy}{x}}{\frac{xy}{xy}}=\frac{y+x}{1}=y+x$$

5. $$\frac{\frac{x}{6}-\frac{1}{3}}{\frac{2}{3x}+\frac{5}{6}}=\frac{\frac{x}{6}(6x)-\frac{1}{3}(6x)}{\frac{2}{3x}(6x)+\frac{5}{6}(6x)}$$
$$=\frac{x^2-2x}{4+5x}$$

7. $$\frac{\frac{7}{5x}-\frac{1}{x}}{\frac{3}{5}+\frac{2}{x}}=\frac{\frac{7}{5x}(5x)-\frac{1}{x}(5x)}{\frac{3}{5}(5x)+\frac{2}{x}(5x)}$$
$$=\frac{7-5}{3x+10}$$
$$=\frac{2}{3x+10}$$

9. $$\frac{\frac{5}{x}+\frac{3}{y}}{3x+5y}=\frac{\frac{5}{x}+\frac{3}{y}}{3x+5y}\cdot\frac{xy}{xy}$$
$$=\frac{\frac{5xy}{x}+\frac{3xy}{y}}{(3x+5y)xy}$$
$$=\frac{5y+3x}{(3x+5y)xy}$$
$$=\frac{1}{xy}$$

11. $$\frac{4-\frac{1}{x^2}}{2+\frac{1}{x}}=\frac{4-\frac{1}{x^2}}{2+\frac{1}{x}}\cdot\frac{x^2}{x^2}$$
$$=\frac{4x^2-\frac{x^2}{x^2}}{2x^2+\frac{x^2}{x}}$$
$$=\frac{4x^2-1}{2x^2+x}$$
$$=\frac{(2x+1)(2x-1)}{x(2x+1)}$$
$$=\frac{2x-1}{x}$$

13. $\dfrac{\frac{10}{x^2-25}}{\frac{3}{x+5}+\frac{2}{x-5}} = \dfrac{\frac{10}{x^2-25}}{\frac{3}{x+5}+\frac{2}{x-5}} \cdot \dfrac{(x+5)(x-5)}{(x+5)(x-5)}$

$= \dfrac{\frac{10(x+5)(x-5)}{(x+5)(x-5)}}{\frac{(x+5)(x-5)}{(x+5)}+\frac{2(x+5)(x-5)}{(x-5)}}$

$= \dfrac{10}{3(x-5)+2(x+5)}$

$= \dfrac{10}{3x-15+2x+10}$

$= \dfrac{10}{5x-5}$

$= \dfrac{10}{5(x-1)}$

$= \dfrac{2}{x-1}$

15. $\dfrac{a+\frac{3}{a}}{\frac{a^2+2}{3a}} = \dfrac{a+\frac{3}{a}}{\frac{a^2+2}{3a}} \cdot \dfrac{3a}{3a}$

$= \dfrac{a(3a)+\left(\frac{3}{a}\right)(3a)}{\left(\frac{a^2+2}{3a}\right)(3a)}$

$= \dfrac{3a^2+9}{a^2+2}$

17. $\dfrac{\frac{3}{x-3}}{\frac{1}{x^2-9}+\frac{2}{x+3}}$

$= \dfrac{\frac{3}{x-3}}{\frac{1}{x^2-9}+\frac{2}{x+3}} \cdot \dfrac{(x+3)(x-3)}{(x+3)(x-3)}$

$= \dfrac{\left(\frac{3}{x-3}\right)\frac{(x+3)(x-3)}{1}}{\frac{1}{(x+3)(x-3)}\frac{(x+3)(x-3)}{1}+\left(\frac{2}{x+3}\right)\frac{(x+3)(x-3)}{1}}$

$= \dfrac{3(x+3)}{1+2(x-3)}$

$= \dfrac{3(x+3)}{1+2x-6}$

$= \dfrac{3x+9}{2x-5}$

19. $\dfrac{\frac{2}{y-1}+2}{\frac{2}{y+1}-2} = \dfrac{\frac{2}{y-1}+2}{\frac{2}{y+1}-2} \cdot \dfrac{(y-1)(y+1)}{(y-1)(y+1)}$

$= \dfrac{\frac{2(y-1)(y+1)}{y-1}+2(y-1)(y+1)}{\frac{2(y-1)(y+1)}{(y+1)}-2(y-1)(y+1)}$

$= \dfrac{2(y+1)[1+(y-1)]}{2(y-1)[1-(y+1)]} = \dfrac{(y+1)(y)}{(y-1)(1-y-1)}$

$= \dfrac{y(y+1)}{(y-1)(-y)} = \dfrac{y+1}{-y+1}$

21. No expression in any denominator can be allowed to be zero, since division by zero is undefined. So -3, 5, and 0 are not allowable replacements for the variable x.

Cumulative Review

23.
$$P = 2(\ell + w)$$
$$P = 2\ell + 2w$$
$$P - 2\ell = 2w$$
$$\frac{P - 2\ell}{2} = w$$

25. $4000 - 125(17) = 1875$

She has yet to pay \$1875.

6.5 Exercises

1.
$$\frac{7}{x} + \frac{3}{4} = \frac{-2}{x}$$
$$4x\left(\frac{7}{x}\right) + 4x\left(\frac{3}{4}\right) = 4x\left(-\frac{2}{x}\right)$$
$$28 + 3x = -8$$
$$3x = -36$$
$$x = -12$$

Check:
$$\frac{7}{-12} + \frac{3}{4} \stackrel{?}{=} \frac{-2}{-12}$$
$$\frac{7}{-12} + \frac{9}{12} \stackrel{?}{=} \frac{1}{6}$$
$$\frac{2}{12} \stackrel{?}{=} \frac{1}{6}$$
$$\frac{1}{6} = \frac{1}{6}$$

3.
$$\frac{-1}{4x} + \frac{3}{2} = \frac{5}{x}$$
$$4x\left(-\frac{1}{4x}\right) + 4x\left(\frac{3}{2}\right) = 4x\left(\frac{5}{x}\right)$$
$$-1 + 6x = 20$$
$$6x = 21$$
$$x = \frac{21}{6}$$
$$x = \frac{7}{2} \text{ or } 3\frac{1}{2}$$

Check:
$$-\frac{1}{4(7/2)} + \frac{3}{2} \stackrel{?}{=} \frac{5}{7/2}$$
$$-\frac{1}{14} + \frac{3}{2} \stackrel{?}{=} \frac{10}{7}$$
$$\frac{1}{14} + \frac{21}{14} \stackrel{?}{=} \frac{10}{7}$$
$$\frac{20}{14} \stackrel{?}{=} \frac{10}{7}$$
$$\frac{10}{7} = \frac{10}{7}$$

5.
$$\frac{5x+3}{3x} = \frac{7}{3} - \frac{9}{x}$$
$$3x\left(\frac{5x+3}{3x}\right) = 3x\left(\frac{7}{3}\right) - 3\left(\frac{9}{x}\right)$$
$$5x + 3 = 7x - 27$$
$$3 = 2x - 27$$
$$30 = 2x$$
$$15 = x$$

Check: $\frac{5(15)+3}{3(15)} \stackrel{?}{=} \frac{3}{\left(-\frac{3}{4}\right)-3}$

$$\frac{78}{45} \stackrel{?}{=} \frac{35}{15} - \frac{9}{15}$$

$$\frac{26}{15} = \frac{26}{15}$$

7. $\frac{x+1}{2x} = \frac{2}{3}$

$$6x\left(\frac{x+1}{2x}\right) = 6x\left(\frac{2}{3}\right)$$

$$3(x+1) = 4x$$

$$3x+3 = 4x$$

$$3 = x$$

Check: $\frac{3+1}{2(3)} \stackrel{?}{=} \frac{2}{3}$

$$\frac{4}{6} \stackrel{?}{=} \frac{2}{3}$$

$$\frac{2}{3} = \frac{2}{3}$$

9. $\frac{6}{3x-5} = \frac{3}{2x}$

$$2x(3x-5)\left(\frac{6}{3x-5}\right) = 2x(3x-5)\left(\frac{3}{2x}\right)$$

$$12x = 9x - 15$$

$$3x = -15$$

$$x = -5$$

Check: $\frac{6}{3(-5)-5} \stackrel{?}{=} \frac{3}{2(-5)}$

$$-\frac{6}{20} \stackrel{?}{=} -\frac{3}{10}$$

$$-\frac{3}{10} = -\frac{3}{10}$$

11. $\frac{2}{2x+5} = \frac{4}{x-4}$

$$(2x+5)(x-4)\left(\frac{2}{2x+5}\right) = (2x+5)(x-4)\left(\frac{4}{x-4}\right)$$

$$2(x-4) = 4(2x+5)$$

$$2x-8 = 8x+20$$

$$-8 = 6x+20$$

$$-28 = 6x$$

$$-\frac{28}{6} = x$$

$$-\frac{14}{3} = x \text{ or } x = -4\frac{2}{3}$$

Check: $\frac{2}{2\left(-\frac{14}{3}\right)+5} \stackrel{?}{=} \frac{4}{-\frac{14}{3}-4}$

$$\frac{2}{-\frac{28}{3}+\frac{15}{3}} \stackrel{?}{=} \frac{4}{-\frac{14}{3}-\frac{12}{3}}$$

$$\frac{2}{-\frac{13}{3}} = \frac{4}{-\frac{26}{3}}$$

$$-\frac{6}{13} = -\frac{6}{13}$$

13. $\frac{2}{x} + \frac{x}{x+1} = 1$

$$x(x+1)\left(\frac{2}{x}\right) + x(x+1)\left(\frac{x}{x+1}\right) = x(x+1)(1)$$

$$(x+1)(2) + x^2 = x^2 + x$$

$$2x+2+x^2 = x^2+x$$

$$2x+2 = x$$

$$2 = -x$$

$$-2 = x$$

Check: $\frac{2}{-2} + \frac{-2}{-2+1} \stackrel{?}{=} 1$

$$-1+2 \stackrel{?}{=} 1$$

$$1 = 1$$

15. $$\frac{85-4x}{x}=7-\frac{3}{x}$$
$$x\left(\frac{85-4x}{x}\right)=7x-x\left(\frac{3}{x}\right)$$
$$85-4x=7x-3$$
$$85=11x-3$$
$$88=11x$$
$$8=x$$

Check: $$\frac{85-4(8)}{8}\overset{?}{=}7-\frac{3}{9}$$
$$\frac{53}{8}\overset{?}{=}\frac{56}{8}-\frac{3}{8}$$
$$\frac{53}{8}=\frac{53}{8}$$

17. $$\frac{3}{x+2}-4=\frac{4x-1}{x+2}$$
$$(x+2)\left(\frac{3}{x+2}\right)-(x+2)(4)=(x+2)\left(\frac{4x-1}{x+2}\right)$$
$$3-4(x+2)=4x-1$$
$$3-4x-8=4x-1$$
$$-4x-5=4x-1$$
$$-5=8x-1$$
$$-4=8x$$
$$-\frac{4}{8}=x$$
$$-\frac{1}{2}=x$$

Check: $$\frac{3}{-\frac{1}{2}+2}-4\overset{?}{=}\frac{4\left(-\frac{1}{2}\right)-1}{-\frac{1}{2}+2}$$
$$2-4\overset{?}{=}-2$$
$$-2=-2$$

19. $$\frac{2}{x-6}-5=\frac{2(x-5)}{x-6}$$
$$(x-6)\left(\frac{2}{x-6}\right)-5(x-6)=(x-6)\left[\frac{2(x-5)}{x-6}\right]$$
$$2-5x+30=2x-10$$
$$-5x+32=2x-10$$
$$42=7x$$
$$6=x$$

Check: $$\frac{2}{6-6}-5\overset{?}{=}\frac{2(6-5)}{6-6}$$
$$\frac{2}{0}-5\overset{?}{=}\frac{2}{0}$$

$x=6$ makes the denominators zero so it is an extraneous solution. There is no solution.

21. $$\frac{2}{x+1}-\frac{1}{x-1}=\frac{2x}{x^2-1}$$
$$\frac{2}{x+1}-\frac{1}{x-1}=\frac{2x}{(x+1)(x-1)}$$
$$(x+1)(x-1)\left(\frac{2}{x+1}\right)-(x+1)(x-1)\left(\frac{1}{x-1}\right)$$
$$=(x+1)(x-1)\left(\frac{2x}{(x+1)(x-1)}\right)$$
$$(x-1)(2)-(x+1)=2x$$
$$2x-2-x-1=2x$$
$$x-3=2x$$
$$-3=x$$

Check: $$\frac{2}{-3+1}-\frac{1}{-3-1}\overset{?}{=}\frac{2(-3)}{(-3)^2-1}$$
$$-1+\frac{1}{4}\overset{?}{=}-\frac{6}{8}$$
$$-\frac{3}{4}=-\frac{3}{4}$$

23. $\frac{y+1}{y^2+2y-3}=\frac{1}{y+3}-\frac{1}{y-1}$

$\frac{y+1}{(y+3)(y-1)}=\frac{1}{y+3}-\frac{1}{y-1}$

$(y+3)(y-1)\left(\frac{y+1}{(y+3)(y-1)}\right)$

$=(y+3)(y-1)\left(\frac{1}{y+3}\right)-(y+3)(y-1)\left(\frac{1}{y-1}\right)$

$y+1=y-1-y-3$

$y+1=-4$

$y=-5$

Check: $\frac{-5+1}{(-5)^2+2(-5)-3}\stackrel{?}{=}\frac{1}{-5+3}-\frac{1}{-5-1}$

$\frac{-4}{12}\stackrel{?}{=}-\frac{1}{2}+\frac{1}{6}$

$-\frac{4}{12}=-\frac{4}{12}$

25. $\frac{2x}{x+4}-\frac{8}{x-4}=\frac{2x^2+32}{x^2-16}$

$\frac{2x}{x+4}-\frac{8}{x-4}=\frac{2x^2+32}{(x+4)(x-4)}$

$(x+4)(x-4)\left(\frac{2x}{x+4}\right)-(x+4)(x-4)\left(\frac{8}{x-4}\right)$

$=(x+4)(x-4)\left(\frac{2x^2+32}{(x+4)(x-4)}\right)$

$(x-4)(2x)-(x+4)(8)=2x^2+32$

$2x^2-8x-8x-32=2x^2+32$

$2x^2-16x-32=2x^2+32$

$-16x-32=32$

$-16x=64$

$x=-4$ Extraneous

No solution.

27. $\frac{4}{x^2-1}+\frac{7}{x+1}=\frac{5}{x-1}$

$\frac{4}{(x+1)(x-1)}+\frac{7}{x+1}=\frac{5}{x-1}$

$(x+1)(x-1)\frac{4}{(x+1)(x-1)}+(x+1)(x-1)$

$\times\frac{7}{x+1}=(x+1)(x-1)\frac{5}{x-1}$

$4+(x-1)(7)=(x+1)(5)$

$4+7x-7=5x+5$

$2x-3=5$

$2x=8$

$x=4$

Check:

$\frac{4}{(4)^2-1}+\frac{7}{4+1}\stackrel{?}{=}\frac{5}{4-1}$

$\frac{4}{15}+\frac{7}{5}\stackrel{?}{=}\frac{5}{3}$

$\frac{25}{15}\stackrel{?}{=}\frac{5}{3}$

$\frac{5}{3}=\frac{5}{3}$

29. $\frac{x+11}{x^2-5x+4}+\frac{3}{x-1}=\frac{5}{x-4}$

$\frac{x+11}{(x-4)(x-1)}+\frac{3}{x-1}=\frac{5}{x-4}$

$(x-4)(x-1)\left(\frac{x+11}{(x-4)(x-1)}\right)$

$+(x-4)(x-1)\left(\frac{3}{x-1}\right)=(x-4)(x-1)\left(\frac{5}{x-4}\right)$

$x+11+(x-4)(3)=(x-1)(5)$

$x+11+3x-12=5x-5$

$4x-1=5x-5$

$-1=x-5$

$4=x$ Extraneous

No solution.

31. $x-2\neq 0$, so $x\neq 2$

$x-4\neq 0$, so $x\neq 4$

$x^2-6x+8\neq 0$

$(x-2)(x-4)\neq 0$

$x-2\neq 0 \qquad x-4\neq 0$

$x\neq 2 \qquad x\neq 4$

Extraneous solutions $x=2$, $x=4$

Cumulative Review

33. $6x^2-x-12=(3x+4)(2x-3)$

35. $x=\text{width}$

$3x-1=\text{length}$

$P=2L+2W$

$54=2(3x-1)+2x$

$54=6x-2+2x$

$56=8x$

$7=x$

$3x-1=3(7)-1-1=20$

Width = 7 meters and length = 20 meters

37. Decrease: $7.3-5.7=1.6$

$\frac{1.6}{7.3}=0.219=21.9\%$

39. Increase: $9.3-8.4=0.9$

$\frac{0.9}{8.4}=0.107$

For 2010 – 2020 the increase is

$0.107(9.3)=0.995$

Expected number $=9.3+0.995$

≈ 10.3 million people

6.6 Exercises

1. $\frac{4}{9}=\frac{8}{x}$

$4x=72$

$x=18$

3. $\frac{x}{17}=\frac{12}{5}$

$5x=204$

$x=\frac{204}{5}$ or $40\frac{4}{5}$

5. $\frac{8}{5} = \frac{x}{7}$

$56 = 5x$

$\frac{56}{5} = x$ or $x = 11\frac{1}{5}$

7. $\frac{7}{x} = \frac{40}{130}$

$40x = 910$

$x = \frac{910}{40} = 22.75$

9. d = distance apart

$\frac{d}{5.5} = \frac{15}{3/4}$

$\frac{3}{4}d = 82.5$

$d = 110$ miles

11. (a) $700 U.S. $\left(\frac{\$1.49 \text{ Australian}}{\$1 \text{ U.S.}}\right)$

= $1043 Australian

(b) $700 U.S. $\left(\frac{\$1.68 \text{ Australian}}{\$1 \text{ U.S.}}\right)$

= $1176 Australian

1176 − 1043 = $133 more Australian

13. x = speed limit in miles per hour

$\frac{x}{90} = \frac{62}{100}$

$100x = 5580$

$x = 55.8 \approx 56$

The limit is 56 miles per hour.

15. x = miles from base of mountain

$\frac{x}{\frac{3}{4}} = \frac{136}{3\frac{1}{2}}$

$\frac{7}{2}x = \frac{3}{4}(136)$

$\frac{7}{2}x = 102$

$x = 29.1$

He is 29 miles away.

17. $x = 20,\quad y = 29,\quad m = 13$

$\frac{n}{m} = \frac{y}{x}$

$\frac{n}{13} = \frac{29}{20}$

$20n = 377$

$n = \frac{377}{20} = 18\frac{17}{20}$ inches

19. $\frac{y}{40} = \frac{175}{35}$

$35y = 7000$

$y = 200$ meters

21. $\frac{d}{k} = \frac{a}{g}$

$\frac{d}{12} = \frac{7}{9}$

$9d = 84$

$d = \frac{84}{9} = 9\frac{1}{3}$ in.

23. $\frac{k}{d} = \frac{h}{b}$

$\frac{k}{32} = \frac{24}{20}$

$k = \frac{768}{20} = 38\frac{2}{5}$ m

25. x = length

$\frac{30}{x} = \frac{5}{8}$

$5x = 8(30)$

$x = \frac{240}{5}$

$x = 48$ inches

27. x = flower height

$\frac{x}{13} = \frac{5}{3}$

$3x = 65$

$x = \frac{65}{3} \approx 22$ inches

Total height $= 13 + 22 = 35$ inches

29. x = amount of accceleration in 11 seconds

$\frac{x}{11} = \frac{3}{2}$

$2x = 33$

$x = 16.5$

$45 + 16.5 = 61.5$

61.5 miles per hour

31.

	D	R	$T = \frac{D}{R}$
Helicopter	1050	s	$\frac{1050}{s}$
Airline	1250	$s+40$	$\frac{1250}{s+40}$

$\frac{1250}{s+40} = \frac{1050}{s}$

$1050s = 1050s + 42,000$

$200s = 42,000$

$s = 210$

$s + 40 = 250$

commuter airline, 250 kilometers/hr; helicopter, 210 kilometers/hr

33. a. $\frac{\$0.79}{7.5 \text{ oz}} = \$0.11/\text{oz}$

b. $\frac{\$1.49}{16 \text{ oz}} = \$0.09$

c. x = price of 40 oz. bucket

$\frac{x}{40} = \frac{1.49}{16}$

$16x = 59.6$

$x = 3.73$

The price would be \$3.73.

35. x = time to mow together

$\frac{1}{4} + \frac{1}{5} = \frac{1}{x}$

$5x + 4x = 20$

$9x = 20$

$x = \frac{20}{9} = 2\frac{2}{9}$ hours

$60\left(\frac{2}{9}\right) = 13.3$

It will take $2\frac{2}{9}$ hours or 2 hours 13 minutes.

37. x = time to rake together

$$\frac{1}{6}+\frac{1}{8}=\frac{1}{x}$$
$$24x\left(\frac{1}{6}\right)+24x\left(\frac{1}{8}\right)=24x\left(\frac{1}{x}\right)$$
$$4x+3x=24$$
$$7x=24$$
$$x=\frac{24}{7}=3\frac{3}{7}$$

It will take $3\frac{3}{7}$ hours or 3 hours 26 minutes.

Cumulative Review

39. $0.000892465 = 8.92465\times10^{-4}$

41. $\frac{x^{-3}y^{-2}}{z^4w^{-8}}=\frac{w^8}{x^3y^2z^4}$

Putting Your Skills to Work

1. $15-12=3$

   $3000

2. $18-16.8=1.2$

   $1200

3. $P=\frac{24x^2+120x}{x^2+8x+15}$

   $P=\frac{24x(x+5)}{(x+3)(x+5)}$

   $P=\frac{24x}{x+3}$

4. $P=\frac{24(11)}{11+3}$

   $P=\frac{264}{14}$

   $P=\$18.857$ thousand

   $P=\$18,857$

5. $$20=\frac{24x}{x+3}$$
   $$20(x+3)=24x$$
   $$20x+60=24x$$
   $$60=4x$$
   $$15=x$$

   $15,000

6. $$21=\frac{24x}{x+3}$$
   $$21(x+3)=24x$$
   $$21x+63=24x$$
   $$63=3x$$
   $$21=x$$

   $21,000

Chapter 6 Review Problems

1. $\dfrac{bx}{bx-by} = \dfrac{bx}{b(x-y)}$
$= \dfrac{x}{x-y}$

2. $\dfrac{4x-4y}{5y-5x} = \dfrac{4(x-y)}{-5(x-y)} = -\dfrac{4}{5}$

3. $\dfrac{2x^2+5x-3}{2x^2-9x+4} = \dfrac{(2x-1)(x+3)}{(2x-1)(x-4)} = \dfrac{x+3}{x-4}$

4. $\dfrac{3x^2+7x+2}{3x^2+13x+4}$
$= \dfrac{(3x+1)(x+2)}{(3x+1)(x+4)}$
$= \dfrac{x+2}{x+4}$

5. $\dfrac{x^2-9}{x^2-10x+21} = \dfrac{(x+3)(x-3)}{(x-7)(x-3)} = \dfrac{x+3}{x-7}$

6. $\dfrac{2x^2+18x+40}{3x+15}$
$= \dfrac{2(x+5)(x+4)}{3(x+5)}$
$= \dfrac{2(x+4)}{3}$

7. $\dfrac{x^3+3x^2}{x^3-2x^2-15x}$
$= \dfrac{x^2(x+3)}{x(x-5)(x+3)}$
$= \dfrac{x}{x-5}$

8. $\dfrac{4x^2+4x-3}{4x^2-2x} = \dfrac{(2x+3)(2x-1)}{2x(2x-1)} = \dfrac{2x+3}{2x}$

9. $\dfrac{2x^2-2xy-24y^2}{2x^2+5xy-3y^2}$
$= \dfrac{2(x-4y)(x+3y)}{(2x-y)(x+3y)}$
$= \dfrac{2(x-4y)}{2x-y}$

10. $\dfrac{4-y^2}{3y^2+5y-2}$
$= \dfrac{(2+y)(2-y)}{(3y-1)(y+2)}$
$= \dfrac{2-y}{3y-1}$

11. $\dfrac{5x^3-10x^2}{25x^4+5x^3-30x^2}$
$= \dfrac{5x^2(x-2)}{5x^2(5x+6)(x-1)}$
$= \dfrac{x-2}{(5x+6)(x-1)}$

12. $\dfrac{16x^2-4y^2}{4x-2y}$

$=\dfrac{4(2x+y)(2x-y)}{2(2x-y)}$

$=2(2x+y)$

$=4x+2y$

13. $\dfrac{3x^2-13x-10}{3x^2+2x}\cdot\dfrac{x^2-25x}{x^2-25}$

$=\dfrac{(3x+2)(x-5)}{x(3x+2)}\cdot\dfrac{x(x-25)}{(x+5)(x-5)}$

$=\dfrac{x-25}{x+5}$

14. $\dfrac{2y^2-18}{3y^2+3y}\div\dfrac{y^2+6y+9}{y^2+4y+3}$

$=\dfrac{2y^2-18}{3y^2+3y}\cdot\dfrac{y^2+4y+3}{y^2+6y+9}$

$=\dfrac{2(y+3)(y-3)}{3y(y+1)}\cdot\dfrac{(y+3)(y+1)}{(y+3)(y+3)}$

$=\dfrac{2(y-3)}{3y}$

15. $\dfrac{2y^2+3y-2}{2y^2+y-1}\div\dfrac{2y^2+y-1}{2y^2-3y-2}$

$=\dfrac{2y^2+3y-2}{2y^2+y-1}\cdot\dfrac{2y^2-3y-2}{2y^2+y-1}$

$=\dfrac{(2y-1)(y+2)}{(2y-1)(y+1)}\cdot\dfrac{(2y+1)(y-2)}{(2y-1)(y+1)}$

$=\dfrac{(y+2)(2y+1)(y-2)}{(y+1)^2(2y-1)}$

16. $\dfrac{6y^2+13y-5}{9y^2+3y}\div\dfrac{4y^2+20y+25}{12y^2}$

$=\dfrac{6y^2+13y-5}{9y^2+3y}\cdot\dfrac{12y^2}{4y^2+20y+25}$

$=\dfrac{(3y-1)(2y+5)}{3y(3y+1)}\cdot\dfrac{12y^2}{(2y+5)(2y+5)}$

$=\dfrac{4y(3y-1)}{(3y+1)(2y+5)}$

17. $\dfrac{3xy^2+12y^2}{2x^2-11x+5}\div\dfrac{2xy+8y}{8x^2+2x-3}$

$=\dfrac{3xy^2+12y^2}{2x^2-11x+5}\cdot\dfrac{8x^2+2x-3}{2xy+8y}$

$=\dfrac{3y^2(x+4)}{(2x-1)(x-5)}\cdot\dfrac{(2x-1)(4x+3)}{2y(x+4)}$

$=\dfrac{3y(4x+3)}{2(x-5)}$

18. $\dfrac{11}{x-2}\cdot\dfrac{2x^2-8}{44}=\dfrac{11}{x-2}\cdot\dfrac{2(x+2)(x-2)}{44}$

$=\dfrac{x+2}{2}$

19. $\dfrac{x^2-5xy-24y^2}{2x^2-2xy-24y^2}\cdot\dfrac{4x^2+4xy-24y^2}{x^2-10xy+16y^2}$

$=\dfrac{(x-8y)(x+3y)}{2(x-4y)(x+3y)}\cdot\dfrac{4(x+3y)(x-2y)}{(x-8y)(x-2y)}$

$=\dfrac{2(x+3y)}{x-4y}$

20. $\dfrac{2x^2+10x+2}{8x-8}\cdot\dfrac{3x-3}{4x^2+20x+4}$
$=\dfrac{2(x^2+5x+1)}{8(x-1)}\cdot\dfrac{3(x-1)}{4(x^2+5x+1)}$
$=\dfrac{3}{16}$

21. $\dfrac{7}{x+1}+\dfrac{4}{2x}=\dfrac{7}{x+1}\cdot\dfrac{2x}{2x}+\dfrac{4}{2x}\cdot\dfrac{x+1}{x+1}$
$=\dfrac{14x+4x+4}{2x(x+1)}$
$=\dfrac{18x+4}{2x(x+1)}$
$=\dfrac{9x+2}{x(x+1)}$

22. $5+\dfrac{1}{x}+\dfrac{1}{x+1}$
$=\dfrac{5}{1}\cdot\dfrac{x(x+1)}{x(x+1)}+\dfrac{1}{x}\cdot\dfrac{x+1}{x+1}+\dfrac{1}{x+1}\cdot\dfrac{x}{x}$
$=\dfrac{5x^2+5x}{x(x+1)}+\dfrac{x+1}{x(x+1)}+\dfrac{x}{x(x+1)}$
$=\dfrac{5x^2+5x+x+1+x}{x(x+1)}$
$=\dfrac{5x^2+7x+1}{x(x+1)}$

23. $\dfrac{7}{x+2}+\dfrac{3}{x-4}$
$=\dfrac{7}{x+2}\cdot\dfrac{x-4}{x-4}+\dfrac{3}{x-4}\cdot\dfrac{x+2}{x+2}$
$=\dfrac{7x-28}{(x+2)(x-4)}+\dfrac{3x+6}{(x+2)(x-4)}$
$=\dfrac{7x-28+3x+6}{(x+2)(x-4)}$
$=\dfrac{10x-22}{(x+2)(x-4)}$

24. $\dfrac{2}{x^2-9}+\dfrac{x}{x+3}$
$=\dfrac{2}{(x+3)(x-3)}+\dfrac{x}{x+3}\cdot\dfrac{x-3}{x-3}$
$=\dfrac{2+x^2-3x}{(x+3)(x-3)}$
$=\dfrac{(x-2)(x-1)}{(x+3)(x-3)}$

25. $\dfrac{x}{y}+\dfrac{3}{2y}+\dfrac{1}{y+2}$
$=\dfrac{x}{y}\cdot\dfrac{2(y+2)}{2(y+2)}+\dfrac{3}{2y}\cdot\dfrac{y+2}{y+2}+\dfrac{1}{y+2}\cdot\dfrac{2y}{2y}$
$=\dfrac{2x(y+2)+3y+6+2y}{2y(y+2)}$
$=\dfrac{2xy+4x+5y+6}{2y(y+2)}$

26. $\frac{4}{a}+\frac{2}{b}+\frac{3}{a+b}$

$= \frac{4}{a}\cdot\frac{b(a+b)}{b(a+b)}+\frac{2}{b}\cdot\frac{a(a+b)}{a(a+b)}+\frac{3}{a+b}\cdot\frac{ab}{ab}$

$= \frac{4ab+4b^2}{ab(a+b)}+\frac{2a^2+2ab}{ab(a+b)}+\frac{3ab}{ab(a+b)}$

$= \frac{4ab+4b^2+2a^2+2ab+3ab}{ab(a+b)}$

$= \frac{2a^2+9ab+4b^2}{ab(a+b)}$

$= \frac{(2a+b)(a+4b)}{ab(a+b)}$

27. $\frac{3x+1}{3x}-\frac{1}{x}=\frac{3x+1}{3x}-\frac{1}{x}\cdot\frac{3}{3}$

$= \frac{3x+1-3}{3x}$

$= \frac{3x-2}{3x}$

28. $\frac{x+4}{x+2}-\frac{1}{2x}$

$= \frac{x+4}{x+2}\cdot\frac{2x}{2x}-\frac{1}{2x}\cdot\frac{x+2}{x+2}$

$= \frac{2x^2+8x}{2x(x+2)}-\frac{x+2}{2x(x+2)}$

$= \frac{2x^2+8x-(x+2)}{2x(x+2)}$

$= \frac{2x^2+8x-x-2}{2x(x+2)}$

$= \frac{2x^2+7x-2}{2x(x+2)}$

29. $\frac{27}{x^2-81}+\frac{3}{2(x+9)}$

$= \frac{27}{(x+9)(x-9)}\cdot\frac{2}{2}+\frac{3}{2(x+9)}\cdot\frac{x-9}{x-9}$

$= \frac{54+3x-27}{2(x+9)(x-9)}$

$= \frac{3x+27}{2(x+9)(x-9)}$

$= \frac{3(x+9)}{2(x+9)(x-9)}$

$= \frac{3}{2(x-9)}$

30. $\frac{1}{x^2+7x+10}-\frac{x}{x+5}$

$= \frac{1}{(x+2)(x+5)}-\frac{x}{x+5}\cdot\frac{x+2}{x+2}$

$= \frac{1-x^2-2x}{(x+2)(x+5)}$

31. $\frac{\frac{3}{2y}-\frac{1}{y}}{\frac{4}{y}+\frac{3}{2y}}=\frac{\frac{3}{2y}-\frac{1}{y}}{\frac{4}{y}+\frac{3}{2y}}\cdot\frac{2y}{2y}$

$= \frac{3-2}{8+3}=\frac{1}{11}$

32. $\frac{\frac{2}{x}+\frac{1}{2x}}{x+\frac{x}{2}}$

$= \frac{\frac{2}{x}+\frac{1}{2x}}{x+\frac{x}{2}}\cdot\frac{2x}{2x}$

$= \frac{4+1}{2x^2+x^2}=\frac{5}{3x^2}$

33. $\frac{w-\frac{4}{w}}{1+\frac{2}{w}} = \frac{w-\frac{4}{w}}{1+\frac{2}{w}} \cdot \frac{w}{w}$

$= \frac{w^2-4}{w+2}$

$= \frac{(w+2)(w-2)}{w+2}$

$= w-2$

34. $\frac{1-\frac{w}{w-1}}{1+\frac{w}{1-w}}$

$= \frac{1-\frac{w}{x-1}}{1+\frac{w}{-(w-1)}}$

$= \frac{1-\frac{w}{w-1}}{1-\frac{w}{w-1}} = 1$

35. $\frac{1+\frac{1}{y^2-1}}{\frac{1}{y+1}-\frac{1}{y-1}} = \frac{1+\frac{1}{(y+1)(y-1)}}{\frac{1}{y+1}-\frac{1}{y-1}} \cdot \frac{(y+1)(y-1)}{(y+1)(y-1)}$

$= \frac{(y+1)(y-1)+1}{y-1-(y+1)}$

$= \frac{y^2}{-2} = -\frac{y^2}{2}$

36. $\frac{\frac{1}{y}+\frac{1}{x+y}}{1+\frac{2}{x+y}}$

$= \frac{\frac{1}{y}+\frac{1}{x+y}}{1+\frac{2}{x+y}} \cdot \frac{y(x+y)}{y(x+y)}$

$= \frac{x+y+y}{y(x+y)+2y}$

$= \frac{x+2y}{xy+y^2+2y}$

$= \frac{x+2y}{y(x+y+2)}$

37. $\frac{\frac{1}{a+b}-\frac{1}{a}}{b} = \frac{\frac{a-(a+b)}{a(a+b)}}{b}$

$= -\frac{1}{a(a+b)}$

38. $\frac{\frac{2}{a+b}-\frac{3}{b}}{\frac{1}{a+b}}$

$= \frac{\frac{2}{a+b}-\frac{3}{b}}{\frac{1}{a+b}} \cdot \frac{b(a+b)}{b(a+b)}$

$= \frac{2b-3(a+b)}{b}$

$= \frac{2b-3a-3b}{b} = \frac{-3a-b}{b}$

39. $\frac{x+5y}{x-6y} \div \left(\frac{1}{5y} - \frac{1}{x+5y}\right)$

$= \frac{x+5y}{x-6y} \div \left(\frac{1}{5y} \cdot \frac{x+5y}{x+5y} - \frac{1}{x+5y} \cdot \frac{5y}{5y}\right)$

$= \frac{x+5y}{x-6y} \div \left[\frac{x+5y}{5y(x+5y)} - \frac{5y}{5y(x+5y)}\right]$

$= \frac{x+5y}{x-6y} \div \frac{x+5y-5y}{5y(x+5y)}$

$= \frac{x+5y}{x-6y} \div \frac{x}{5y(x+5y)}$

$= \frac{x+5y}{x-6y} \cdot \frac{5y(x+5y)}{x}$

$= \frac{5y(x+5y)^2}{x(x-6y)}$

40. $\left(\frac{1}{x+2y}-\frac{1}{x-y}\right)\div\frac{2x-4y}{x^2-3xy+2y^2}$
$=\left(\frac{1}{x+2y}-\frac{1}{x-y}\right)\cdot\frac{x^2-3xy+2y^2}{2x-4y}$
$=\frac{x-y-(x+2y)}{(x+2y)(x-y)}\cdot\frac{(x-y)(x-2y)}{2(x-2y)}$
$=\frac{-3y}{2(x+2y)}$

41.
$$\frac{8a-1}{6a+8}=\frac{3}{4}$$
$$\frac{8a-1}{2(3a+4)}=\frac{3}{4}$$
$$4(3a+4)\left[\frac{8a-1}{2(3a+4)}\right]=4(3a+4)\left(\frac{3}{4}\right)$$
$$2(8a-1)=3(3a+4)$$
$$16a-2=9a+12$$
$$7a-2=12$$

42.
$$\frac{8}{a-3}=\frac{12}{a+3}$$
$$(a-3)(a+3)\frac{8}{a-3}=(a-3)(a+3)\frac{12}{a+3}$$
$$8a+24=12a-36$$
$$60=4a$$
$$15=a$$

43.
$$\frac{2x-1}{x}-\frac{1}{2}=-2$$
$$2x\left(\frac{2x-1}{x}\right)-2x\left(\frac{1}{2}\right)=-2(2x)$$
$$4x-2-x=-4x$$
$$7x=2$$
$$x=\frac{2}{7}$$

44.
$$\frac{5-x}{x}-\frac{7}{x}=-\frac{3}{4}$$
$$4x\left(\frac{5-x}{x}\right)-4x\left(\frac{7}{x}\right)=4x\left(-\frac{3}{4}\right)$$
$$4(5-x)-28=-3x$$
$$20-4x-28=-3x$$
$$-4x-8=-3x$$
$$-8=x$$

45.
$$\frac{5}{2}-\frac{2y+7}{y+6}=3$$
$$2(y+6)\left(\frac{5}{2}\right)-2(y+6)\left(\frac{2y+7}{y+6}\right)=2(y+6)(3)$$
$$5y+30-4y-14=6y+36$$
$$-20=5y$$
$$-4=y$$

46.
$$\frac{5}{4}-\frac{1}{2x}=\frac{1}{x}+2$$
$$4x\left(\frac{5}{4}\right)-4x\left(\frac{1}{2x}\right)=4x\left(\frac{1}{x}\right)+4x(2)$$

$$5x-2=4+8x$$
$$-2=4+3x$$
$$-6=3x$$
$$-2=x$$

47. $$\frac{7}{8x}-\frac{3}{4}=\frac{1}{4x}+\frac{1}{2}$$
$$8x\left(\frac{7}{8x}\right)-8x\left(\frac{3}{4}\right)=8x\left(\frac{1}{4x}\right)+8x\left(\frac{1}{2}\right)$$
$$7-6x=2+4x$$
$$7=2+10x$$
$$5=10x$$
$$\frac{1}{2}=x$$

48. $$\frac{1}{3x}+2=\frac{5}{6x}-\frac{1}{2}$$
$$6x\left(\frac{1}{3x}\right)+6x(2)=6x\left(\frac{5}{6x}\right)-6x\left(\frac{1}{2}\right)$$
$$2+12x=5-3x$$
$$15x=3$$
$$x=\frac{1}{5}$$

49. $$\frac{3}{y-3}=\frac{3}{2}+\frac{y}{y-3}$$
$$2(y-3)\left(\frac{3}{y-3}\right)$$
$$=2(y-3)\left(\frac{3}{2}\right)+2(y-3)\left(\frac{y}{y-3}\right)$$
$$6=3(y-3)+2y$$
$$6=3y-9+2y$$
$$6=5y-9$$
$$15=5y$$
$$3=y$$

Since $y=3$ causes a denominator in the original equation to equal 0, there is no solution.

50. $$\frac{x\ \ 8}{x-2}=\frac{2x}{x+2}-2$$
$$(x+2)(x-2)\left(\frac{x-8}{x-2}\right)$$
$$=(x+2)(x-2)\left(\frac{2x}{x+2}\right)-(x+2)(x-2)(2)$$
$$x^2-6x-16=2x^2-4x-2x^2+8$$
$$x^2-2x-24=0$$
$$(x-6)(x+4)=0$$

$x-6=0$ $\quad$ $x+4=0$

$x=6$ $\quad$ $x=-4$

51. $$\frac{9}{2}-\frac{7y-4}{y+2}=-\frac{1}{4}$$
$$4(y+2)\left(\frac{9}{2}\right)-4(y+2)\left(\frac{7y-4}{y+2}\right)$$
$$=4(y+2)\left(-\frac{1}{4}\right)$$
$$18(y+2)-4(7y-4)=-(y+2)$$
$$18y+36-28y+16=-y-2$$
$$-10y+52=-y-2$$
$$52=9y-2$$
$$54=9y$$
$$6=y$$

52. $\frac{3y-1}{3y}-\frac{6}{5y}=\frac{1}{y}-\frac{4}{15}$

$$15y\left(\frac{3y-1}{3y}\right)-15y\left(\frac{6}{5y}\right)=15y\left(\frac{1}{y}\right)-15y\left(\frac{4}{15}\right)$$

$$15y-5-18=15-4y$$
$$19y=38$$
$$y=2$$

53. $\frac{y+18}{y^2-16}=\frac{y}{y+4}-\frac{y}{y-4}$

$$\frac{y+18}{(y+4)(y-4)}=\frac{y}{y+4}-\frac{y}{y-4}$$
$$y+18=(y-4)(y)-(y+4)(y)$$
$$y+18=y^2-4y-y^2-4y$$
$$y+18=-8y$$
$$18=-9y$$
$$-2=y$$

54. $\frac{4}{x^2-1}=\frac{2}{x-1}+\frac{2}{x+1}$

$$4=(x+1)(2)+(x-1)(2)$$
$$4=2x+2+2x-2$$
$$4=4x$$
$$1=x$$

$x=1$ is not an allowed solution.

There is no solution.

55. $\frac{9y-3}{y^2+2y}-\frac{5}{y+2}=\frac{3}{y}$

$$y(y+2)\left[\frac{9y-3}{y(y+2)}\right]-y(y+2)\left(\frac{5}{y+2}\right)=y(y+2)\left(\frac{3}{y}\right)$$
$$9y-3-5y=3y+6$$
$$y=9$$

56. $\frac{2}{3-3y}+\frac{2}{2y-1}=\frac{4}{3y-3}$

$$\frac{2}{-3(y-1)}+\frac{2}{2y-1}=\frac{4}{3(y-1)}$$
$$\frac{-2}{3(y-1)}+\frac{2}{2y-1}=\frac{4}{3(y-1)}$$
$$(2y-1)(-2)+3(y-1)(2)=(2y-1)(4)$$
$$-2(2y-1)+6(y-1)=4(2y-1)$$
$$-4y+2+6y-6=8y-4$$
$$2y-4=8y-4$$
$$-4=6y-4$$
$$0=6y$$
$$0=y$$

57. $\frac{x}{4}=\frac{12}{17}$

$$17x=(4)(12)$$
$$17x=48$$
$$\frac{17x}{17}=\frac{48}{17}$$
$$x=2.8$$

58. $\frac{8}{5} = \frac{2}{x}$

$8x = 10$

$x = \frac{5}{4} = 1.25 = 1.3$

59. $\frac{33}{10} = \frac{x}{8}$

$264 = 10x$

$26.4 = x$

60. $\frac{5}{x} = \frac{22}{9}$

$22x = (5)(9)$

$22x = 45$

$\frac{22x}{22} = \frac{45}{22}$

$x = 2.0$

61. $\frac{13.5}{0.6} = \frac{360}{x}$

$13.5x = 216$

$x = 16$

62. $\frac{2\frac{1}{2}}{3\frac{1}{4}} = \frac{7}{x}$

$\frac{\frac{5}{2}}{\frac{13}{4}} = \frac{7}{x}$

$\frac{5}{2}x = \frac{91}{4}$

$x = 9.1$

63. x = gallons to cover 400 square feet

$\frac{x}{400} = \frac{5}{240}$

$240x = 2000$

$x = \frac{2000}{240} = 8\frac{1}{3}$

8.3 gallons of paint are needed.

64. $\frac{3}{100} = \frac{5}{x}$

$3x = 5(100)$

$3x = 500$

$\frac{3x}{3} = \frac{500}{3}$

$x = \frac{500}{3} = 166\frac{2}{3}$

167 cookies

65. x = gallons to get to Denver

$\frac{x}{1300} = \frac{7}{200}$

$200x = 9100$

$x = \frac{91}{2} = 45\frac{1}{2}$

He will use 46 gallons of gas.

66. $\frac{4}{122} = \frac{3}{x}$

$4x = (122)(3)$

$4x = 366$

$\frac{4x}{4} = \frac{366}{4}$

$x = 91.5$ miles

67.

	D	R	$T=\frac{D}{R}$
Train	180	$s+20$	$\frac{180}{s+20}$
Car	120	s	$\frac{120}{s}$

$$\frac{180}{s+20}=\frac{120}{s}$$
$$180s=120s+2400$$
$$60s=2400$$
$$s=40$$
$$s+20=60$$

Car's speed is 40 mph.

Trains speed is 60 mph.

68.

	Time for whole job	Part of job done in 1 hr
Professional	5	1/5
John	8	1/8
Together	x	$1/x$

$$\frac{1}{5}+\frac{1}{8}=\frac{1}{x}$$
$$40x\left(\frac{1}{5}\right)+40x\left(\frac{1}{8}\right)=40x\left(\frac{1}{x}\right)$$
$$8x+5x=40$$
$$13x=40$$
$$\frac{13x}{13}=\frac{40}{13}$$
$$x=3\frac{1}{13}$$
$$\frac{1}{13}\times\frac{60}{1}=5$$

3 hours, 5 minutes

69.
$$\frac{5.75}{3}=\frac{x}{95}$$
$$(5.75)(95)=3x$$
$$546.25=3x$$
$$\frac{546.25}{3}=\frac{3x}{3}$$
$$182.1=x$$

182 feet tall

70. x = height of building
$$\frac{x}{450}=\frac{8}{3}$$
$$3x=3600$$
$$x=1200$$

Office building is 1200 feet tall.

71. x = time to paint together
$$\frac{1}{5}+\frac{1}{10}=\frac{1}{x}$$
$$10x\left(\frac{1}{5}\right)+10x\left(\frac{1}{10}\right)=10x\left(\frac{1}{x}\right)$$
$$2x+x=10$$
$$3x=10$$
$$x=3\frac{1}{3}$$

$3\frac{1}{3}$ hours or 3 hours, 20 minutes

72. x = time to plow together

$$\frac{1}{20}+\frac{1}{30}=\frac{1}{x}$$

$$60x\left(\frac{1}{20}\right)+60x\left(\frac{1}{30}\right)=60x\left(\frac{1}{x}\right)$$

$$3x+2x=60$$

$$5x=60$$

$$x=12 \text{ hours}$$

73. $$\frac{a^2+a-12}{48a^2-16a^3}=\frac{(a+4)(a-3)}{16a^2(3-a)}$$
$$=\frac{(a+4)(a-3)}{-16a^2(a-3)}$$
$$=-\frac{a+4}{16a^2}$$

74. $$\frac{6b^3-24b^2}{b^2+b-20}=\frac{6b^2(b-4)}{(b+5)(b-4)}$$
$$=\frac{6b^2}{b+5}$$

75. $$\frac{x^2-y^2}{x^2+4xy+3y^2}\cdot\frac{x^2+xy-6y^2}{x^2+xy-2y^2}$$
$$=\frac{(x+y)(x-y)}{(x+y)(x+3y)}\cdot\frac{(x+3y)(x-2y)}{(x+2y)(x-y)}$$
$$=\frac{x-2y}{x+2y}$$

76. $$\frac{x^2-6xy-16y^2}{x^2+4xy-21y^2}\cdot\frac{x^2-8xy+15y^2}{x^2=3xy-10y^2}$$
$$=\frac{(x-8y)(x+2y)}{(x+7y)(x-3y)}\cdot\frac{(x-3y)(x-5y)}{(x-5y)(x+2y)}$$
$$=\frac{x-8y}{x+7y}$$

77. $$\frac{x}{x+3}+\frac{9x+18}{x^2+3x}=\frac{x}{x+3}\cdot\frac{x}{x}+\frac{9x+18}{x(x+3)}$$
$$=\frac{x^2+9x+18}{x(x+3)}$$
$$=\frac{(x+6)(x+3)}{x(x+3)}$$
$$=\frac{x+6}{x}$$

78. $$\frac{x-30}{x^2-5x}+\frac{x}{x-5}=\frac{x-30}{x(x-5)}+\frac{x}{x-5}\cdot\frac{x}{x}$$
$$=\frac{x-30+x^2}{x(x-5)}$$
$$=\frac{x^2+x-30}{x(x-5)}$$
$$=\frac{(x+6)(x-5)}{x(x-5)}$$
$$=\frac{x+6}{x}$$

79. $\frac{a+b}{ax+ay}-\frac{a+b}{bx+by}$

$=\frac{a+b}{a(x+y)}\cdot\frac{b}{b}-\frac{a+b}{b(x+y)}\cdot\frac{a}{a}$

$=\frac{b(a+b)-a(a+b)}{ab(x+y)}$

$=\frac{ab+b^2-a^2-ab}{ab(x+y)}$

$=\frac{b^2-a^2}{ab(x+y)}$

80. $\frac{\frac{5}{3x}+\frac{2}{9x}}{\frac{3}{x}+\frac{8}{3x}}=\frac{\frac{5}{3x}+\frac{2}{9x}}{\frac{3}{x}+\frac{8}{3x}}\cdot\frac{9x}{9x}$

$=\frac{\frac{5(9x)}{3x}+\frac{2(9x)}{9x}}{\frac{3(9x)}{x}+\frac{8(9x)}{3x}}$

$=\frac{15+2}{27+24}$

$=\frac{17}{51}$

$=\frac{1}{3}$

81. $\frac{\frac{4}{5y}-\frac{8}{xy}}{y+\frac{y}{5}}=\frac{\frac{4}{5y}-\frac{8}{y}}{y+\frac{y}{5}}\cdot\frac{5y}{5y}$

$=\frac{\frac{4(5y)}{5y}-\frac{8(5y)}{y}}{y(5y)+\frac{y(5y)}{5}}$

$=\frac{4-40}{5y^2+y^2}$

$=\frac{-36}{6y^2}$

$=-\frac{6}{y^2}$

82. $\frac{x-3y}{x+2y}\div\frac{2}{y}-\frac{12}{x+3y}$

$=\frac{\frac{x-3y}{x+2y}}{\frac{2}{y}-\frac{12}{x+3y}}$

$=\frac{\frac{x-3y}{x+2y}}{\frac{2}{y}-\frac{12}{x+3y}}\cdot\frac{y(x+2y)(x+3y)}{y(x+2y)(x+3y)}$

$=\frac{y(x-3y)(x+3y)}{2(x+2y)(x+3y)-12y(x+2y)}$

$=\frac{y(x-3y)(x+3y)}{2(x+2y)[x+3y-6y]}$

$=\frac{y(-3y)(x+3y)}{2(x+2y)(x-3y)}$

$=\frac{y(x+3y)}{2(x+2y)}$

83. $\frac{10}{x-3}=\frac{9}{x-5}$

$10(x-5)=9(x-3)$

$10x-50=9x-27$

$x=23$

84. $\frac{7x-4}{6x-5}=\frac{8}{7}$

$7(7x-4)=8(6x-5)$

$49x-28=48x-40$

$x-28=-40$

$x=-12$

85. $2+\frac{4}{b-1}=\frac{4}{b^2-b}$

$2b(b-1)+b(b-1)\left(\frac{4}{b-1}\right)=b(b-1)\left(\frac{4}{b(b-1)}\right)$

$2b^2-2b+4b=4$

$2b^2+2b=4$

$2b^2+2b-4-0$

$2(b^2+b-2)=0$

$b^2+b-2=0$

$b+2=0 \qquad b-1=0$

$b=-2 \qquad b=1$

Check: $2+\frac{4}{1-1}\overset{?}{=}\frac{4}{1^2-1}$

$2+\frac{4}{0}\overset{?}{=}\frac{4}{0}$

$b=1$ does not check

Solution: $b=-2$

86. x = weight of 40 m of wire

$\frac{x}{40}=\frac{8}{30}$

$30x=8(40)$

$30x=320$

$x=10.7$

It weighs 10.7 kg

87. x = distance driven in 7 hours

$\frac{x}{7}=\frac{270}{4.5}$

$4.5x=7(270)$

$4.5x=1890$

$x=420$

She drove 420 miles

How Am I Doing? Chapter 6 Test

1. $\frac{2ac+2ad}{3a^2c+3a^2d}=\frac{2a(c+d)}{3a^2(c+d)}=\frac{2}{3a}$

2. $\frac{8x^2-2x^2y^2}{y^2+4y+4}$

$=\frac{2x^2(2-y)(2+y)}{(y+2)^2}$

$=\frac{2x^2(2-y)}{y+2}$

3. $\frac{x^2+2x}{2x-1}\cdot\frac{10x^2-5x}{12x^3+24x^2}$

$=\frac{x(x+2)}{2x-1}\cdot\frac{5x(2x-1)}{12x^2(x+2)}$

$=\frac{5}{12}$

4. $\frac{x+2y}{12y^2}\cdot\frac{4y}{x^2+xy-2y^2}$

$=\frac{x+2y}{12y^2}\cdot\frac{4y}{(x+2y)(x-y)}$

$=\frac{1}{3y(x-y)}$

5. $\frac{2a^2-3a-2}{a^2+5a+6}\div\frac{a^2-5a+6}{a^2-9}$

$=\frac{2a^2-3a-2}{a^2+5a+6}\cdot\frac{a^2-9}{a^2-5a+6}$

$=\frac{(2a+1)(a-2)}{(a+2)(a+3)}\cdot\frac{(a+3)(a-3)}{(a-2)(a-3)}$

$=\frac{2a+1}{a+2}$

6. $$\frac{1}{a^2-a-2}+\frac{3}{a-2}$$
$$=\frac{1}{(a-2)(a+1)}+\frac{3}{a-2}\cdot\frac{a+1}{a+1}$$
$$=\frac{1+3(a+1)}{(a-2)(a+1)}$$
$$=\frac{1+3a+3}{(a-2)(a+1)}$$
$$=\frac{3a+4}{(a-2)(a+1)}$$

7. $$\frac{x-y}{xy}-\frac{a-y}{ay}=\frac{x-y}{xy}\cdot\frac{a}{a}-\frac{a-y}{ay}\cdot\frac{x}{x}$$
$$=\frac{ax-ay-ax+xy}{axy}$$
$$=\frac{xy-ay}{axy}$$
$$=\frac{y(x-a)}{axy}$$
$$=\frac{x-a}{ax}$$

8. $$\frac{3x}{x^2-3x-18}-\frac{x-4}{x-6}$$
$$=\frac{3x}{(x-6)(x+3)}-\frac{x-4}{x-6}$$
$$=\frac{3x}{(x-6)(x+3)}-\frac{x-4}{x-6}\cdot\frac{x+3}{x+3}$$
$$=\frac{3x}{(x-6)(x+3)}-\frac{x^2-x-12}{(x-6)(x+3)}$$
$$=\frac{3x-(x^2-x-12)}{(x-6)(x+3)}$$
$$=\frac{3x-x^2+x+12}{(x-6)(x+3)}$$
$$=\frac{-x^2+4x+12}{(x-6)(x+3)}$$
$$=\frac{-(x-6)(x+2)}{(x-6)(x+3)}$$
$$=-\frac{x+2}{x+3}$$

9. $$\frac{\frac{x}{3y}-\frac{1}{2}}{\frac{4}{3y}-\frac{2}{x}}=\frac{\frac{x}{3y}-\frac{1}{2}}{\frac{4}{3y}-\frac{2}{x}}\cdot\frac{6xy}{6xy}$$
$$=\frac{2x^2-3xy}{8x-12y}$$
$$=\frac{x(2x-3y)}{4(2x-3y)}$$
$$=\frac{x}{4}$$

10. $$\frac{\frac{6}{b}-4}{\frac{5}{bx}-\frac{10}{3x}}=\frac{\frac{6}{b}-4}{\frac{5}{bx}-\frac{10}{3x}}\cdot\frac{3bx}{3bx}$$
$$=\frac{\frac{6(3bx)}{b}-4(3bx)}{\frac{5(3bx)}{bx}-\frac{10(3bx)}{3x}}$$
$$=\frac{18x-12bx}{15-10b}$$
$$=\frac{6x(3-2b)}{5(3-2b)}$$
$$=\frac{6x}{5}$$

11. $\dfrac{2x^2+3xy-9y^2}{4x^2+13xy+3y^2}$

$=\dfrac{(2x-3y)(x+3y)}{(4x+y)(x+3y)}$

$=\dfrac{2x-3y}{4x+y}$

12. $\dfrac{1}{x+4}-\dfrac{2}{x^2+6x+8}$

$=\dfrac{1}{x+4}-\dfrac{2}{(x+4)(x+2)}$

$=\dfrac{1}{x+4}\cdot\dfrac{x+2}{x+2}-\dfrac{2}{(x+4)(x+2)}$

$=\dfrac{x+2}{(x+4)(x+2)}-\dfrac{2}{(x+4)(x+2)}$

$=\dfrac{x+2-2}{(x+4)(x+2)}$

$=\dfrac{x}{(x+4)(x+2)}$

13. $\dfrac{4}{3x}-\dfrac{5}{2x}=5-\dfrac{1}{6x}$

$6x\left(\dfrac{4}{3x}\right)-6x\left(\dfrac{5}{2x}\right)=6x(5)-6x\left(\dfrac{1}{6x}\right)$

$8-15=30x-1$

$-7=30x-1$

$-6=30x$

$\dfrac{-6}{30}=x$

$-\dfrac{1}{5}=x$

14. $\dfrac{x-3}{x-2}=\dfrac{2x^2-15}{x^2+x-6}-\dfrac{x+1}{x+3}$

$\dfrac{x-3}{x-2}=\dfrac{2x^2-15}{(x+3)}-\dfrac{x+1}{x+3}$

$(x-2)(x+3)\left(\dfrac{x-3}{x-2}\right)$

$=(x-2)(x+3)\left[\dfrac{2x^2-15}{(x+3)(x-2)}\right]$

$-(x-2)(x+3)\left(\dfrac{x+1}{x+3}\right)$

$(x+3)(x-3)=2x^2-15-(x-2)(x+1)$

$x^2-9=2x^2-15-(x^2-x-2)$

$x^2-9=2x^2-15-x^2+x+2$

$x^2-9=x^2+x-13$

$-9=x-13$

$4=x$

Check: $\dfrac{4-3}{4-2}\stackrel{?}{=}\dfrac{2(4)^2-15}{4^2+4-6}-\dfrac{4+1}{4+3}$

$\dfrac{1}{2}\stackrel{?}{=}\dfrac{17}{14}-\dfrac{5}{7}$

$\dfrac{7}{14}\stackrel{?}{=}\dfrac{17}{14}-\dfrac{10}{14}$

$\dfrac{7}{14}=\dfrac{7}{14}$

15. $3-\dfrac{7}{x+3}=\dfrac{x-4}{x+3}$

$(x+3)(3)-(x+3)\left(\dfrac{7}{x+3}\right)=(x+3)\left(\dfrac{x-4}{x+3}\right)$

$3x+9-7=x-4$

$2x=-6$

$x=-3$

$x = -3$ is not allowed.

There is no solution.

16. $\frac{3}{3x-5} = \frac{7}{5x+4}$

$3(5x+4) = 7(3x-5)$

$15x + 12 = 21x - 35$

$12 = 6x - 35$

$47 = 6x$

$\frac{47}{6} = x$

Check:

$\frac{3}{3\left(\frac{47}{6}\right)-5} \stackrel{?}{=} \frac{7}{5\left(\frac{47}{6}\right)+4}$

$\frac{3}{\frac{111}{6}} \stackrel{?}{=} \frac{7}{\frac{259}{6}}$

$\frac{18}{111} \stackrel{?}{=} \frac{42}{259}$

$\frac{6}{37} = \frac{6}{37}$

17. $\frac{9}{x} = \frac{13}{5}$

$45 = 13x$

$\frac{45}{13} = x$

18. $\frac{9.3}{2.5} = \frac{x}{10}$

$\frac{10(9.3)}{2.5} = x$

$37.2 = x$

19. x = on time flights

$\frac{x}{200} = \frac{113}{150}$

$150x = 113(200)$

$150x = 22,600$

$x = 151$ flights

20. x = Cost of wood for 92 days

$\frac{x}{92} = \frac{100}{25}$

$25x = 9200$

$x = 368$

It will cost \$368.

21. $\frac{\text{height}}{\text{shadow}}: \frac{6}{7} = \frac{87}{x}$

$6x = 609$

$x = 101.5$

102 feet

Cumulative Test for Chapters 0 - 6

1. 0.25% of $2.57 = (0.0025)(2.57)$

$= 0.006$ centimeters

2. $\text{Percent} = \frac{252}{720} = 0.35 = 35\%$

3. $\text{Difference} = 6\% - 5.5\% = 0.5\%$

$\text{Savings} = 0.005(18,500) = \$92.50$

4. $5(x-3)-2(4-2x)=7(x-1)-(x-2)$
$$5x-15-8+4x=7x-7-x+2$$
$$9x-23=6x-5$$
$$3x-23=-5$$
$$3x=18$$
$$x=6$$

5. $A=\pi r^2 h$
$$\frac{A}{\pi r^2}=\frac{\pi r^2 h}{\pi r^2}$$
$$\frac{A}{\pi r^2}=h$$

6. $4(2-x)<3$
$$8-4x<3$$
$$-4x<-5$$
$$x>\frac{5}{4}$$

7. $\frac{3}{5}x-2\le\frac{1}{2}x+\frac{3}{5}$
$$10\left(\frac{3}{5}x\right)-10(2)\le 10\left(\frac{1}{2}x\right)+10\left(\frac{3}{5}\right)$$
$$6x-20\le 5x+6$$
$$x-20\le 6$$
$$x\le 26$$

8. $3ax+3bx-2ay-2by$
$$=3x(a+b)-2y(a+b)$$
$$=(a+b)(3x-2y)$$

9. $8a^3-38a^2b-10ab^2$
$$=2a(4a^2-19ab-5b^2)$$
$$=2a(4a+b)(a-5b)$$

10. $\frac{-12x^{-2}y^4}{3x^{-5}y^{-8}}=-4x^3y^{12}$

11. $\frac{4x^2-25}{2x^2+9x-35}=\frac{(2x+5)(2x-5)}{(2x-5)(x+7)}$
$$=\frac{2x+5}{x+7}$$

12. $\frac{x^2-4}{x^2-25}\cdot\frac{3x^2-14x-5}{3x^2+6x}$
$$=\frac{(x+2)(x-2)}{(x+5)(x-5)}\cdot\frac{(3x+1)(x-5)}{3x(x+2)}$$
$$=\frac{(x-2)(3x+1)}{3x(x+5)}$$

13. $\frac{2x^2-9x+9}{8x-12}\div\frac{x^2-3x}{2x}$
$$=\frac{2x^2-9x+9}{8x-12}\cdot\frac{2x}{x^2-3x}$$
$$=\frac{(2x-3)(x-3)}{4(2x-3)}\cdot\frac{2x}{x(x-3)}$$
$$=\frac{1}{2}$$

14. $\frac{5}{2x+4}+\frac{3}{x-3}$

$= \frac{5}{2(x+2)}\cdot\frac{x-3}{x-3}+\frac{3}{x-3}\cdot\frac{2(x+2)}{2(x+2)}$

$= \frac{5x-15+6x+12}{2(x+2)(x-3)}$

$= \frac{11x-3}{2(x+2)(x-3)}$

15. $\frac{8}{c^2-4}-\frac{2}{c^2-5c+6}$

$= \frac{8}{(c+2)(c-2)}\cdot\frac{c-3}{c-3}-\frac{2}{(c-2)(c-3)}\cdot\frac{c+2}{c+2}$

$= \frac{8(c-3)}{(c+2)(c-2)(c-3)}-\frac{2(c+2)}{(c-2)(c-3)(c+2)}$

$= \frac{8c-24-(2c+4)}{(c+2)(c-2)(c-3)}$

$= \frac{6c-28}{(c+2)(c-2)(c-3)}$

16. $\frac{3x-2}{3x+2}=2$

$(3x+2)\left(\frac{3x-2}{3x+2}\right)=(3x+2)(2)$

$3x-2=6x+4$

$-6=3x$

$-2=x$

17. $\frac{x-3}{x}=\frac{x+2}{x+3}$

$x(x+2)=(x-3)(x+3)$

$x^2+2x=x^2-9$

$2x=-9$

$x=-\frac{9}{2}$

18. $\frac{\frac{1}{x-3}+\frac{5}{x^2-9}}{\frac{6x}{x+3}} = \frac{\frac{1}{x-3}+\frac{5}{(x+3)(x-3)}}{\frac{6x}{x+3}}\cdot\frac{(x+3)(x-3)}{(x+3)(x-3)}$

$= \frac{x+3+5}{6x(x-3)}$

$= \frac{x+8}{6x(x-3)}$

19. $\frac{\frac{3}{a}+\frac{2}{b}}{\frac{5}{a^2}-\frac{2}{b^2}} = \frac{\frac{3}{a}+\frac{2}{b}}{\frac{5}{a^2}-\frac{2}{b^2}}\cdot\frac{a^2b^2}{a^2b^2}$

$= \frac{3ab^2+2a^2b}{5b^2-2a^2}$

20. $\frac{2x+1}{3}=\frac{4}{5}$

$5(2x+1)=3(4)$

$10x+5=12$

$10x=7$

$x=\frac{7}{10}$

21. $\frac{\text{mi}}{\text{in.}}: \frac{130}{2\frac{1}{2}}=\frac{x}{4}$

$\left(2\frac{1}{2}\right)x=520$

$\frac{5}{2}x=520$

$x=208$

22. x = phone calls to make 110 sales

$\frac{x}{110}=\frac{22}{5}$

$5x=2420$

$x=484$

He must make 484 calls.

Chapter 7

7.1 Exercises

1. The x – coordinate of the origin is 0.

3. $(5,\ 1)$ is an ordered pair because the order is important. The graphs of $(5,\ 1)$ and $(1,\ 5)$ are different.

5.

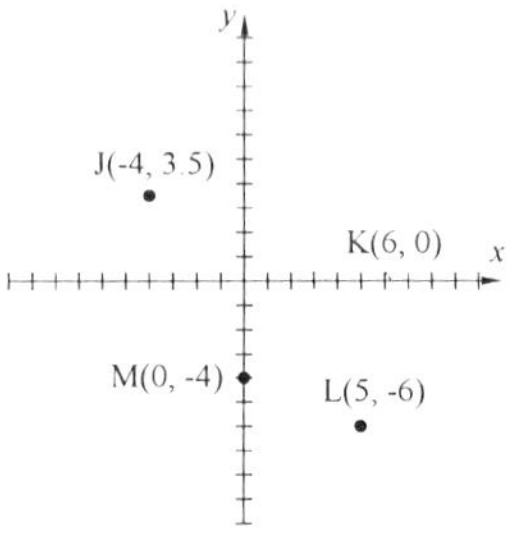

7. R: $(-3,\ -5)$

   S: $\left(-4\frac{1}{2},\ 0\right)$

   X: $(3,\ -5)$

   Y: $\left(2\frac{1}{2},\ 6\right)$

9. $(-4,\ -1), (-3,\ -2), (-2,\ -3)$
   $(-1,\ -5), (0,\ -3), (2,\ -1)$

11. $y = 4x + 7$

   (a) $(0,\ ),\ y = 4(0) + 7 = 7$
   $(0,\ 7)$

   (b) $(2,\ ),\ y = 4(2) + 7 = 15$
   $(2,\ 15)$

13. $y = -4x + 2$

   (a) $(-5,\ ),\ y = -4(-5) + 2 = 22$
   $(5,\ 22)$

   (b) $(4,\ ),\ y = -4(4) + 2 = -14$
   $(4,\ -14)$

15. $3x - 4y = 11$

   (a)
   $$\begin{aligned} 3(-3) - 4y &= 11 \\ -9 - 4y &= 11 \\ -4y &= 20 \\ y &= -5 \end{aligned}$$
   $(-3,\ -5)$

   (b)
   $$\begin{aligned} 3x - 4(1) &= 11 \\ 3x - 4 &= 11 \\ 3x &= 15 \\ x &= 5 \end{aligned}$$
   $(5,\ 1)$

17. $2y+3x=-6$

(a) $2y+3(-2)=-6$

$2y-6=-6$

$2y=0$

$y=0$

$(-2, 0)$

(b) $2(3)+3x=-6$

$6+3x=-6$

$3x=-12$

$x=-4$

$(-4, 3)$

19. $2x+\frac{1}{5}y=6$

(a) $2x+\frac{1}{5}(20)=6$

$2x+4=6$

$2x=2$

$x=1$

$(1, 20)$

19. (b) $2\left(\frac{9}{5}\right)+\frac{1}{5}y=6$

$\frac{18}{5}+\frac{1}{5}y=6$

$18+y=30$

$y=12$

$\left(\frac{9}{5}, 12\right)$

21. B5

23. E1

25. D3

27. (a)

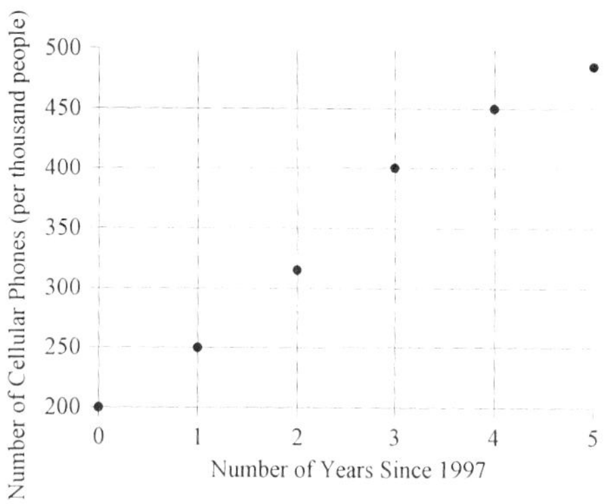

(b) The number of cellular users has increased significantly in the 5 year period. The rate of increase is smaller as time goes on.

29. (a)

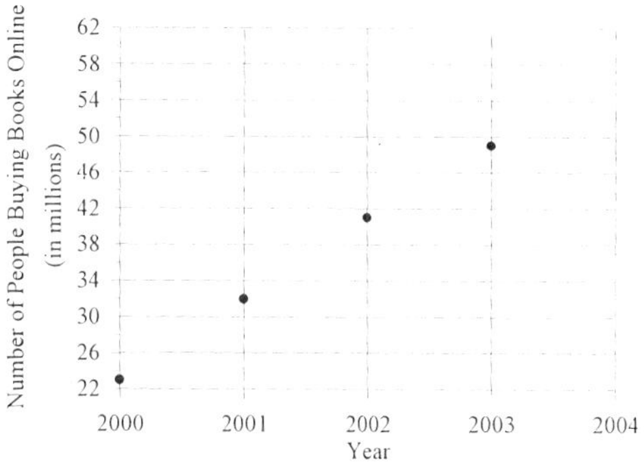

(b) An estimated 57 to 58 million people would buy books online in 2004.

Cumulative Review

31. $\frac{8x}{x+y}+\frac{8y}{x+y}=\frac{8x+8y}{x+y}=\frac{8(x+y)}{x+y}=8$

33. $A=\pi r^2,\ r=19$

$A=3.14(19)^2$

$=1133.54$ square yards

35. Decrease $=6{,}109{,}005-358{,}205$

$=5{,}750{,}800$

% Decrease $=\frac{5{,}750{,}800}{6{,}109{,}005}\times 100\%$

$=94.1\%$

There was a 94.1% decrease in readership.

7.2 Exercises

1. Replacing x by -2 and y by 5 in $2x+5y=0$ does not result in a true statement so $(-2, 5)$ is not a solution.

3. The $x-$intercept of a line is the point where the line crosses the $\underline{x\text{-axis}}$.

5. $y=-2x+1$

$(0, 1), (-2, 5), (1, -1)$

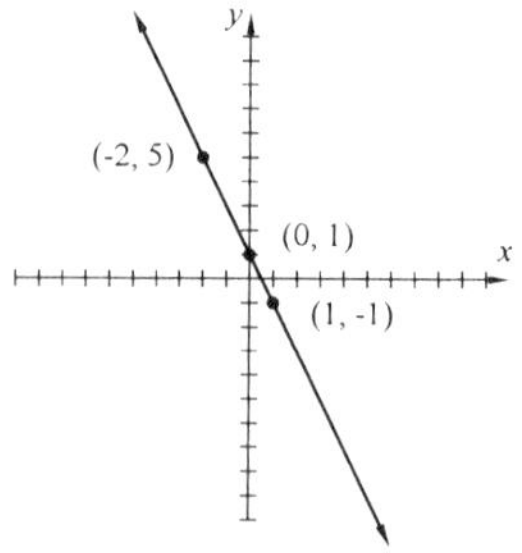

7. $y=x-4$

$(0, -4), (2, -2), (4, 0)$

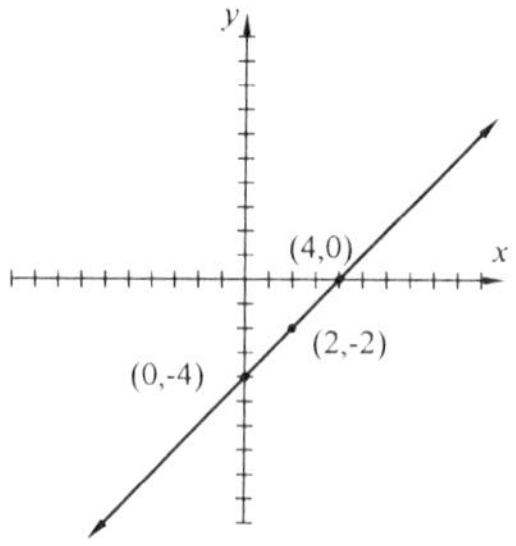

9. $y=-2x+3$

x	y
2	-1
4	-5
0	3

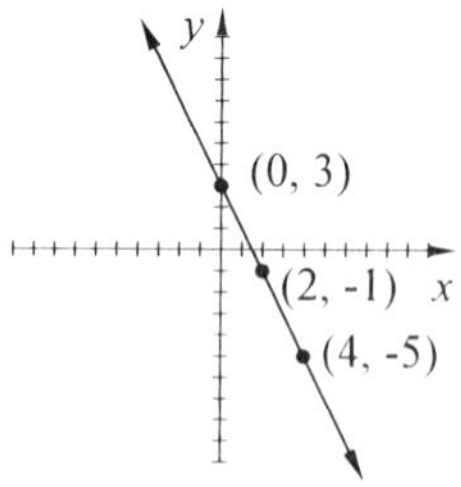

11. $y=3x+2$

x	y
-1	-1
0	2
1	5

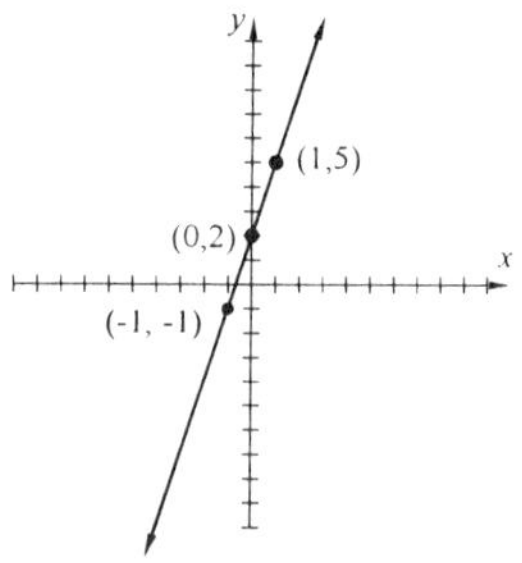

13. $3x - 2y = 0$

x	y
-2	-3
0	0
2	3

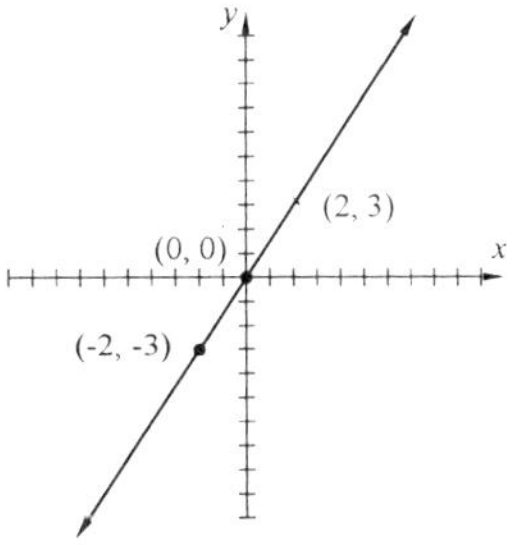

15. $y = -\frac{3}{4}x + 3$

x	y
-4	6
0	3
4	0

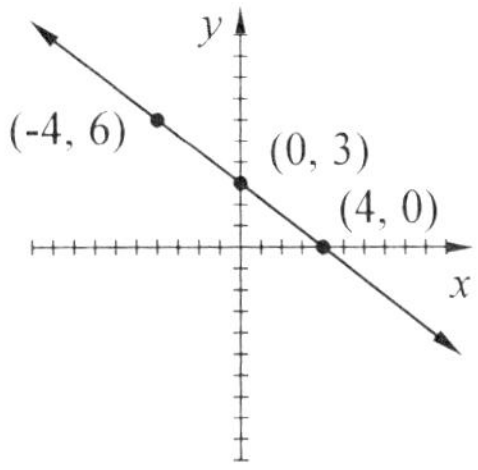

17. $4x + 3y = 12$

x	y
0	4
3	0
6	-4

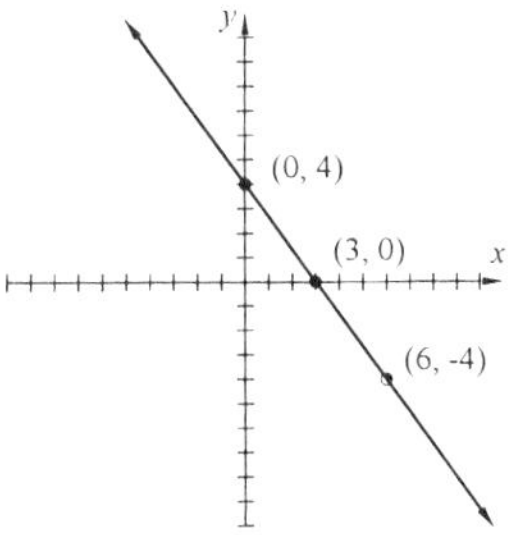

19. $y = 6 - 2x$

x	y
0	6
1	4
3	0

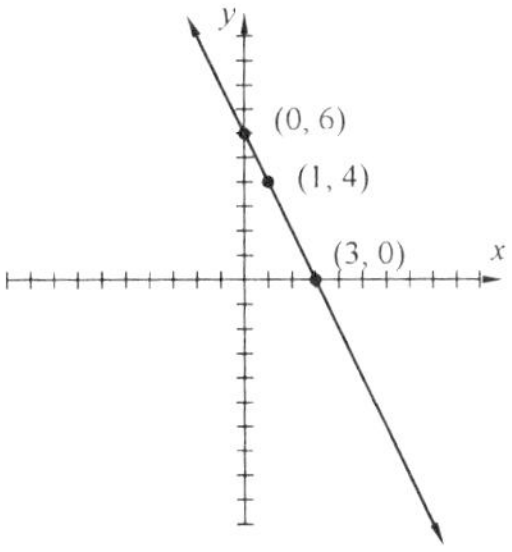

21. $x + 3 = 6y$

x	y
-3	0
0	$\frac{1}{2}$
3	1

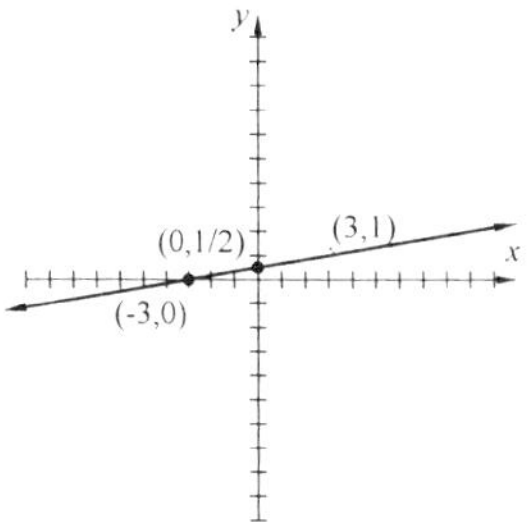

23. $y - 2 = 3y$

$-2 = 2y$

$-1 = y$

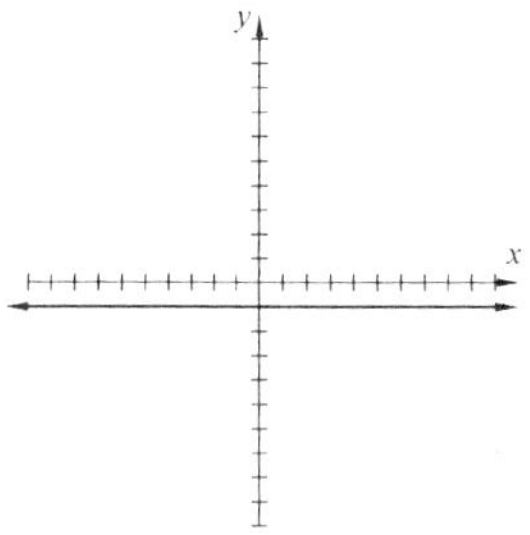

25. $2x + 9 = 5x$

$-3x + 9 = 0$

$-3x = -9$

$x = 3$

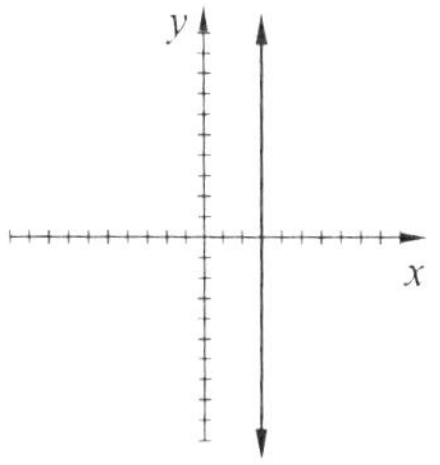

27. $2x + 5y - 2 = -12$

$2x + 5y = -10$

x	y
-5	0
0	-2
5	-4

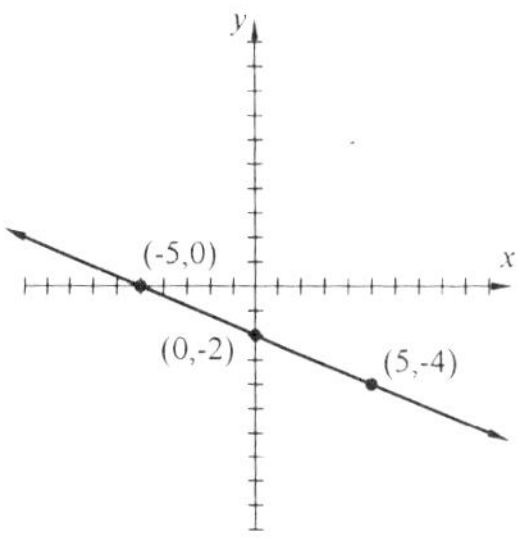

29. $C = 8m$

m	0	15	30	45	60	75
C	0	120	240	360	480	600

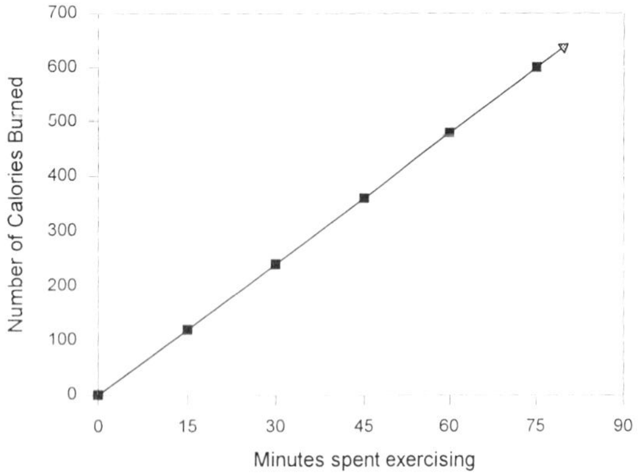

31. $S = 12t + 280$

t	S
0	280
6	352
15	460
21	532

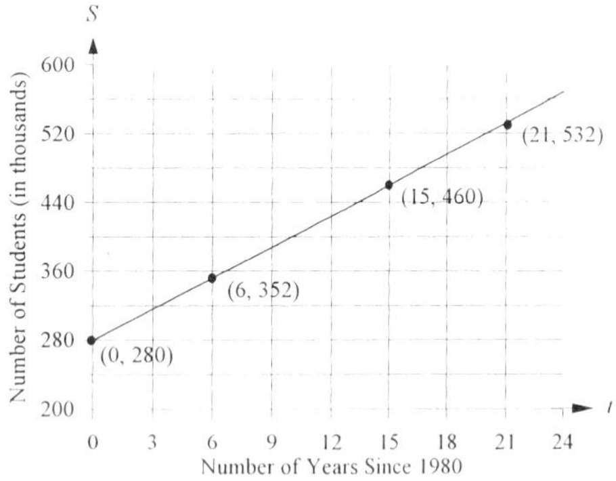

Cumulative Review

33.
$$\frac{x}{x+4} + \frac{x-1}{x-4} = 2$$
$$(x-4)(x) + (x+4)(x-1) = 2(x+4)(x-4)$$
$$x^2 - 4x + x^2 + 3x - 4 = 2x^2 - 32$$
$$-x = -28$$
$$x = 28$$

35. $x = \text{length}$

$$\frac{x}{2} + 1 = \text{width}$$
$$p = 2L + 2w$$
$$53 = 2x + 2\left(\frac{x}{2} + 1\right)$$
$$53 = 2x + x + 2$$
$$51 = 3x$$
$$17 = x$$
$$\frac{17}{2} + 1 = 9\frac{1}{2}$$

width $= 9\frac{1}{2}$ meters

Length = 17 meters

37. $x = \text{cost of mailing from Japan}$

$$\frac{1}{2}x - 2 = \text{cost of mailing from U.S.}$$
$$x + \frac{1}{2}x - 2 = 184$$
$$2x + x - 4 = 368$$
$$3x - 4 = 368$$
$$3x = 372$$
$$x = 124$$

The cost is \$1.24

39. $x =$ length of one base

$2x - 3 =$ length of other base

$$A = \frac{1}{2}(B + b)h$$

$$63 = \frac{1}{2}(2x - 2 + x)(14)$$

$$126 = (3x - 3)(14)$$

$$126 = 42x - 42$$

$$168 = 42x$$

$$4 = x$$

$$2x - 3 = 2(4) - 3 = 5$$

7.3 Exercises

Use $m = \dfrac{y_2 - y_1}{x_2 - x_1}$ in Exercises 1-11.

1. You cannot find the slope of the line passing through $(5, -12)$ and $(5, -6)$ because division by zero is impossible.

3. $(4, 1)$ and $(6, 7)$

$$m = \frac{7 - 1}{6 - 4} = \frac{6}{2} = 3$$

5. $(6, 6)$ and $(9, 3)$

$$m = \frac{3 - 6}{9 - 6} = \frac{-3}{3} = -1$$

7. $(-2, 1)$ and $(3, 4)$

$$m = \frac{4 - 1}{3 - (-2)} = \frac{3}{5}$$

9. $(-6, -5)$ and $(2, -7)$

$$m = \frac{-7 - (-5)}{2 - (-6)} = \frac{-2}{8} = -\frac{1}{4}$$

11. $(-3, 0)$ and $(0, -4)$

$$m = \frac{-4 - 0}{0 - (-3)} = \frac{-4}{3} = -\frac{4}{3}$$

Use $y = mx + b$ in Exercises 13-41.

13. $\left(\frac{3}{4}, -4\right)$ and $(2, -8)$

$$m = \frac{-8 - (-4)}{2 - \frac{3}{4}} = \frac{-4}{\frac{5}{4}} = -\frac{16}{5}$$

15. $y = 8x + 9;\ m = 8,\ b = 9$

17. $y = -3x + 4,\ m = -3,\ b = 4$

19. $y = -\frac{8}{7}x + \frac{3}{4};\ m = -\frac{8}{7},\ b = -\frac{3}{4}$

21. $y = -6x;\ m = -6,\ b = 0$

23. $6x + y = \frac{4}{5}$

$$y = -6x + \frac{4}{5};\ m = -6,\ b = \frac{4}{5}$$

25. $5x + 2y = 3$

$$y = -\frac{5}{2}x + \frac{3}{2};\ m = -\frac{5}{2},\ b = \frac{3}{2}$$

27. $7x - 3y = 4$

$$y = \frac{7}{3}x - \frac{4}{3}; \quad m = \frac{7}{3}, \quad b = -\frac{4}{3}$$

29. $m = \frac{3}{5}$, y-intercept $(0, 3)$

$$y = \frac{3}{5}x + 3$$

31. $m = 6$, y-intercept $(0, -3)$

$$y = 6x - 3$$

33. $m = -\frac{5}{4}$, y-intercept $\left(0, -\frac{3}{4}\right)$

$$y = -\frac{5}{4}x - \frac{3}{4}$$

35. $m = \frac{3}{4}$, $b = -4$

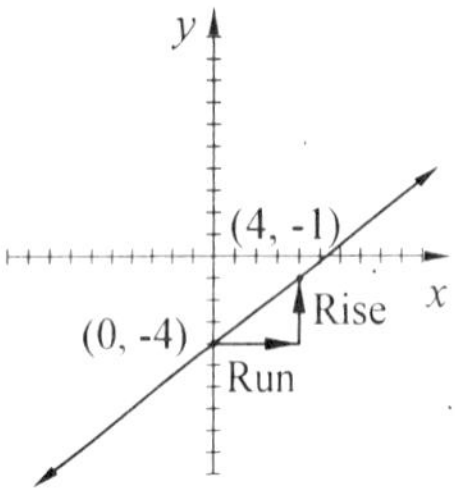

37. $m = -\frac{5}{3}$, $b = 2$

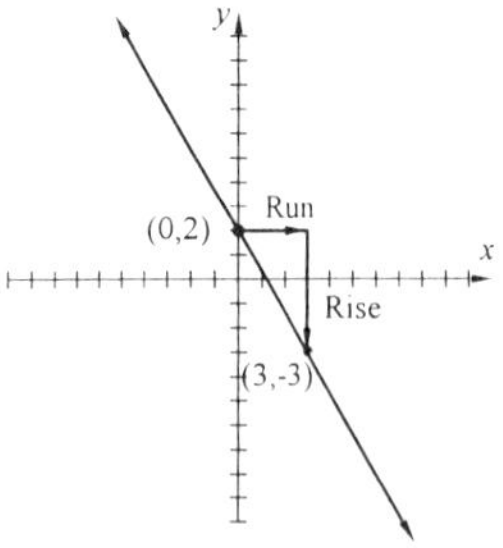

39. $y = \frac{2}{3}x + 2$

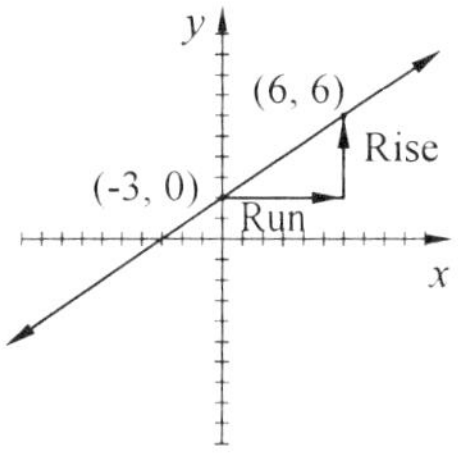

41. $y + 2x = 3$

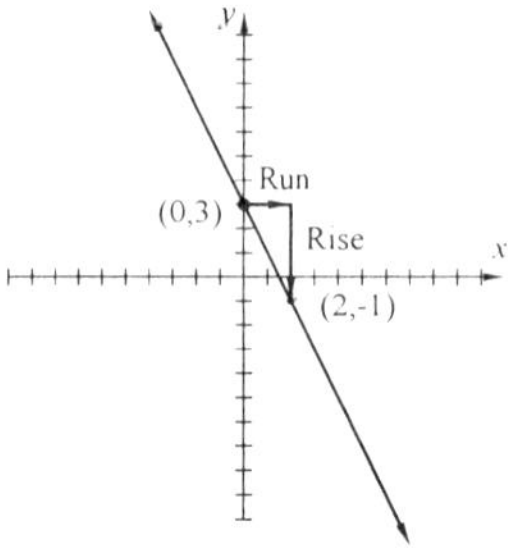

43. $y = 2x$

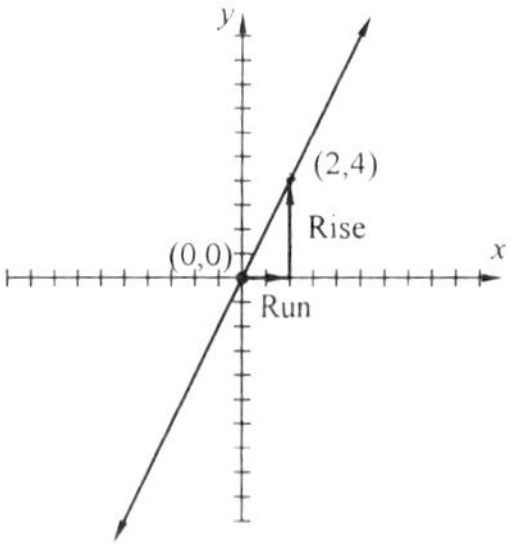

45. $m = \frac{5}{6}$

(a) $m_{\parallel} = m = \frac{5}{6}$

(b) $m_{\perp} = -\frac{1}{m} = -\frac{6}{5}$

47. $m_1 = 6$

(a) $m_2 = m_2 = 6$

(b) $m_2 = -\frac{1}{m_1} = -\frac{1}{6}$

49. $y = \frac{1}{3}x + 2$

$m = \frac{1}{3}$

(a) $m_{11} = m = \frac{1}{3}$

(b) $m_{\perp} = -\frac{1}{m} = -3$

51.

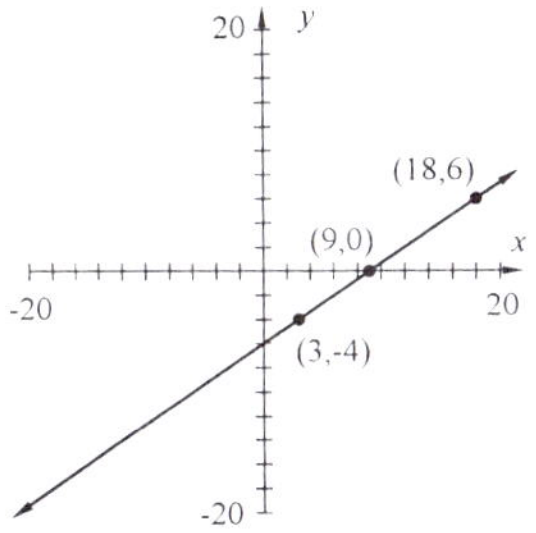

Yes

$m = \frac{0-(-4)}{9-3} = \frac{4}{6} = \frac{2}{3}$

From the graph, $b = -6$

$y = \frac{2}{3}x - 6$

53. $y = 14(4x + 35)$

(a) $y = 562x + 490$

(b) $m = 56,\ b = 490$

(c) The slope is the increase in the United States Federal Budget in billions of dollars per year.

Cumulative Review

55. $3x + 8 > 2x + 12$

$3x - 2x + 8 > 2x - 2x + 12$

$x + 8 > 12$

$x > 4$

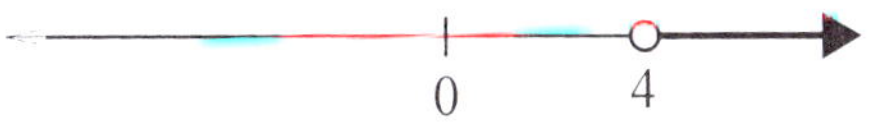

57. $\frac{1}{2}(x+2) \le \frac{1}{3}x + 5$

$3(x+2) \le 2x + 30$

$3x + 6 \le 2x + 30$

$x \le 24$

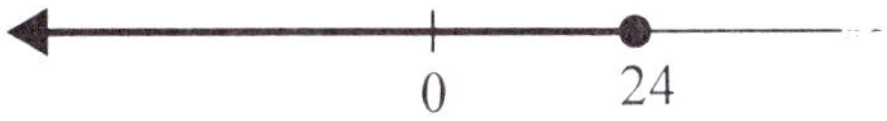

59. Decrease in range = $0.62 \times$ cruising range

$= 0.62(390) = 241.8$

$390 - 241.8 = 148.2$

Range at maximum speed = 148 miles

61. 1990-1999: $\frac{240-206}{206} = 0.165 = 16.5\%$

1999-2008: $\frac{288-240}{240} = 0.20 = 20\%$

Difference = $20.0\% - 16.5\% = 3.5\%$

How Am I Doing? Sections 7.1-7.3

1.

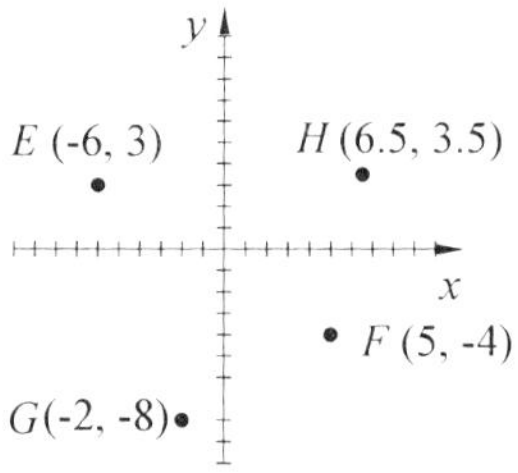

2. $A\left(3,\ 9\right)$

$B\left(7,\ -7\right)$

$C\left(-7, -8\right)$

$D\left(-4,\ 3\right)$

3. $y=-7x+3$

$y=-7\left(4\right)+3=-25,\ \left(4,\ -25\right)$

$y=-7\left(0\right)+3=3,\ \left(0,\ 3\right)$

$y=-7\left(-2\right)+3=17,\ \left(-2,\ 17\right)$

4. $6x-5y=-30$

x	y
0	6
−5	0
−10	−6

$0-5y=-30 \Rightarrow y=6$

$6x-5\left(0\right)-30 \Rightarrow x=-5$

$6x-5\left(-6\right)=-30 \Rightarrow x=-10$

5. $y=-4x-3$

x	y
−2	5
−1	1
0	−3

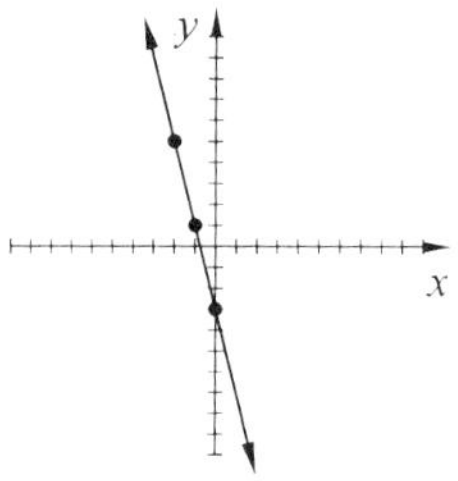

6. $y=\frac{3}{4}x-1$

x	y
−4	−4
0	−1
4	2

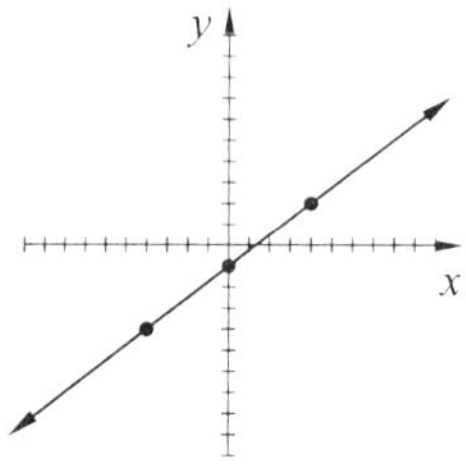

7. $y=\frac{3}{5}x$

x	y
−5	−3
0	0
5	3

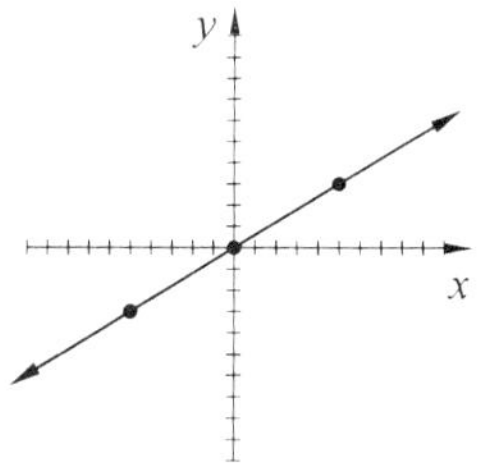

8. $-5x+2=y+3$

$-5x-1=y$

x	y
−1	4
0	−1
1	−6

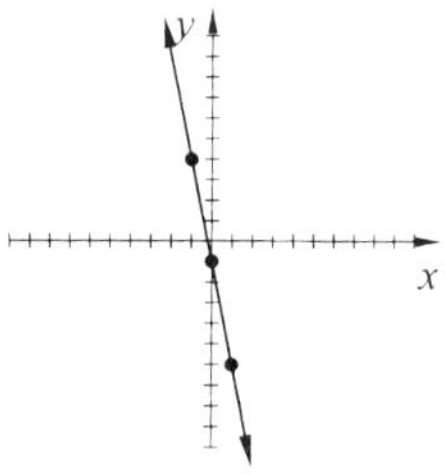

9. $\left(-6,\ -8\right)$ and $\left(2,\ -3\right)$

$$y=\frac{-3-\left(-8\right)}{2-\left(-6\right)}=\frac{5}{8}$$

10. $9x-3y-5=0$

$$-3y-5=-9x$$
$$-3y=-9x+5$$
$$y=3x-\frac{5}{3}$$
$$m=3,\ y\text{-intercept}\left(0,-\frac{5}{3}\right)$$

11. $m=\frac{3}{4}$

$$m_{\perp}=-\frac{1}{m}=-\frac{4}{3}$$

12. $3y+11x+7=0$

$$3y=-11x-7$$
$$y=-\frac{11}{3}x-\frac{7}{3}$$
$$m=-\frac{11}{3}$$
$$m_{11}=-\frac{11}{3}$$

7.4 Exercises

Use $m=\frac{y_2-y_1}{x_2-x_1}$ and $y=mx+b$ as needed in Exercises 1-31.

1. $m=4,\ (-3,0)$

$$0=4(-3)+b\Rightarrow b=12$$
$$y=4x+12$$

3. $m=-3,\ (2,5)$

$$5=-3(2)+b\Rightarrow b=11$$
$$y=-3x+11$$

5. $m=-3,\ \left(\frac{1}{2},2\right)$

$$2=-3\left(\frac{1}{2}\right)+b\Rightarrow b=\frac{7}{2}$$
$$y=-3x+\frac{7}{2}$$

7. $m=-\frac{2}{5},\ (5,-3)$

$$-3=-\frac{2}{5}(5)+b\Rightarrow b=-1$$
$$y=-\frac{2}{5}x-1$$

9. $(3,-12)$ and $(-4,2)$

$$m=\frac{2-(-12)}{-4-3}=\frac{14}{-7}=-2$$
$$2=-2(-4)+b\Rightarrow b=-6$$
$$y=-2x-6$$

11. $(2,-6)$ and $(-1,6)$

$$m=\frac{6-(-6)}{-1-2}=\frac{12}{-3}=-4$$
$$-6=-4(-1)+b\Rightarrow b=2$$
$$y=-4x+2$$

13. $(3, 5)$ and $(-1, -15)$

$m = \frac{-15-5}{-1-3} = \frac{-20}{-4} = 5$

$5 = 5(3) + b \Rightarrow b = -10$

$y = 5x - 10$

15. $\left(1, \frac{5}{6}\right)$ and $\left(3, \frac{3}{2}\right)$

$m = \frac{\frac{3}{2} - \frac{5}{6}}{3-1} = \frac{\frac{4}{6}}{2} = \frac{1}{3}$

$\frac{5}{6} = \frac{1}{3}(1) + b \Rightarrow b = \frac{1}{2}$

$y = \frac{1}{3}x + \frac{1}{2}$

17. $m = -2,\ (4, 3)$

$3 = -2(4) + b \Rightarrow b = 11$

$y = -2x + 11$

19. $(2, -3)$ and $(-1, 6)$

$m = \frac{6-(-3)}{-1-2} = \frac{9}{-3} = -3$

$-3 = -3(2) + b \Rightarrow b = 3$

$y = -3x + 3$

21. $b = 1,\ m = -\frac{2}{3}$

$y = -\frac{2}{3}x + 1$

23. $b = -4,\ m = \frac{2}{3}$

$y = \frac{2}{3}x - 4$

25. $b = 0,\ m = -\frac{2}{3}$

$y = -\frac{2}{3}x$

27. $b = -2,\ m = 0$

$y = -2$

29. $m = 0,\ (7, -2)$

The line is horizontal so

$y = -2$

31. $(4, -6)$ perpendicular to x – axis so m is undefined.

$x = 4$ for all values of y.

$x = 4$

33. $(0, -4) \Rightarrow b = -4$

Parallel to $y = \frac{3}{4}x + 2 \Rightarrow m_1 = \frac{3}{4}$

$m_2 = m_1 = \frac{3}{4}$

$y = \frac{3}{4}x - 4$

35. Perpendicular to $y = 2x - 9 \Rightarrow m_1 = 2$

$$m_2 = -\frac{1}{m_1} = -\frac{1}{2}$$

$$y = mx + b, \ (2, 3)$$

$$3 = -\frac{1}{2}(2) + b \Rightarrow b = 4$$

$$y = -\frac{1}{2}x + 4$$

37. $(0, 227)$ and $(10, 251)$

$$m = \frac{251 - 227}{10 - 0} = \frac{24}{10} = 2.4, \ b = 227$$

$$y = 2.4x + 227$$

Cumulative Review

39.

$$\frac{3}{t} - \frac{2}{t-1} = \frac{4}{t}$$

$$t(t-1)\left(\frac{3}{t}\right) - t(t-1)\left(\frac{2}{t-1}\right) = t(t-1)\left(\frac{4}{t}\right)$$

$$3(t-1) - 2t = 4(t-1)$$

$$3t - 3 - 2t = 4t - 4$$

$$t - 3 = 4t - 4$$

$$1 = 3t$$

$$\frac{1}{3} = t$$

41. 2nd week:

Price $= 80 - 0.15(80) = \$68$

3rd week:

Price $= 68 - 0.1(68) = \$61.20$

43. Number shipped $= 35(60) = 2100$

Percent defective $= \frac{104}{2100} \cdot 100\% \approx 4.95\%$

45. $x =$ width

$2x + 70 =$ length

$$P = 2L + 2W$$

$$3440 = 2(2x + 70) + 2x$$

$$3440 = 4x + 140 + 2x$$

$$3440 = 6x + 140$$

$$3300 = 6x$$

$$550 = x$$

$$2x + 70 = 2(550) + 70 = 1170$$

width $= 550$ feet

length $= 1170$ feet

7.5 Exercises

1. Any convenient point not on the boundary line may be used as a test point.

3. $y > 2 - 3x$

Test point: $(0, 0)$

$0 > 2 - 3(0)$

$0 > 2$　False

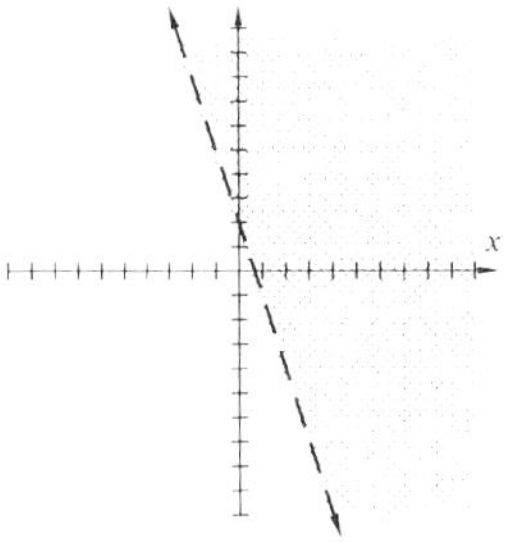

5. $2x - 3y < 6$

   Test point: $(0, 0)$

   $2(0) - 3(0) < 6$

   $0 < 6$ True

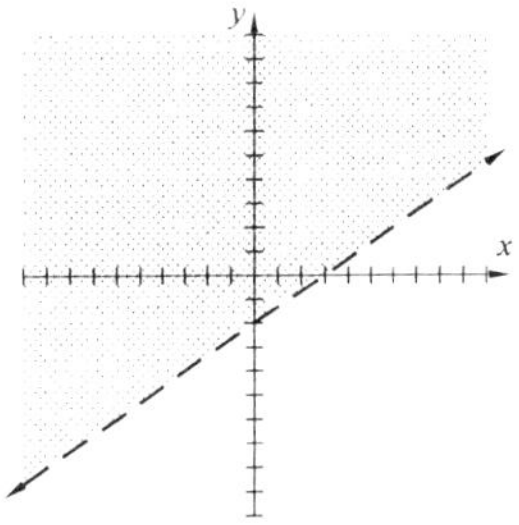

7. $2x - y \geq 3$

   Test point: $(0, 0)$

   $2(0) - (0) \geq 3$

   $0 \geq 3$ False

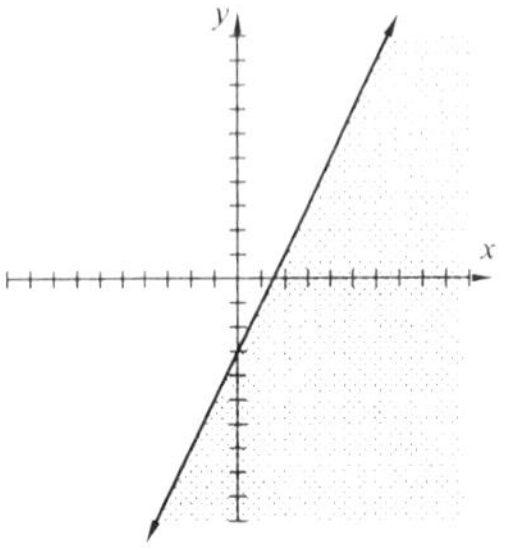

9. $y \geq 4x$

   Test point: $(1, 0)$

   $0 \geq 4(1)$

   $0 \geq 4$ False

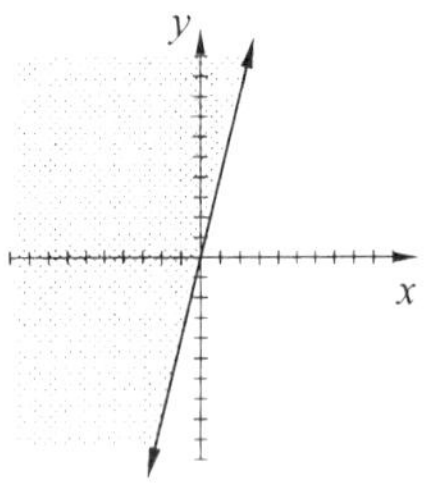

11. $y < -\frac{1}{2}x$

    Test point: $(1, 0)$

    $0 < -\frac{1}{2}(1)$

    $0 < -\frac{1}{2}$ False

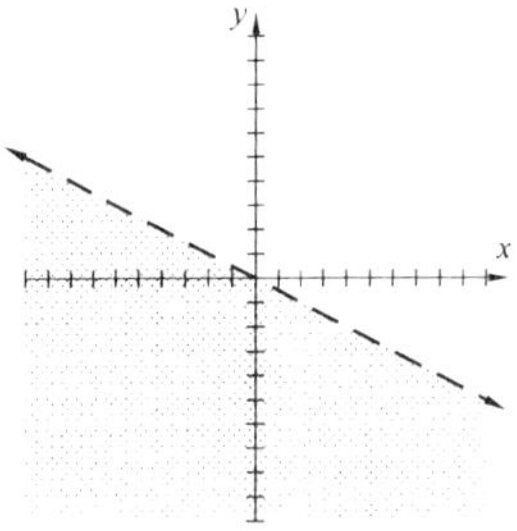

13. $x \geq 2$

    Test point: $(0, 0)$

    $0 \geq 2$ False

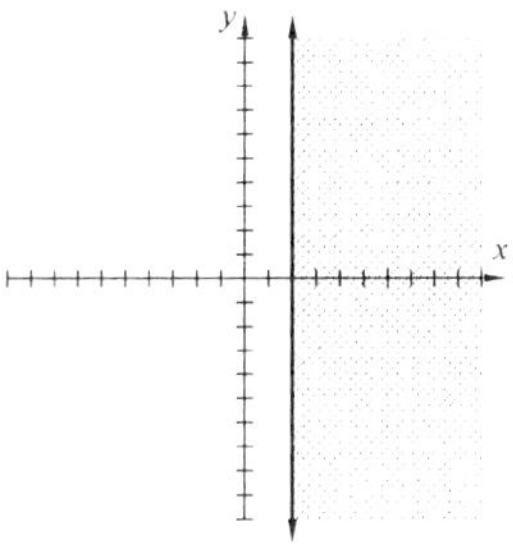

15. $2x - 3y + 6 \geq 0$

    Test point: $(0, 0)$

    $2(0) - 3(0) + 6 \geq 0$

    $6 \geq 0$ True

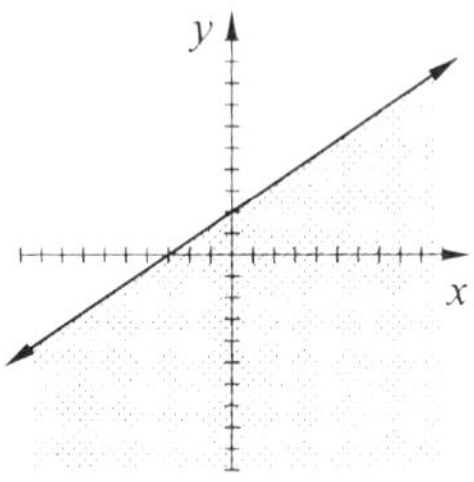

17. $2x > -3y$

    Test point: $(1, 1)$

    $2(1) > -3(1)$

    $2 > -3$ True

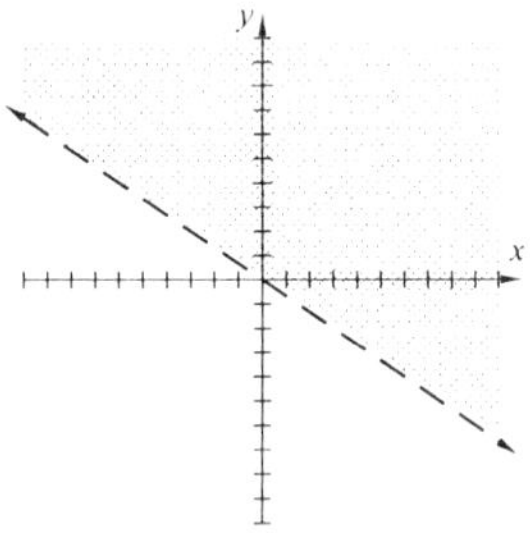

19. $2x > 3 - y$

    Test point: $(0, 0)$

    $2(0) > 3 - (0)$

    $0 > 3$ False

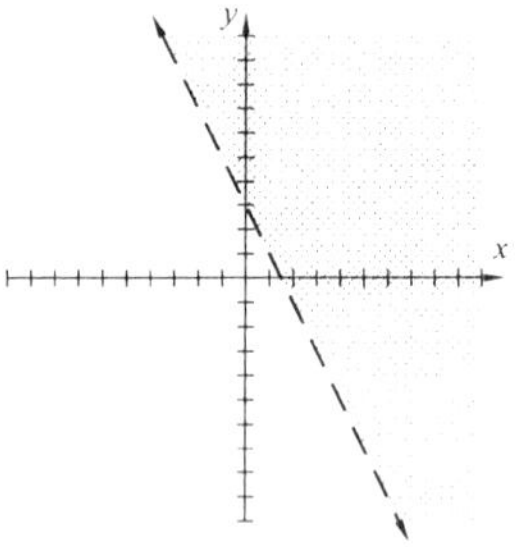

21. $x > -2y$

Test point: $(1, 0)$

$(1) > -2(0)$

$1 > 0$ True

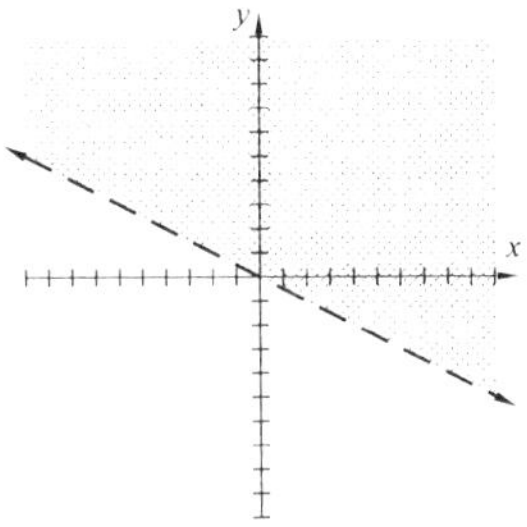

23. $3x + 3 > 5x - 3$

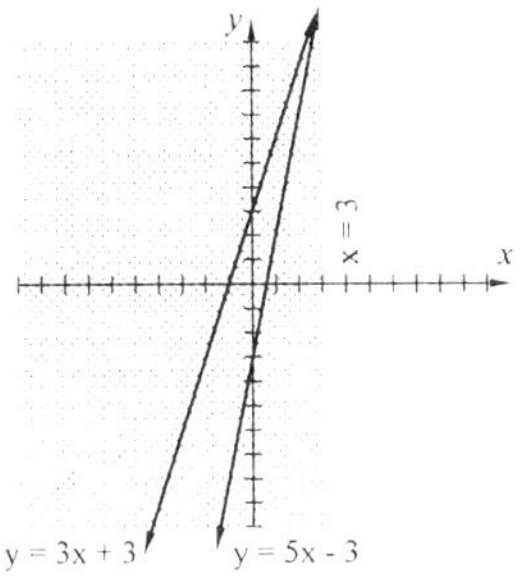

(a) $x < 3$

(b) $3x + 3 > 5x - 3$

$-2x > -6$

$x < 3$

(c) $3x + 3$ will be greater than $5x - 3$ for those values of x where the graph of $y = 3x + 3$ lies above the graph of $y = 5x - 3$.

Cumulative Review

25. $5x^2 + 12x - 9 = 0$

$(6x - 3)(x + 3) = 0$

$5x - 3 = 0 \qquad x + 3 = 0$

$5x = 3 \qquad x = -3$

$x = \frac{3}{5}$

27. x = the number of miles over 36,000

Total cost = cost for months + cost for excess miles.

$195(36) + 0.15x \le 8400$

$7020 + 0.15x \le 8400$

$0.15x \le 1380$

$x \le 9200$

She may drive 9200 miles above the limit.

29. x = amount of zinc

$\frac{4}{3}x$ = amount of copper

$x + \frac{4}{3}x = 210$

$3x + 4x = 630$

$7x = 630$

$x = 90$ ounces of zinc

7.6 Exercises

1. You can describe a function using a table of values, an algebraic equation, or a graph.

3. The domain of a function is the set of possible values of the independent variable.

5. If a vertical line can intersect the graph more than once, the relation is not a function.

7. $\left\{\left(\frac{2}{5}, 4\right), \left(3, \frac{2}{5}\right), \left(-3, \frac{2}{5}\right), \left(\frac{2}{5}, -1\right)\right\}$

Domain $= \left\{-3, \frac{2}{5}, 3\right\}$

Range $= \left\{-1, \frac{2}{5}, 4\right\}$

Not a function

9. $\{(7, 3.1), (5, 0), (7, 2.3)\}$

Domain $= \{5, 7\}$

Range $= \{0, 2.3, 3.1\}$

Not a function

11. $\{(12, 1), (14, 3), (1, 12), (9, 12)\}$

Domain $= \{1, 9, 12, 14\}$

Range $= \{1, 3, 12\}$

Function

13. $\{(3, 75), (5, 95), (3, 85), (7, 100)\}$

Domain $= \{3, 5, 7\}$

Range $= \{75, 85, 95, 100\}$

Not a function

15. $y = x^2 + 3$

x	y
−2	7
−1	4
0	3
1	4
2	7

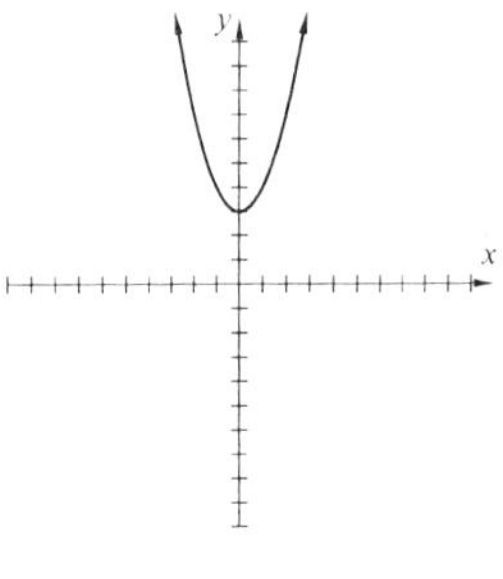

17. $y = 2x^2$

x	y
−2	8
−1	2
0	0
1	2
2	8

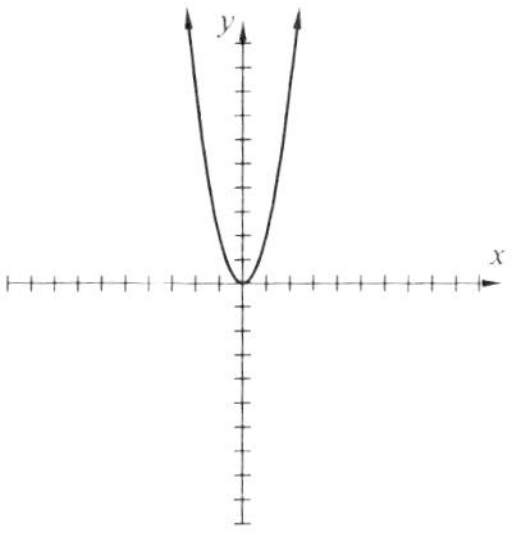

19. $x = -2y^2$

x	y
−8	−2
−2	−1
0	0
−2	1
−8	2

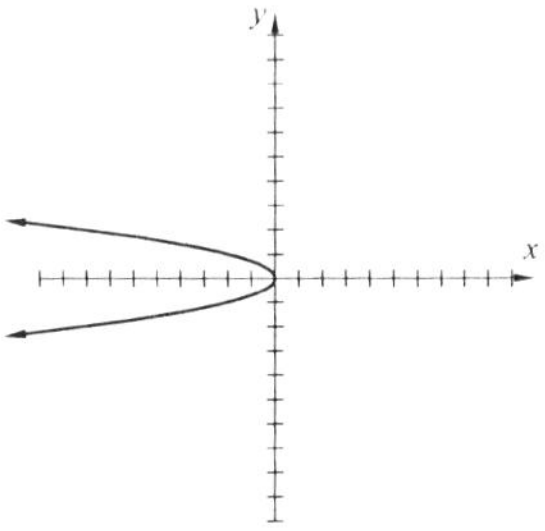

21. $x = y^2 - 4$

x	y
0	−2
−3	−1
−4	0
−3	1
0	2

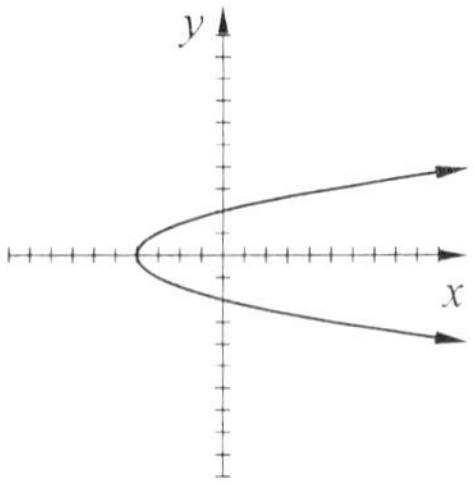

23. $y = \frac{2}{x}$

x	y
-4	$-\frac{1}{2}$
-2	-1
-1	-2
1	?
2	1
4	$\frac{1}{2}$

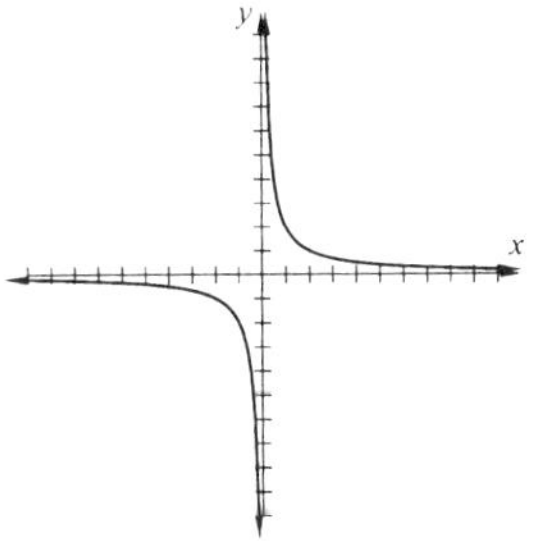

25. $y = \frac{4}{x^2}$

x	y
-4	$\frac{1}{4}$
-2	1
-1	4
1	4
2	1
4	$\frac{1}{4}$

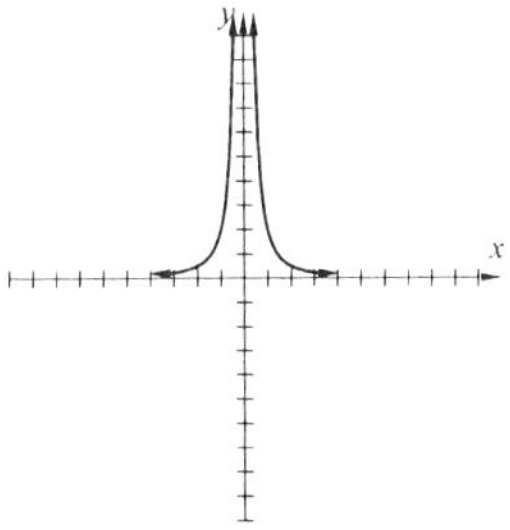

27. $x = (y+1)^2$

x	y
4	-3
1	-2
0	-1
1	0
4	1

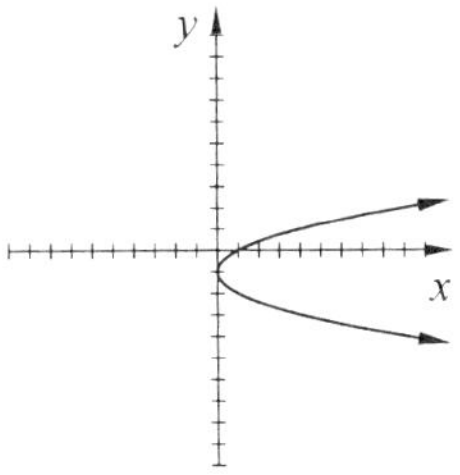

29. $y = \frac{4}{x-2}$

x	y
0	-2
1	-4
$\frac{3}{2}$	-8
$\frac{5}{2}$	8
3	4
4	2

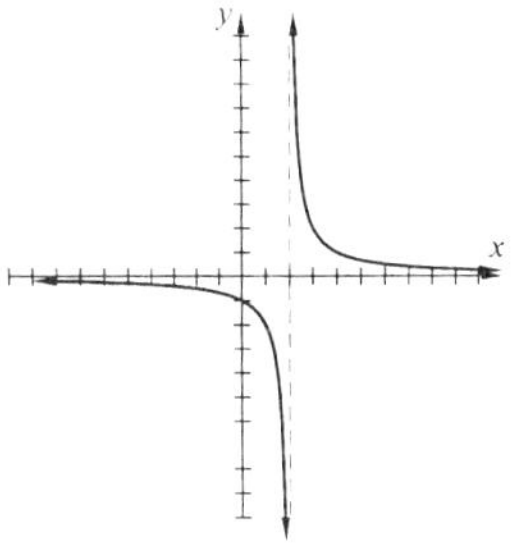

31. Function, passes the vertical line test

33. Not a function, fails the vertical line test

35. Function, passes the vertical line test

37. Not a function, fails the vertical line test

39. $f(x) = 2 - 3x$

(a) $f(-8) = 2 - 3(-8) = 26$

(b) $f(0) = 2 - 3(0) = 2$

(c) $f(2) = 2 - 3(2) = -4$

41. $f(x) = 2x^2 - x + 3$

(a) $f(0) = 2(0)^2 - 0 + 3 = 3$

(b) $f(-3) = 2(-3)^2 - (-3) + 3 = 24$

(c) $f(2) = 2(2)^2 - (2) + 3 = 9$

43. $f(x) = \frac{3}{x+2}$

(a) $f(3) = \frac{3}{3+2} = \frac{3}{5}$

(b) $f(0) = \frac{3}{0+2} = \frac{3}{2}$

(c) $f(-5) = \frac{3}{-5+2} = -1$

45. Yes

47. $f(1980) = 45\%$

$x = 1980$

49. $f(x) = 2x^2 + 20x + 800$

$f(0) = 2(0)^2 + 20(0) + 800 = 800$

$f(3) = 2(3)^2 + 20(3) + 800 = 878$

$f(6) = 2(6)^2 + 20(6) + 800 = 992$

$f(10) = 2(10)^2 + 20(10) + 800 = 1200$

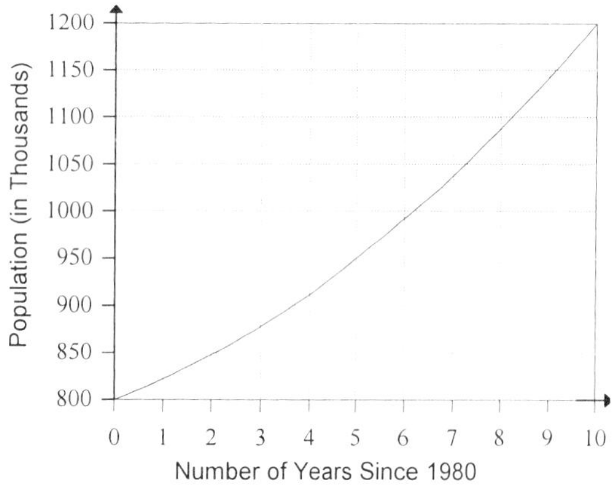

The curve slopes upward more steeply for larger values of x. The growth rate is increasing as x gets larger.

Cumulative Review

51. $(5x^2 - 6x + 2) + (-7x^2 - 8x + 4)$

$= 5x^2 - 6x + 2 - 7x^2 - 8x + 4$

$= (5-7)x^2 + (-6-8)x + 2 + 4$

$= -2x^2 - 14 + 6$

53. $(4x^2 - 2x + 3)(x + 2)$

$= 4x^3 - 2x^2 + 3x + 8x^2 - 4x + 6$

$= 4x^3 + (-2+8)x^2 + (3-4)x + 6$

$= 4x^3 + 6x^2 - x + 6$

Putting Your Skills to Work

1. $2.47(60)(60)(24) = 213,408$ acres/day

$213{,}408(365) \approx 78,000,000$ acres/year

2. $1756(2.47) = 4337$ million acres

$= 4,337,000,000$ acres

3. $1992: 4337 - 2(78) = 4181$ million acres

$1995: 4337 - 5(78) = 3947$ million acres

4. t = number of years after 1990

$(0, 4337)$ and $(5, 3947)$

$m = \frac{3947 - 4337}{5 - 0} = -78, \ b = 4337$

$r(t) = -78t + 4337$

5.

t	0	5	10	15	20
r	4337	3947	3557	3167	2777

6.

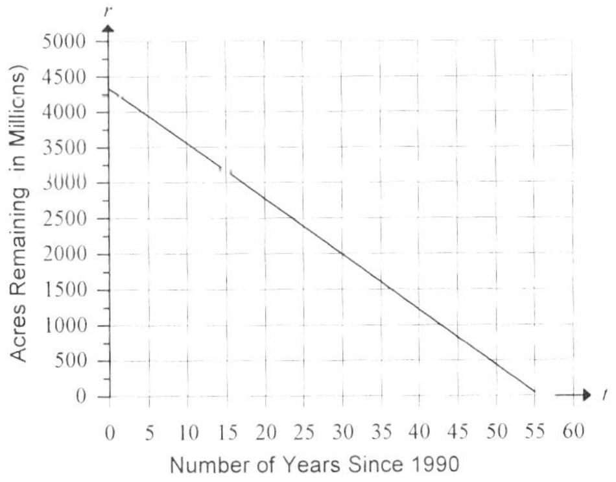

7. $t = 2035 - 1990 = 45$

$r(45) \approx 827$ million acres

8. $r(55) = 0$

There will be no rainforests after $1990 + 55 = 2045$.

Chapter 7 Review Problems

1. A: $(2, -3)$, B: $(-1, 0)$, C: $(3, 2)$, D: $(-2, -3)$

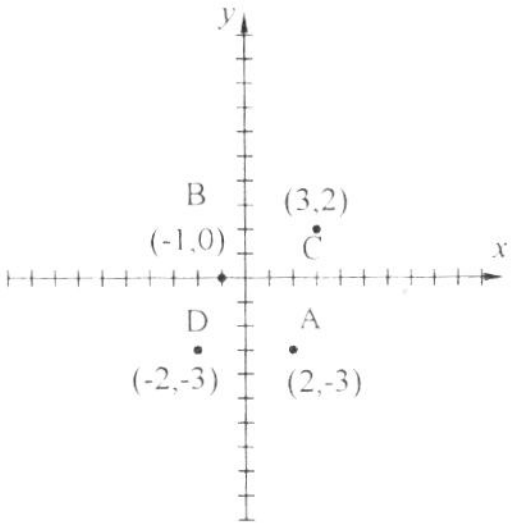

2. $E: (4, 4)$, $F: (0, 3)$, $G: (1, -4)$, $H: (-4, -1)$

3. $y = 7 - 3x$

(a) $y = 7 - 3x$

(b) $10 = 7 - 3x \rightarrow x = -1, (-1, 10)$

4. $2x + 5y = 12$

(a) $(1, \ ): 2(1) + 5y = 12$

$5y = 10$

$y = 2: (1, 2)$

(b) $(\ , 4): 2x + 5(4) = 12$

$2x = -8$

$x = -4: (-4, 4)$

5. $x = 6$

(a) $(\ , -1)$: $x = 6$: $(6, -1)$

(b) $(\ , 3)$: $x = 6$: $(6, 3)$

6. $3y = 2x + 6$

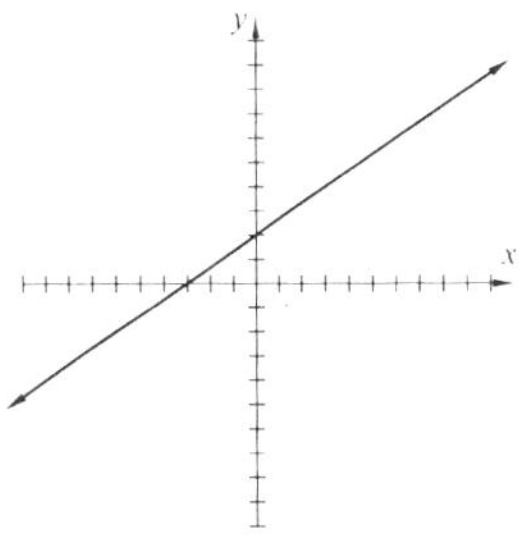

7. $5y + x = -15$

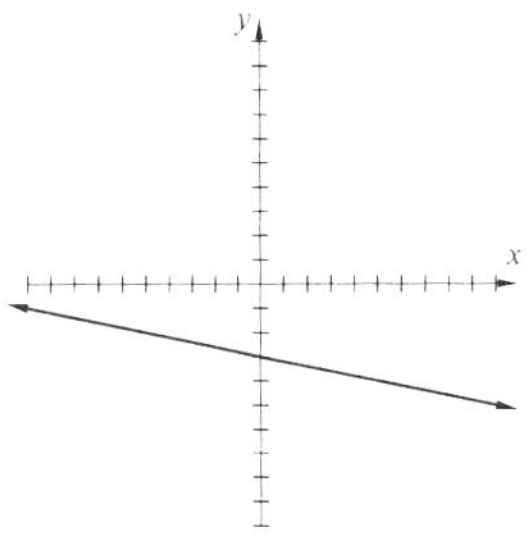

8. $2y+4x=-8+2y$

$4x=-8$

$x=-2$

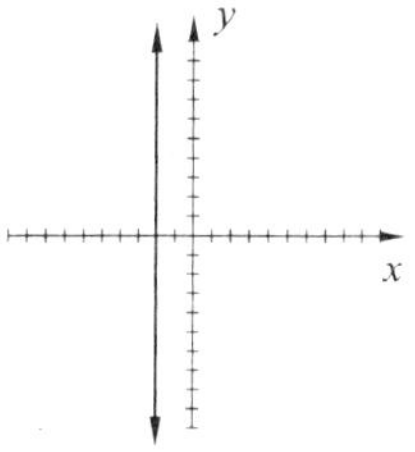

9. $(5,-3)$ and $\left(2,-\frac{1}{2}\right)$

$m=\frac{y_2-y_1}{x_2-x_1}=\frac{-\frac{1}{2}-(-3)}{2-5}=-\frac{5}{6}$

10. $9x-11y+15=0$

$9x+15=11y$

$\frac{9}{11}x+\frac{15}{11}=y$

$m=\frac{9}{11},\quad b=\frac{15}{11}$

11. $m=-\frac{1}{2},\quad b=3$

$y=mx+b$

$y=-\frac{1}{2}x+3$

12. $y=\frac{3}{5}x-2,\ m=\frac{3}{5}$

$m_\perp=-\frac{1}{m}=-\frac{5}{3}$

13. $y=-\frac{1}{2}x+3$

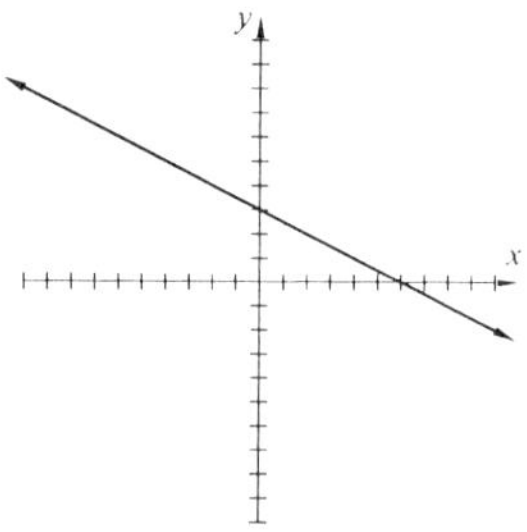

14. $2x-3y=-12$

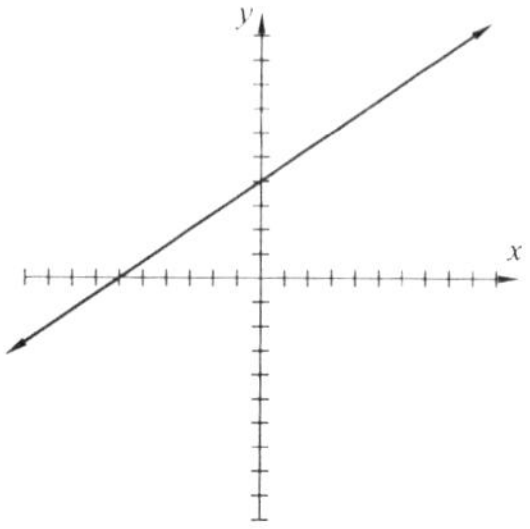

15. $5x+2y=20+2y$

$5x=20$

$x=4$

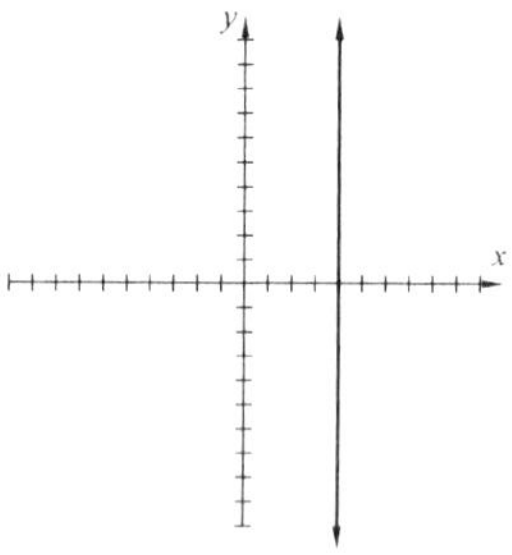

16. $(5,6),\quad m=2$

$y=mx+b$

$6=2(5)+b\Rightarrow b=-4$

$y=2x-4$

17. $(3, -4)$, $m = -6$

$$y = mx + b$$
$$-4 = -6(3) + b \Rightarrow b = 14$$
$$y = -6x + 14$$

18. $(-4, 3)$, $m = -\frac{3}{2}$

$$y = mx + b$$
$$3 = -\frac{3}{2}(-4) + b \Rightarrow b = -3$$
$$y = -\frac{3}{2}x - 3$$

19. $(3, 7)$ and $(-6, 7)$

$$m = \frac{y_2 - y_1}{x_2 - x_1} = \frac{7-7}{-6-3} = 0$$

Horizontal line with $y = 7$

$$y = 7$$

20. $y = -\frac{2}{3}x + 4$, $m = -\frac{2}{3}$

$$m_{11} = m = -\frac{2}{3}$$

21. $-3x + 4y = 8$

$$4y = 3x + 8$$
$$y = \frac{3}{4}x + 1$$
$$m = \frac{3}{4},\ m_{\perp} = -\frac{1}{m} = -\frac{4}{3}$$

22. $(0, -3)$ and $(3, -1)$

$$m = \frac{-3-(-1)}{0-3} = \frac{2}{3},\ b = -3$$
$$y = \frac{2}{3}x - 3$$

23. $(0, 1)$ and $(1, -2)$

$$m = \frac{-2-1}{1-0} = -3,\ b = 1$$
$$y = -3x + 1$$

24. $x = 5$

25. $y < \frac{1}{3}x + 2$

Test point: $(0, 0)$

$$0 < \frac{1}{3}(0) + 2$$

$0 < 2$ True

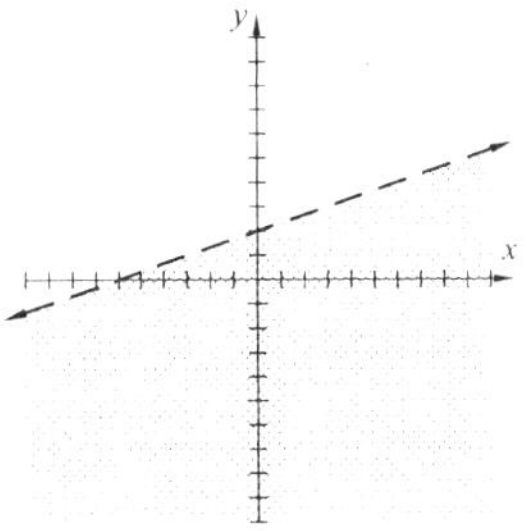

26. $3y + 2x \geq 12$

Test point: $(0, 0)$

$$3(0) + 2(0) \geq 12$$

$0 \geq 12$ False

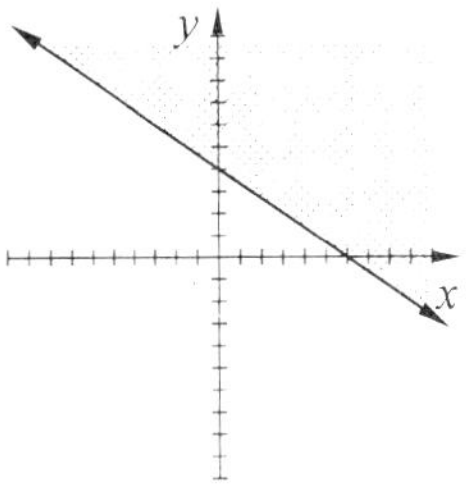

27. $x \leq 2$

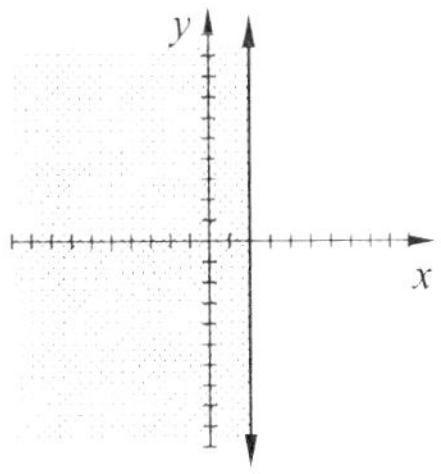

28. $\{(5, -6), (-6, 5), (-5, 5), (-6, -6)\}$
    Domain $\{-6, -5, 5\}$
    Range $\{-6, 5\}$
    Not a function

29. $\{(2, -3), (5, -3), (6, 4), (-2, 4)\}$
    Domain $\{-2, 2, 5, 6\}$
    Range $\{-3, 4\}$
    Function

30. Function, passes the vertical line test

31. Not a function, fails the vertical line test

32. Function, passes the vertical line test

33. $y = x^2 - 5$

x	y
−2	−1
−1	−4
0	−5
1	−4
2	−1

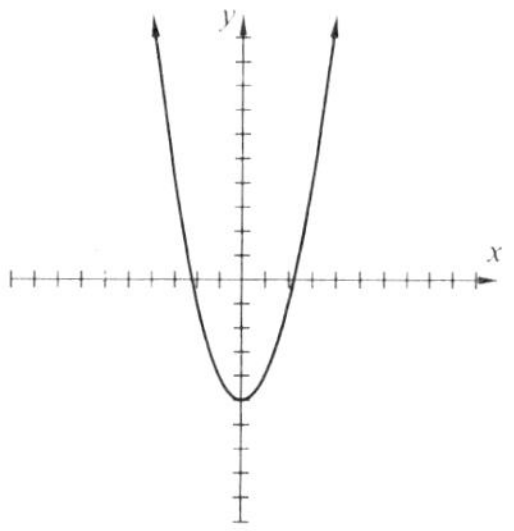

34. $x = y^2 + 3$

x	y
7	−2
4	−1
3	0
4	1
7	2

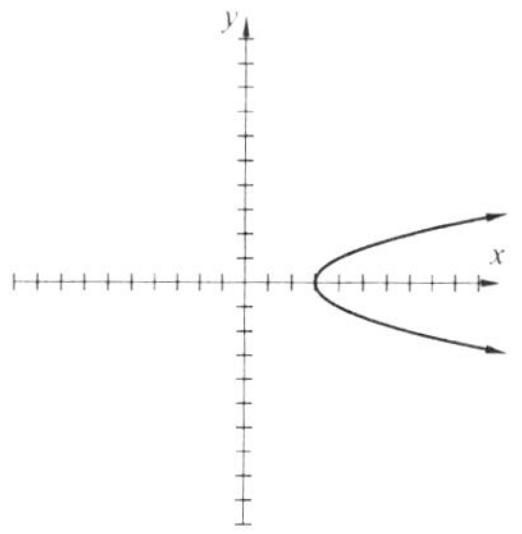

35. $y = (x-3)^2$

x	y
1	4
2	1
3	0
4	1
5	4

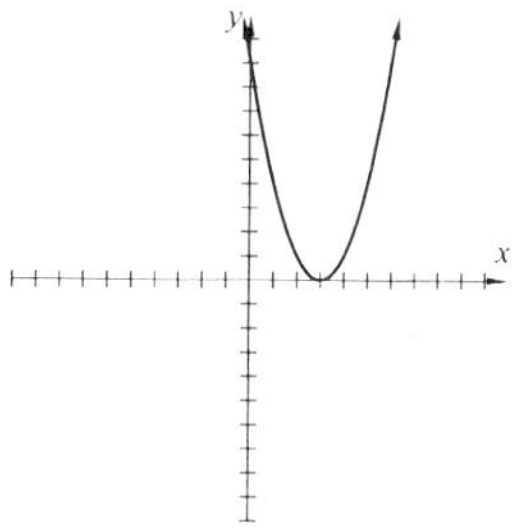

36. $f(x) = 7 - 6x$
    (a) $f(0) = 7 - 6(0) = 7$
    (b) $f(-4) = 7 - 6(-4) = 31$

37. $g(x) = -2x^2 + 3x + 4$
    (a) $g(-1) = -2(-1)^2 + 3(-1) + 4 = -1$
    (b) $g(3) = -2(3)^2 + 3(3) + 4 = -5$

38. $f(x) = 8x - 3$
    (a) $f(-1) = 8(-1) - 3 = -11$
    (b) $f\left(\frac{1}{2}\right) = 8\left(\frac{1}{2}\right) - 3 = 1$

39. $f(x) = \frac{2}{x+4}$
    (a) $f(-2) = \frac{2}{-2+4} = 1$
    (b) $f(6) = \frac{2}{6+4} = \frac{1}{5}$

40. $f(x) = x^2 - 2x + \frac{3}{x}$

(a) $f(-1) = (-1)^2 - 2(-1) + \frac{3}{(-1)} = 0$

(b) $f(3) = (3)^2 - 2(3) + \frac{3}{3} = 4$

Use $y = 30 + 0.09x$ in Exercises 41-46.

41. $y = 30 + 0.09(2000) = \$210$

42. $y = 30 + 0.09(1600) = \$174$

43. $y = 0.09x + 30$

$y = mx + b,\ b = 30$

It tells us that if no electricity is used, the cost is a minimum of \$30.

44. $y = 0.09x + 30$

$y = mx + b,\ m = 0.09$

It tells us that the bill increases \$0.09 for each kilowatt-hour of use.

45. $147 = 0.09x + 30$

$117 = 0.09x$

$x = 1300$ kilowatt-hours

46. $246 = 0.09x + 30$

$216 = 0.09x$

$x = 2400$ kilowatt-hours

Use $y = 0.4x + 23.6$ in Exercises 47-52.

47. $x = 2008 - 1960 = 48$

$y = 0.4(48) + 23.6$

$y = 42.8$ quadrillion Btu

48. $x - 2010 - 1960 = 50$

$y - 0.4(50) + 23.6$

$y = 43.6$ quadrillion Btu

49. The slope is 0.4. The significance of the slope is that it tells us the rate of change per year. In this case it tells us that for each additional year the U.S. consumes 0.4 quadrillion Btu of petroleum.

50. The y-intercept is $(0,\ 23.6)$. This tells us that when $x = 0$, which is the year 1960, a total of 23.6 quadrillion Btu of petroleum were used.

51. $y = 47.6$

$47.6 = 0.4x + 23.6$

$24 = 0.4x$

$60 = x$

$\text{Year} = 1960 + 60 = 2020$

52. $y = 51.6$

$51.6 = 0.4x + 23.6$

$28 = 0.4x$

$70 = x$

$\text{Year} = 1960 + 70 = 2030$

53. $5x+3y=-15$

x	y
0	−5
−3	0

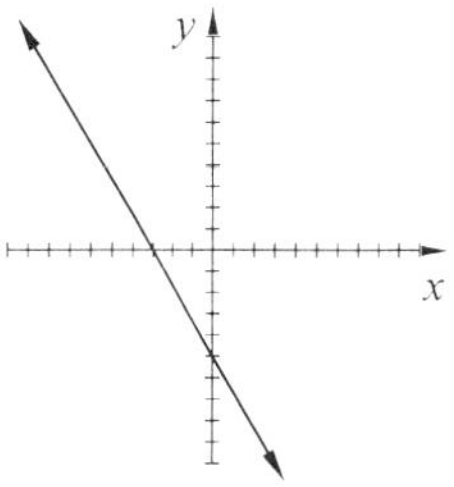

54. $y=\frac{3}{4}x-3$

x	y
0	−3
4	0

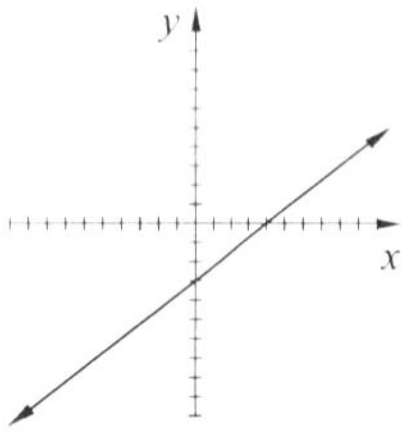

55. $(2,-7)$ and $(-3,-5)$

$$y=\frac{-5-(-7)}{-3-2}=\frac{2}{-5}=-\frac{2}{5}$$

56. $7x+6y-10=0$

$$6y-10=-7x$$
$$6y=-7x+10$$
$$y=-\frac{7}{6}x+\frac{5}{3}$$

57. $m=\frac{2}{3}$, $(3,-5)$

$$-5=\frac{2}{3}(3)+b\Rightarrow b=-7$$
$$y=\frac{2}{3}x-7$$

58. $y<-2x+1$

Test point: $(0, 0)$

$0<-2(0)+1$

$0<1$ True

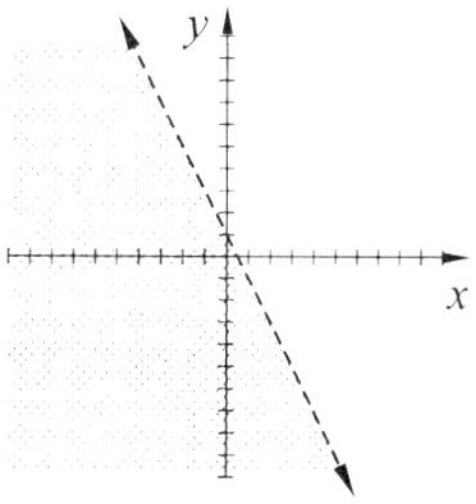

Use $y=-130x+22,000$ in Exercises 59-64.

59. 1980:

$x=1980-1980=0$

$y=-130(0)+22,000$

$y=22,000$

22,000,000 people

1990:

$x=1990-1980=10$

$y=-130(10)+22,000$

$y=20,700$

20,700,000 people

2000:

$x=2000-1980=20$

$y=-130(20)+22,000$

$y=19,400$

19,400,000 people

60.

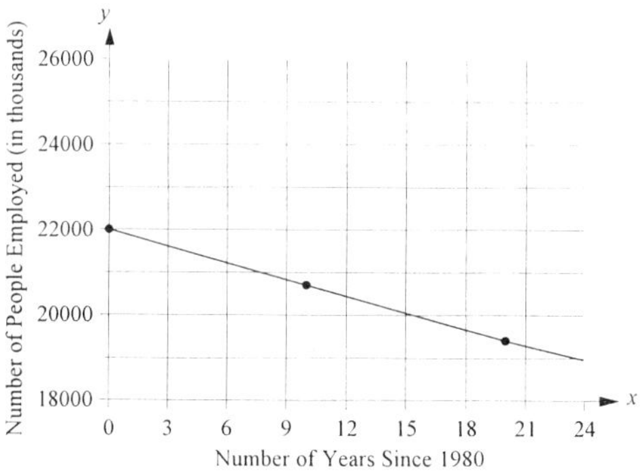

61. The slope is -130. The slope tells us that the number of people employed in manufacturing decreases each year by 130 thousand.

62. The y-intercept is $(0, 22,000)$. This tells us that in the year 1980 the number of manufacturing jobs was 22,000,000.

63. $y = 18,100$
$18,100 = -130x + 22,000$
$-3900 = -130x$
$30 = x$
$\text{Year} = 1980 + 30 = 2010$

64. $y = 16,800$
$16,800 = -130x + 22,000$
$-5200 = -130x$
$40 = x$
$\text{Year} = 1980 + 40 = 2020$

How Am I Doing? Chapter 7 Test

1. $B = (6, 1)$ $C = (-4, -3)$
$D = (-3, 0)$ $E = (5, -2)$

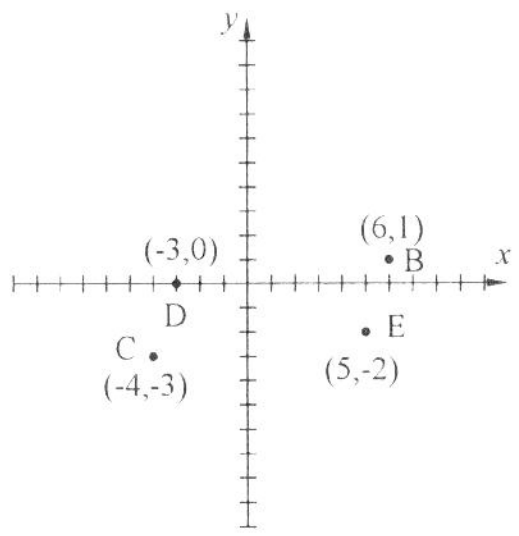

2. $6x - 3 = 5x - 2y$
$x + 2y = 3$

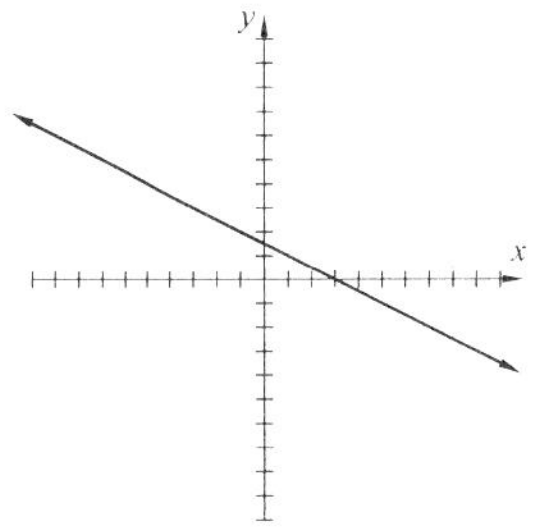

3. $8x + 2y = -4$

x	y
0	-2
$-1/2$	0
-1	2

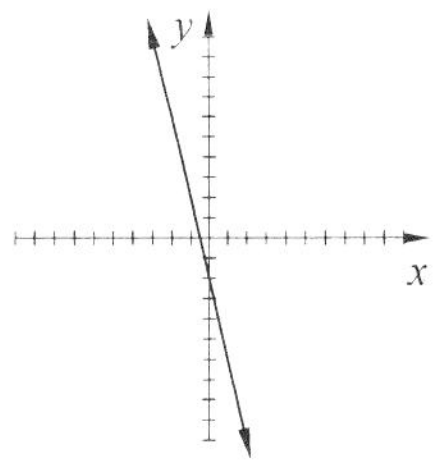

4. $y = \frac{2}{3}x - 4$

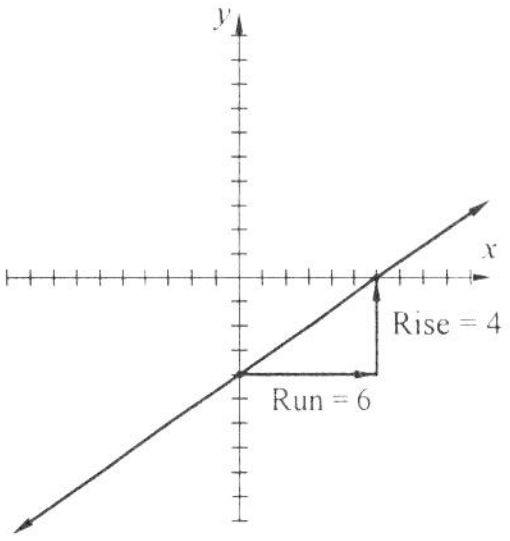

5. $3x + 2y - 5 = 0$
$2y = -3x + 5$
$y = -\frac{3}{2}x + \frac{5}{2}$
$m = -\frac{3}{2},\ b = \frac{5}{2}$

6. $(8, 6)$ and $(-3, -5)$
$m = \frac{-5-6}{-3-8} = \frac{-11}{-11} = 1$

7. $(4, -2),\ m = \frac{1}{2}$

$$y = mx + b$$

$$-2 = \frac{1}{2}(4) + b \Rightarrow b = -4$$

$$y = \frac{1}{2}x - 4 \text{ or } x - 2y = 8$$

8. $(-3, 11)$ and $(-6, 11)$

$$m = \frac{11-11}{-6-(-3)} = 0$$

9. $(5, -4)$ and $(-3, 8)$

$$m = \frac{y_2 - y_1}{x_2 - x_1} = \frac{8-(-4)}{-3-5} = -\frac{3}{2}$$

$$y = mx + b$$

$$-4 = \left(-\frac{3}{2}\right)(5) + b$$

$$-8 = -15 + 2b$$

$$\frac{7}{2} = b$$

$$y = -\frac{3}{2}x + \frac{7}{2}$$

10. $(2, 7)$ and $(2, -2)$

$$m = \frac{y_2 - y_1}{x_2 - x_1} = \frac{-2-7}{2-2} = \frac{-9}{0} \text{ undefined}$$

Vertical line: $x = 2$

11. $4y \le 3x$

Test point: $(1,1)$

$4(1) \le 3(1)$

$4 \le 3$ False

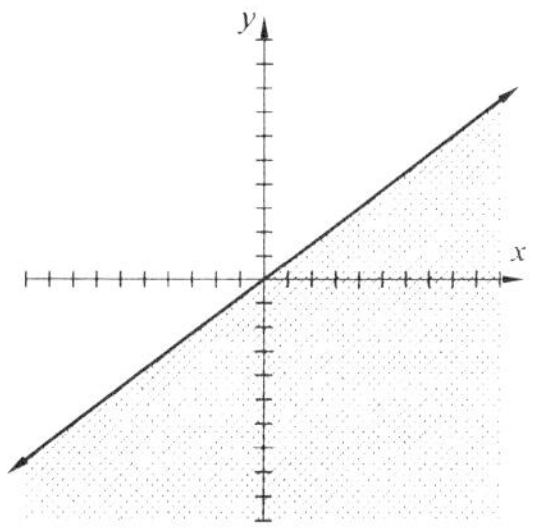

12. $-3x - 2y > 10.$

Test point: $(0, 0)$

$-3(0) - 2(0) > 10$

$0 > 10$ False

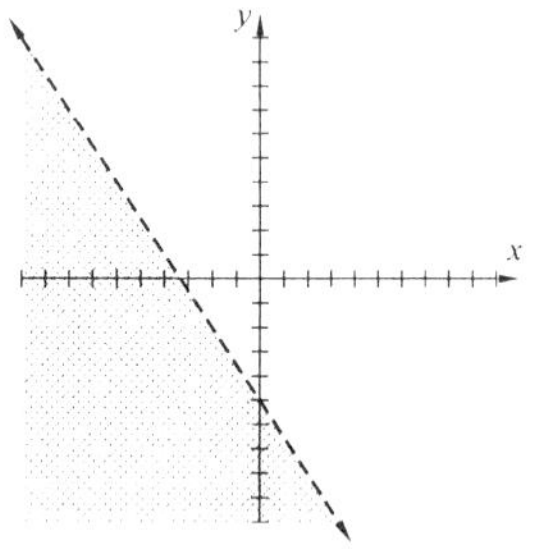

13. $\{(2, -8), (3, -7), (2, 5)\}$

no; two different ordered pairs have the same first coordinate.

14. Function, passes the vertical line test

15. $y = 2x^2 - 3$

x	y
-2	5
-1	-1
0	-3
1	-1
2	5

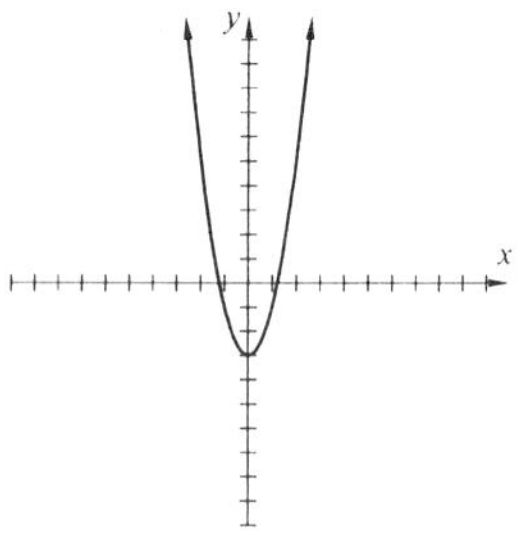

16. $f(x) = -x^2 - 2x - 3$

(a) $f(0) = -(0)^2 - 2(0) - 3 = -3$

(b) $f(-2) = -(-2)^2 - 2(-2) - 3 = -3$

17. $g(x) = \frac{3}{x-4}$

(a) $g(3) = \frac{3}{3-4} = -3$

(b) $g(-11) = \frac{3}{-11-4} = \frac{3}{-15} = -\frac{1}{5}$

Cumulative Test for Chapters 0-7

1. $\left(-3x^3y^4z\right)^3 = -27x^9y^{12}z^3$

2. $\frac{-16x^{-3}y^4}{24x^5y^{-6}} = \frac{-2y^{10}}{3x^8}$

3. $2 - 3(4 - x) = x - (3 - x)$

$2 - 12 + 3x = x - 3 + x$

$x = 7$

4. $$\frac{600}{R} = \frac{600}{R+50} + \frac{500(R+20)}{R^2+50R}$$

$$R(R+50)\left(\frac{600}{R}\right) = R(R+50)\left[\frac{600}{R+50} + \frac{500(R+20)}{R^2+50R}\right]$$

$$600(R+50) = 600R + 500(R+20)$$

$$600R + 30,000 = 600R + 500R + 10,000$$

$$20,000 = 500R$$

$$40 = R$$

5. $x = 2^3 - 6 \div 3 - 1$

$= 8 - 6 \div 3 - 1$

$= 8 - 2 - 1$

$= 5$

6. $A = lw + lh$

$A - lh = lw$

$\frac{A - lh}{l} = w$

7. $50a^2 - 98b^2$

$= 2\left(25a^2 - 49b^2\right)$

$= 2(5a + 7b)(5a - 7b)$

8. $4x^2 + 4x - 15 = (2x - 3)(2x + 5)$

9. $$\frac{x - \frac{1}{x}}{\frac{1}{2} + \frac{1}{2x}} = \frac{x - \frac{1}{x}}{\frac{1}{2} + \frac{1}{2x}} \cdot \frac{2x}{2x}$$

$$= \frac{2x^2 - 2}{x + 1}$$

$$= \frac{2(x+1)(x-1)}{x+1}$$

$$= 2(x - 1)$$

10. $(6, 8)$ and $(7, 11)$

$m = \frac{y_2 - y_1}{x_2 - x_1} = \frac{11 - 8}{7 - 6} = \frac{3}{1} = 3$

$y = mx + b$

$8 = 3(6) + b \Rightarrow b = -10$

$y = 3x - 10$

11. $(7, -4)$ and slope is undefined

$x = 7$

12. $(3, -5), \quad m = -\frac{2}{3}$

$$y = mx + b$$
$$-5 = \left(-\frac{2}{3}\right)(3) + b$$
$$-5 = -2 + b$$
$$-3 = b$$
$$y = -\frac{2}{3}x - 3$$

13. $(-8, -3)$ and $(11, -3)$

$$m = \frac{y_2 - y_1}{x_2 - x_1} = \frac{-3-(-3)}{11-(-8)} = 0$$

14. $3x - 7y = -2$

$$-7y = -3x - 2$$
$$y = \frac{3}{7}x + \frac{2}{7}$$
$$y = mx + b$$
$$m = \frac{3}{7}$$

15. $y = \frac{2}{3}x - 4$

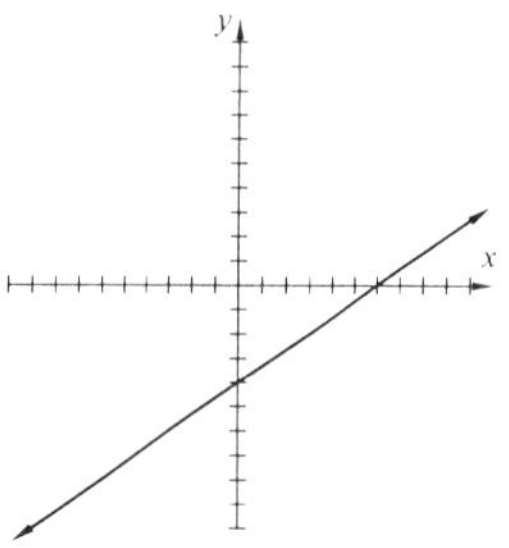

16. $3x + 8 = 5x$

$$8 = 2x$$
$$4 = x$$

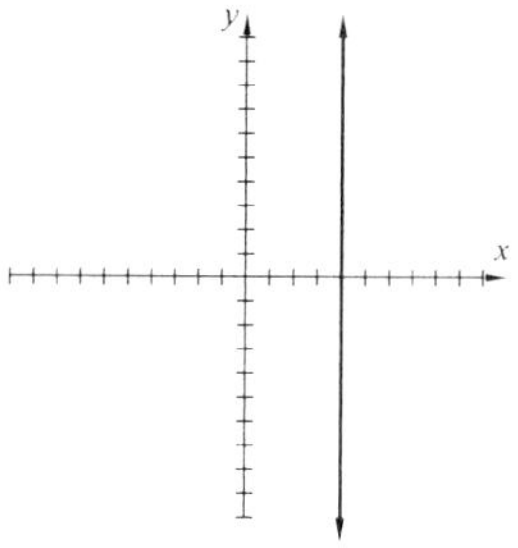

17. $2x + 5y \le -10$

Test point: $(0, 0)$

$2(0) + 5(0) \le -10$

$0 \le -10$ False

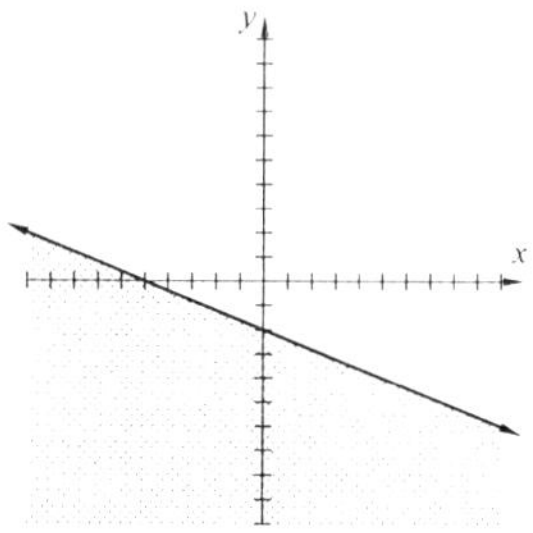

18. $x = y^2 + 3$

x	y
7	−2
4	−1
3	0
4	1
7	2

Not a Function

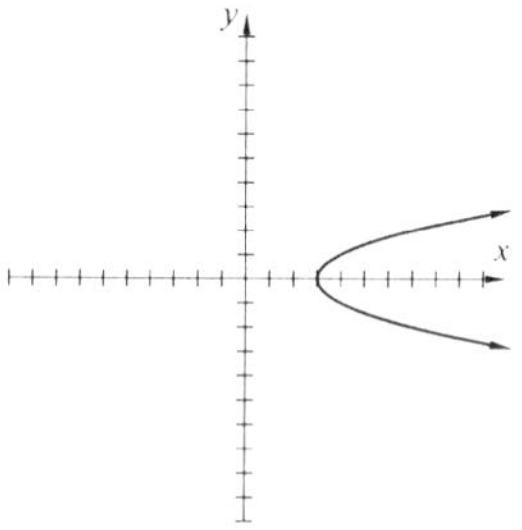

19. $\{(3, 10), (10, -3), (0, -3), (5, 10)\}$

This is a function

20. $f(x) = 3x^2 - 2x + \frac{4}{x}$

(a) $f(-1) = 3(-1)^2 - 2(-1) + \frac{4}{-1} = 1$

(b) $f(2) = 3(2)^2 - 2(2) + \frac{4}{2} = 10$

Chapter 8

8.1 Exercises

1. The lines $y = 2x - 5$ and $y = 2x + 6$ are parallel with y – intercepts at -5 and 6 respectively. There is no solution. The system is inconsistent.

3. If two lines have different slopes you can conclude they <u>intersect</u> and the system has <u>one</u> solution.

5. If two lines have different slopes but the same y – intercept, the lines intersect at the y – intercept which is the solution to the system.

7. $x - y = 3$
   $x + y = 5$
   $(4,1)$

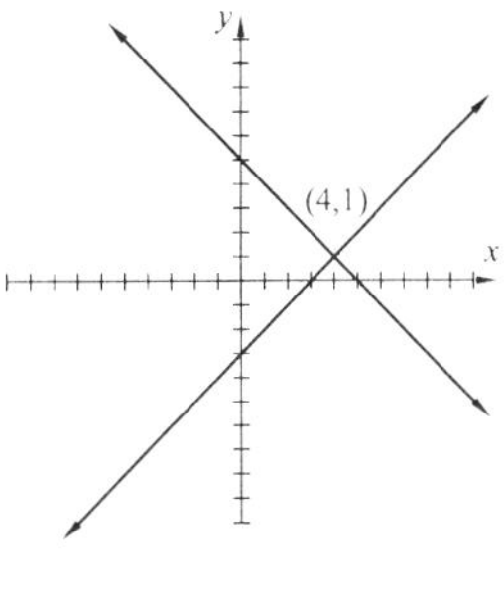

9. $2x + y = 0$
   $-3x + y = 5$
   $(-1, 2)$

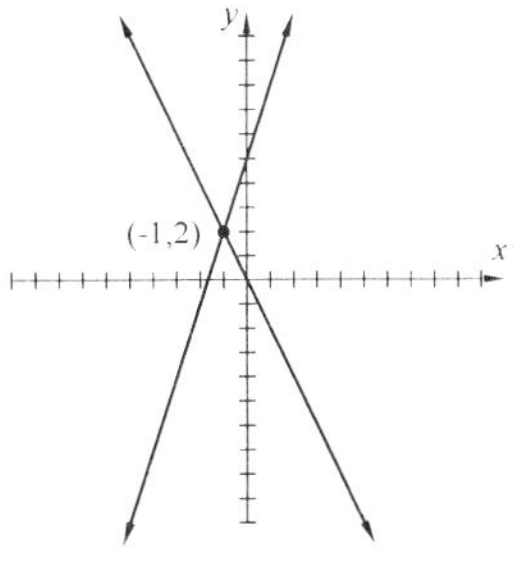

11. $4x + y = 5$
    $3x - 2y = 12$
    $(2, -3)$

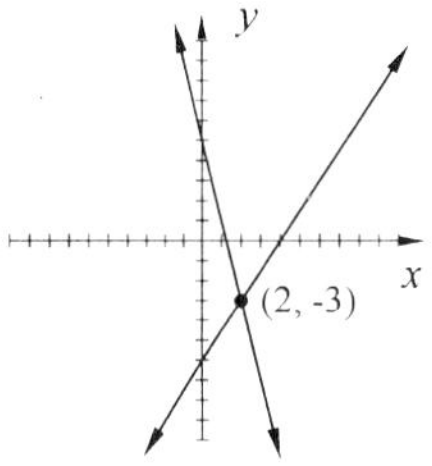

13. $-2x + y - 3 = 0$
    $4x + y + 3 = 0$
    $(-1, 1)$

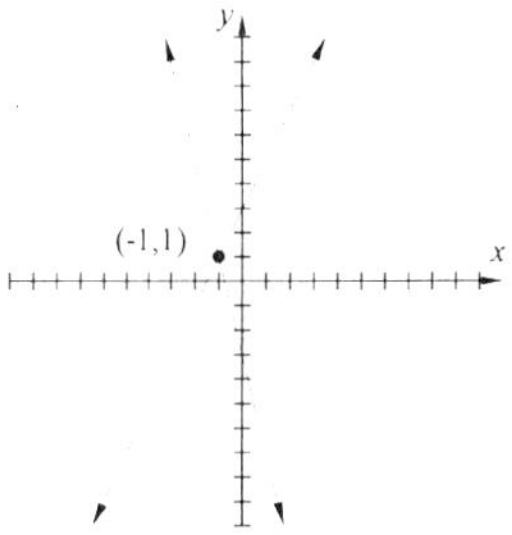

15. $3x - 2y = -18$
    $2x + 3y = 14$
    $(-2, 6)$

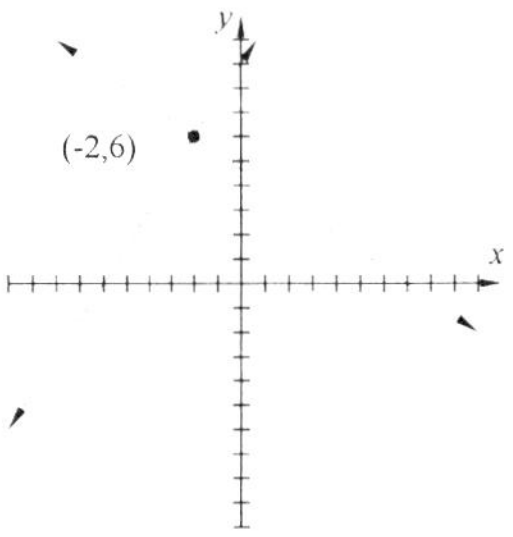

17. $y = \dfrac{3}{4}x + 7$
    $y = -\dfrac{1}{2}x + 2$
    $(-4, 4)$

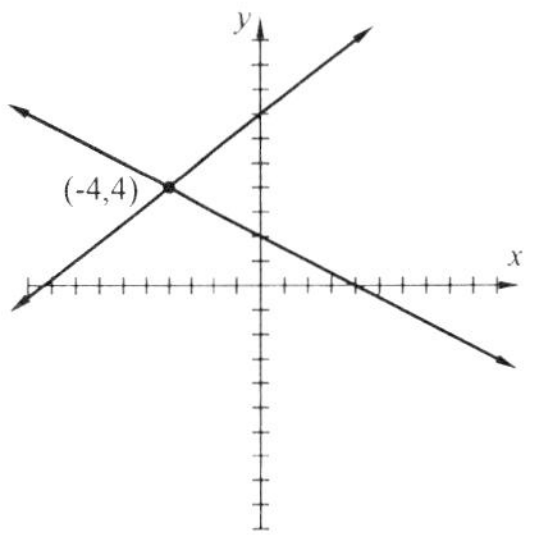

19. $3x - 2y = -4$

$-9x + 6y = -9$

No solution

Inconsistent system.

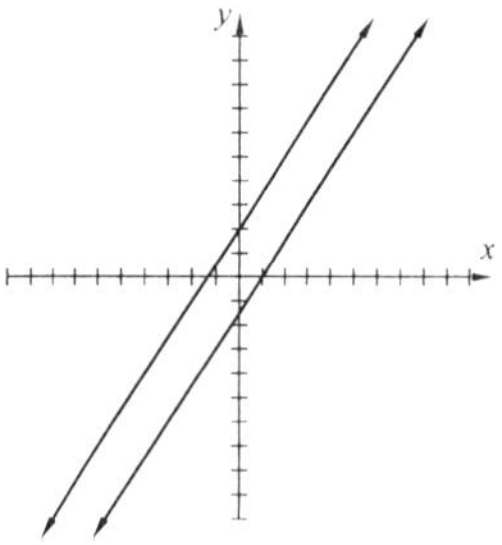

21. $y - 2x - 6 = 0$

$\frac{1}{2}y - 3 = x$

Infinite number of solutions, dependent equations

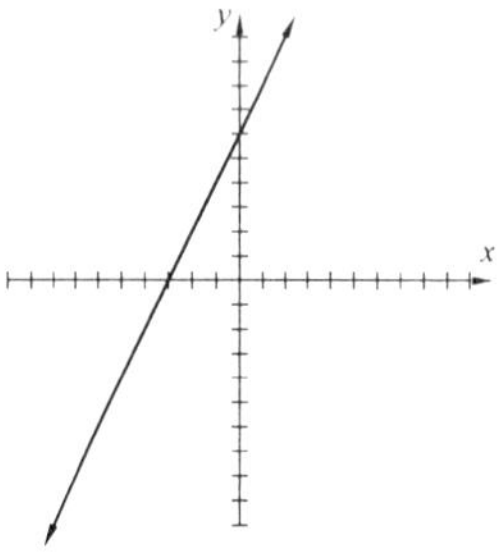

23. $y = \frac{1}{2}x - 2$

$y = \frac{2}{3}x - 1$

$(-6, -5)$

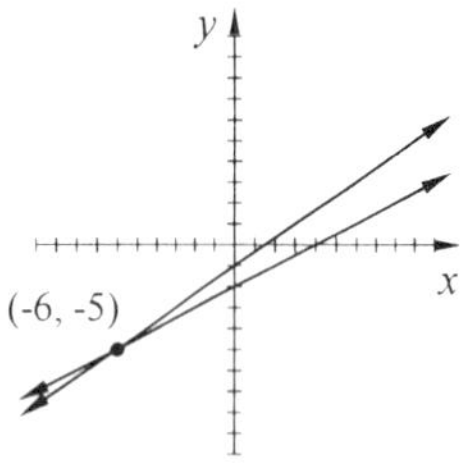

25. (a) Newtown: $y = 7x$

Tryus: $y = 3x + 36$

(b) $y = 7x$

x	y
1	7
4	28
8	56
12	84

$y = 3x + 36$

x	y
1	39
4	48
8	60
12	72

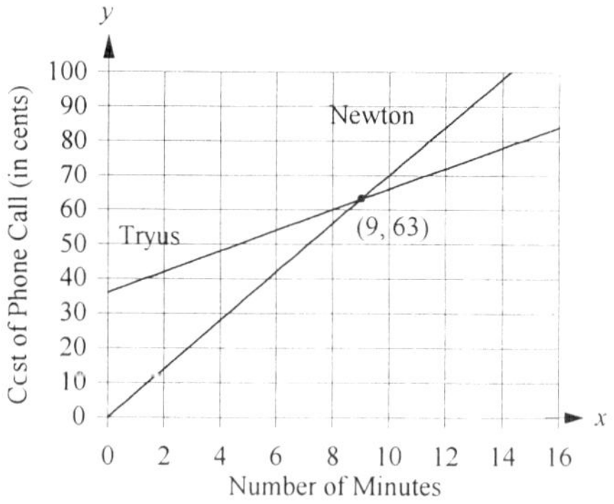

(c) 9 minutes

(b) Tyrus Telephone Company

27. $y = 56x + 1808$

$y = -62x - 2086$

$(-33, -40)$

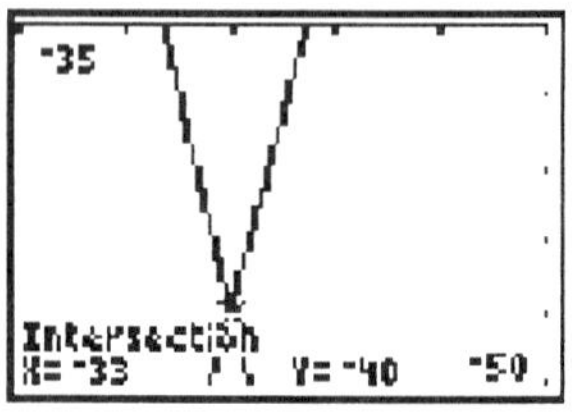

29. $88x + 57y = 683.10$

$95x - 48y = 7460.64$

$(47.52, -61.38)$

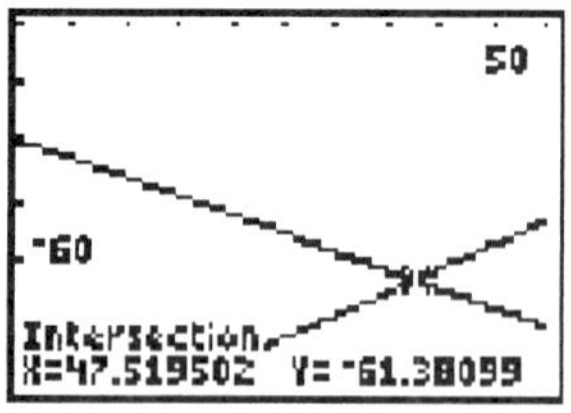

Cumulative Review

31. $7-3(4-x)\le 11x-6$

$$7-12+3x\le 11x-6$$
$$-5+3x\le 11x-6$$
$$-5-8x\le -6$$
$$-8x\le -1$$
$$x\ge \frac{1}{8}$$

33. $(3x-7)^2=(3x-7)(3x-7)$

$$=9x^2-21x-21x+49$$
$$=9x^2-42x+49$$

35. x = the number of people in 1991

$4x=360,000$

$x=90,000$ people

37. $1500(365)=547,500$ people

8.2 Exercises

1. $4x+3y=9$ (1)

$x-3y=6$ (2)

Solve (2) for x

$x=3y+6$ (3)

Substitute $3y+6$ for x in (1)

$$4(3y+6)+3y=9$$
$$12y+24+3y=9$$
$$15y+24=9$$
$$15y=-15$$
$$y=-1$$

Substitute -1 for y in (3)

$x=3(-1)+6=3$

$(3,-1)$

Check: $4(3)+3(-1)\stackrel{?}{=}9$

$9=9$

3. $2x+y=4$ (1)

$2x-y=0$ (2)

Solve (1) for y

$y=-2x+4$ (3)

Substitute $-2x+4$ for y in (2)

$$2x-(-2x+4)=0$$
$$2x+2x-4=0$$
$$4x-4=0$$
$$4x=4$$
$$x=1$$

Substitute 1 for x in (3)

$y=-2(1)+4=2$

$(1,2)$

Check: $2(1)-2\stackrel{?}{=}0$

$0=0$

5. $5x + 2y = 5 \quad (1)$

$3x + y = 4 \quad (2)$

Solve (2) for y

$y = -3x + 4 \quad (3)$

Substitute $-3x + 4$ for y in (1)

$$5x + 2(-3x + 4) = 5$$
$$5x - 6x + 8 = 5$$
$$-x = -3$$
$$x = 3$$

Substitute 3 for x in (3)

$y = -3(3) + 4 = -5$

$(3, -5)$

Check: $5(3) + 2(-5) \stackrel{?}{=} 5$

$5 = 5$

7. $3x - 2y = -8 \quad (1)$

$x + y = 4 \quad (2)$

Solve (2) for y

$y = -x + 4 \quad (3)$

Substitute $-x + 4$ for y in (1)

$$3x - 2(-x + 4) = -8$$
$$3x + 2x - 8 = -8$$
$$5x = 0$$
$$x = 0$$

Substitute 0 for x in (3)

$y = -0 + 4 = 4$

$(0, 4)$

Check: $3(0) - 2(4) \stackrel{?}{=} -8$

$-8 = -8$

9. $3a - 5b = 2 \quad (1)$

$3a + b = 32 \quad (2)$

Solve (2) for b

$b = -3a + 32 \quad (3)$

Substitute $-3a + 32$ for b in (1)

$$3a - 5(-3a + 32) = 2$$
$$3a + 15a - 160 = 2$$
$$18a - 160 = 2$$
$$18a = 162$$
$$a = 9$$

11. $3x - y = 3 \quad (1)$

$x + 3y = 11 \quad (2)$

Solve (2) for x

$x = -3y + 11 \quad (3)$

Substitute $-3y + 11$ for x in (1)

$$3(-3y + 11) - y = 3$$
$$-9y + 33 - y = 3$$
$$-10y + 33 = 3$$
$$-10y = -30$$
$$y = 3$$

Substitute 3 for y in (3)

$x = -3(3) + 11$

$x = 2$

$(2, 3)$

Check: $3(2) - 3 \stackrel{?}{=} 6$

$3 = 3$

13. $p+2q-4=0 \quad (1)$

$7p-q-3=0 \quad (2)$

Solve (1) for p

$p=4-2q \quad (3)$

Substitute $4-2q$ for p in (2)

$$7(4-2q)-q-3=0$$
$$28-14q-q-3=0$$
$$-15q=-25$$
$$q=\frac{5}{3}$$

Substitute $\frac{5}{3}$ for q in (3)

$$p=4-2\left(\frac{5}{3}\right)=\frac{12}{3}-\frac{10}{3}=\frac{2}{3}$$

$\left(\frac{2}{3}, \frac{5}{3}\right)$

Check: $\frac{2}{3}+2\left(\frac{5}{3}\right)-4\stackrel{?}{=}0$

$$\frac{2}{3}+\frac{10}{3}-\frac{12}{3}\stackrel{?}{=}0$$
$$0=0$$

15. $3x-y-9=0 \quad (1)$

$8x+5y-1=0 \quad (2)$

Solve (1) for y

$3x-9=y \quad (3)$

Substitute $3x-9$ for y in (2)

$$8x+5(3x-9)-1=0$$
$$8x+15x-45-1=0$$
$$23x-46=0$$
$$x=2$$

Substitute 2 for x in (3)

$y=3(2)-9=-3$

$(2, -3)$

Check: $8(2)+5(-3)-1\stackrel{?}{=}0$

$$0=0$$

17. $\frac{5}{3}x+\frac{1}{3}y=-3 \quad (1)$

$-2x+3y=24 \quad (2)$

Multiply (1) by 3

$5x+y=-9 \quad (3)$

Solve (3) for y

$y=-5x-9 \quad (4)$

Substitute $-5x-9$ for y in (2)

$$-2x+3(-5x-9)=24$$
$$-2x-15x-27=24$$
$$-17x=51$$
$$x=-3$$

Substitute -3 for x in (4)

$y=-5(-3)-9=6$

$(-3, 6)$

19. $4x+5y=2 \quad (1)$

$\frac{1}{5}x+y=\frac{-7}{5} \quad (2)$

Multiply (2) by 5

$x+5y=-7$

Solve for x

$x=-5y-7 \quad (3)$

Substitute $-5y-7$ for x in (1)

$4(-5y-7)+5y=2$

$-20y-28+5y=2$

$-15y=30$

$y=-2$

Substitute -2 for y in (3)

$x=-5(-2)-7=3$

$(3, -2)$

21. $\frac{4}{7}x-\frac{2}{7}y=2$

$4x+2y=14 \quad (1)$

$3x+y=13 \quad (2)$

Solve (2) for y

$y=-3x+13 \quad (3)$

Substitute $-3x+13$ for y in (1)

$4x+2(-3x+13)=14$

$4x-6x+26=14$

$-2x+26=14$

$-2x=-12$

$x=6$

Substitute 6 for x in (3)

$y=-3(6)+13$

$y=-5$

$(6, -5)$

23. $3x+y=6 \quad (1)$

$3x-2y=-3 \quad (2)$

Solve (1) for y

$y=-3x+6 \quad (3)$

Substitute $-3x+6$ for y in (2)

$3x-2(-3x+6)=-3$

$3x+6x-12=-3$

$9x-12=-3$

$9x=9$

$x=1$

Substitute 1 for x in (3)

$y=-3(1)+6$

$y=3$

$(1, 3)$

25. $3x-4y=36 \quad (1)$

$y+3=3(x+1)$

$y+3=3x+3$

$y=3x \quad (2)$

Substitute $3x$ for y in (1)

$3x-4(3x)=36$

$3x-12x=36$

$-9x=36$

$x=-4$

Substitute -4 for x in (2)

$y=3(-4)=-12$

$(-4, -12)$

27. $2x = 4(2y+2)$

$3(x-3y)+2 = 17$

$2x = 8y+8$

$3x-9y+2=17$

$x = 4y+4 \quad (1)$

$3x-9y=15 \quad (2)$

Substitute $4y+4$ for x in (2)

$3(4y+4)-9y=15$

$12y+12-9y=15$

$3y=3$

$y=1$

Substitute 1 for y in (1)

$x=4(1)+4=8$

$(8, 1)$

Check: $3[8-3(1)]+2 \stackrel{?}{=} 17$

$17 = 17$

29. You would need 3 (7) equations to solve for 3(7) unknowns. Substitution can reduce the system to one equation with one unknown. Then substituting back successively gives another unknown, thus requiring one equation for each unknown.

31. The solution to a system must satisfy both equations.

33. If the graphs are parallel, the system has no solutions because there are no points in common.

35. (a) LEP: $y = 90x+50$

AP: $y = 80x+100$

(b) $90x+50 = 80x+100$

$10x = 50$

$x - 5$

$y = 90(5)+50 = 500$

It would take 5 times and cost $500

(c) $90x+50-(80x+100) = 80$

$10x-50 = 80$

$10x = 130$

$x = 13$

They expect plowing 13 times and he should hire Adirondack Plowing.

Cumulative Review Problems

37. $6x+3y \ge 9$

Test point: $(0, 0)$

$6(0)+3(0) \ge 9$

$0 \ge 9$ False

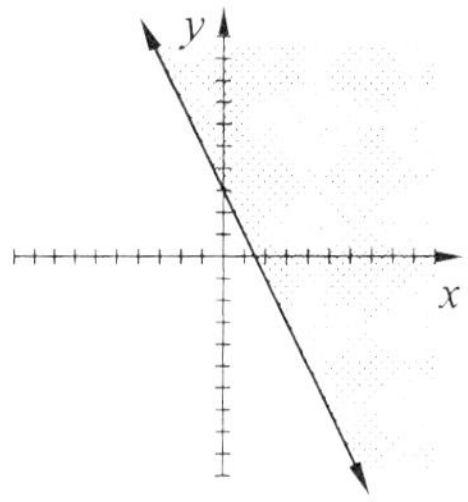

39. total cost = basic contract + bonus

bonus = $0.03 \times$ basic contract

total cost = $3.8 + 0.03(3.8)$

$= 3.914$

$= \$3{,}914{,}000$

41. $\frac{(5\text{trips})(360 \text{ miles/trip})(\$1.80/\text{gal})}{20 \text{ miles/gal}}$

$= \frac{5(360)(1.8)}{20}$

$= 162$

The cost of gasoline for family trips for one year is \$162.

8.3 Exercises

1. $3x + 2y = 5$ (1)

$5x - \ \ y = 3$ (2)

Multiply (2) by 2 to eliminate y

3. $4x - 3y = 10$ (1)

$5x + 4y = 0$ (2)

Multiply (1) by 4 and (2) by 3 to eliminate y

5. $-x + y = -3$ (1)

$\underline{-2x - y = 6}$ (2)

$-3x = 3$

$x = -1$

Substitute -1 for x in (1)

$-(-1) + y = -3$

$y = -4$

$(-1, -4)$

Check: $-2(-1) - (-4) \stackrel{?}{=} 6$

$6 = 6$

7. $2x + 3y = 1$ (1)

$x - 2y = 4$ (2)

Multiply (2) by -2

$2x + 3y = 1$

$\underline{-2x + 4y = -8}$

$7y = -7$

$y = -1$

Substitute -1 for y in (2)

$x - 2(-1) = 4$

$x = 2$

$(2, -1)$

9. $4x + -2y = -2$ (1)

$4x + 3y = -12$ (2)

Multiply (2) by -1

$4x - 2y = -2$

$\underline{-4x - 3y = 12}$

$-5y = 10$

$y = -2$

Substitute -2 for y in (1)

$4x - 2(-2) = -2$

$4x + 4 = -2$

$4x = -6$

$x = -\frac{3}{2}$

$\left(-\frac{3}{2}, -2\right)$

Check: $4\left(-\frac{3}{2}\right) + 3(-2) \stackrel{?}{=} -12$

$-12 = -12$

11. $5x-15y=9 \quad (1)$
$-x+10y=1 \quad (2)$
Multiply (2) by 5
$5x-15y=9$
$\underline{-5x+50y=5}$
$35y=14$
$y=\frac{2}{5}$
Substitute $\frac{2}{5}$ for y in (2)
$-x+10\left(\frac{2}{5}\right)=1$
$-x=-3$
$x=3$
$\left(3, \frac{2}{5}\right)$

Check: $5(3)-15\left(\frac{2}{5}\right)\stackrel{?}{=}9$
$9=9$

13. $2x+5y=2 \quad (1)$
$3x+\ \ y=3 \quad (2)$
Multiply (2) by -5
$2x+5y=2$
$\underline{-15x-5y=-15}$
$-13x \quad =-13$
$x=1$
Substitute 1 for x in (2)
$3(1)+y=3$
$y=0$
$(1, 0)$
Check: $2(1)+5(0)\stackrel{?}{=}2$
$2=2$

Substitute 4 for x in (2)
$2(4)-y=6$
$y=-2$
$y=2$
$(4, 2)$
Check: $5(4)-3(2)\stackrel{?}{=}14$
$14=14$

15. $8x+6y=-2 \quad (1)$
$10x-9y=-8 \quad (2)$
Multiply (1) by 3 and (2) by 2
$24x+18y=-6$
$\underline{20x-18y=-16}$
$44x \quad =-22$
$x=-\frac{1}{2}$
Substitute $-\frac{1}{2}$ for x in (1)
$8\left(-\frac{1}{2}\right)+6y=-2$
$6y=2$
$y=\frac{1}{3}$
$\left(-\frac{1}{2}, \frac{1}{3}\right)$

Check: $10\left(-\frac{1}{2}\right)-9\left(\frac{1}{3}\right)\stackrel{?}{=}-8$

$-8=-8$

17. $2x+3y=-8$ (1)

$5x+4y=-34$ (2)

Multiply (1) by -5 and (2) by 2

$-10x-15y=40$

$\underline{10x+8y=-68}$

$-7y=-28$

$y=4$

Substitute 4 for y in (1)

$2x+3(4)=-8$

$2x=-20$

$x=-10$

$(-10, 4)$

Check: $5(-10)+4(4)\stackrel{?}{=}-34$

$-34=-34$

19. $2x-7y=8$ (1)

$3x-4y=-1$ (2)

Multiply (1) by 3 and (2) by -2

$6x-21y=24$

$\underline{-6x+8y=2}$

$-13y=26$

$y=-2$

Substitute -2 for y in (2)

$3x-4(-2)=-1$

$3x+8=-1$

$3x=-9$

$x=-3$

$(-3, -2)$

Check: $2(-3)-7(-2)\stackrel{?}{=}8$

$8=8$

21. $12x-6y=-2$ (1)

$9x-7y=-10$ (2)

Multiply (1) by 3 and (2) by 4

$36x-18y=-6$

$\underline{-36x-28y=-40}$

$-46y=-46$

$y=1$

Substitute 1 for y in (1)

$12x-6(1)=-2$

$12x=4$

$x=\frac{1}{3}$

$\left(\frac{1}{3}, 1\right)$

Check: $-9\left(\frac{1}{3}\right)-7(1)\stackrel{?}{=}-10$

$-10=-10$

23. $\frac{1}{4}x-\frac{3}{4}y=2$ (1)

$2x+3y=1$ (2)

Multiply equation (1) by 4

25. $\frac{1}{2}x+\frac{2}{3}y=\frac{1}{3}$ (1)

$\frac{3}{4}x-\frac{4}{5}y=2$ (2)

Multiply (1) by 6 and (2) by 20.

27. $x+\frac{5}{4}y=\frac{9}{4}$ (1)

$\frac{2}{5}x-y=\frac{3}{5}$ (2)

Multiply (1) by 4 and (2) by 5

$4x+5y=9$

$2x-5y=3$

$6x \quad = 12$

$x=2$

Substitute 2 for x in (1)

$2+\frac{5}{4}y=\frac{9}{4}$

$\frac{5}{4}y=\frac{1}{4}$

$y=\frac{1}{5}$

$\left(2, \frac{1}{5}\right)$

Check: $\frac{2}{5}(2)-\frac{1}{5}\stackrel{?}{=}\frac{3}{5}$

$\frac{3}{5}=\frac{3}{5}$

29. $\frac{x}{6}+\frac{y}{2}=-\frac{1}{2}$ (1)

$x-9y=21$ (2)

Multiply (1) by -6

$-x-3y=3$

$x-9y=21$

$-12y=24$

$y=-2$

Substitute -2 for y in (2)

$x-9(-2)=21$

$x=3$

$(3,-2)$

Check: $\frac{3}{6}+\frac{-2}{2}\stackrel{?}{=}-\frac{1}{2}$

$\frac{1}{2}-\frac{2}{2}\stackrel{?}{=}-\frac{1}{2}$

$-\frac{1}{2}=-\frac{1}{2}$

31. $\frac{5}{6}x+y=-\frac{1}{3}$ (1)

$-8x+9y=28$ (2)

Multiply (1) by 18 and (2) by -2

$15x+18y=-6$

$16x-18y=-56$

$31x \quad =-62$

$x=-2$

Substitute -2 for x in (2)

$-8(-2)+9y=28$

$9y=12$

$y=\frac{4}{3}$

$\left(-2, \frac{4}{3}\right)$

Check: $\frac{5}{6}(-2)+\frac{4}{3}\stackrel{?}{=}-\frac{1}{3}$

$-\frac{1}{3}=-\frac{1}{3}$

33. $\frac{2}{3}x+\frac{3}{5}y=-\frac{1}{5}$ (1)

$\frac{1}{4}x+\frac{1}{3}y=\frac{1}{4}$ (2)

Multiply (1) by 15 and (2) by 12

$10x+9y=-3$ (3)

$3x+4y=3$ (4)

Multiply (3) by 3 and (4) by -10

$30x+27y=-9$

$-30x-40y=-30$

$-13y=-39$

$y=3$

Substitute 3 for y in (2)

$\frac{1}{4}x+\frac{1}{3}(3)=\frac{1}{4}$

$\frac{1}{4}x+1=\frac{1}{4}$

$x+4=1$

$x=-3$

$(-3, 3)$

Check: $\frac{2}{3}(-3)+\frac{3}{5}(3)\stackrel{?}{=}-\frac{1}{5}$

$-2+\frac{9}{5}\stackrel{?}{=}-\frac{1}{5}$

$-\frac{1}{5}=-\frac{1}{5}$

35. $0.5x-0.3y=0.1$ (1)

$5x+3y=6$ (2)

Multiply (1) by 10

$5x-3y=1$

$5x+3y=6$

37. $4x+0.5y=9$ (1)

$0.2x-0.05y=1$ (2)

Multiply (1) by 10 and (2) by 100

$40x+5y=90$

$20x-5y=100$

39. $0.5x-0.2y=0.4$ (1)

$-0.3x+0.4y=0.6$ (2)

Multiply (1) and (2) by 10

$5x-2y=4$ (3)

$-3x+4y=6$ (4)

Multiply (3) by 2

$10x-4y=8$

$-3x+4y=6$

$7x=14$

$x=2$

Substitute 2 for x (1)

$0.5(2)-0.2y=0.4$

$-0.2y+1=0.4$

$-0.2y=-0.6$

$y=3$

$(2, 3)$

41. $0.02x-0.04y=0.26$ (1)

$0.07x-0.09y=0.66$ (2)

Multiply (1) and (2) by 100

$2x-4y=26$ (3)

$7x-9y=66$ (4)

Multiply (3) by 7 and (4) by -2

$14x-28y=182$

$-14x+18y=-132$

$-10y=50$

$y=-5$

Substitute -5 for y in (3)

$2x-4(-5)=26$

$2x=6$

$x=3$

$(3, -5)$

43. $0.4x-\ \ 5y=-1.2$ (1)

$-0.03x+0.5y=0.14$ (2)

Multiply (1) by 10 and (2) by 100

$4x-50y=-12$

$-3x+50y=14$

$x=2$

Substitute 2 for x in (1)

$0.4(2)-5y=-1.2$

$-5y=-2$

$y=\frac{2}{5}$

$\left(2, \frac{2}{5}\right)$

45. $2x+y=11$ (1)

$x+3y=18$ (2)

Multiply (1) by -3

$-6x-3y=-33$

$x+3y=\ \ 18$

$-5x \quad =-15$

$x=3$

Substitute 3 for x in (1)

$2(3)+y=11$

$6+y=11$

$y=5$

$(3, 5)$

47. $\frac{5}{4}x+y=16$ (1)

$x-\frac{5}{3}y=-2$ (2)

Multiply (1) by 4 and (2) by 3

$5x+4y=64$ (3)

$3x-5y=-6$ (4)

Multiply (3) by 5 and (4) by 4

$25x+20y=320$

$12x-20y=-24$

$37x \quad = 296$

$x=8$

Substitute 8 for x in (1)

$\frac{5}{4}(8)+y=16$

$10+y=16$

$y=6$

$(8, 6)$

49. $0.2x+0.1y=1.1$ (1)

$0.01x+0.03y=0.18$ (2)

Multiply (1) by 10 and (2) by 100

$2x+y=11$ (3)

$x+3y=18$ (4)

Multiply (4) by -2

$2x+\ \ y=11$

$-2x-6y=-36$

$-5y=-25$

$y=5$

Substitute 5 for y in (4)

$x+3(5)=18$

$x+15=18$

$x=3$

$(3, 5)$

51. $5(3x-y)=2-(y-x)$

$9(2x-2)=y-3x$

Expand both equations

$15x-5y=2-y+x$

$18x-18=y-3x$

Rearrange

$14x-4y=2 \quad (1)$

$21x-\ y=18 \quad (2)$

Multiply (2) by -4

$14x-4y=2$

$-84x+4y=-72$

$-70x \quad =-70$

$x=1$

Substitute 1 for x in (2)

$21(1)-y=18$

$-y=-3$

$y=3$

$(1, 3)$

Cumulative Review

53. $100\%-89\%=11\%$

$0.11(5000)=550$

$0.11(7000)=770$

Between 550 and 770 airplanes

55. $d=rt \Rightarrow t=\frac{d}{r}$

Bobby's time for 100 laps

< Carlos' time for 96 laps

$\frac{100\times 4.75}{r}<\frac{96\times 4.75}{95}$

$\frac{475}{r}<4.8$

$475<4.8r$

$98.96<r$

Bobby will have to average more than 99mph.

57. $75x^2-3=3(25x^2-1)$

$=3\left[(5x)^2-(1)^2\right]$

$=3(5x+1)(5x-1)$

How Am I Doing? Sections 8.1-8.3

1.

$2x-y=-8$

$-x+2y=10$

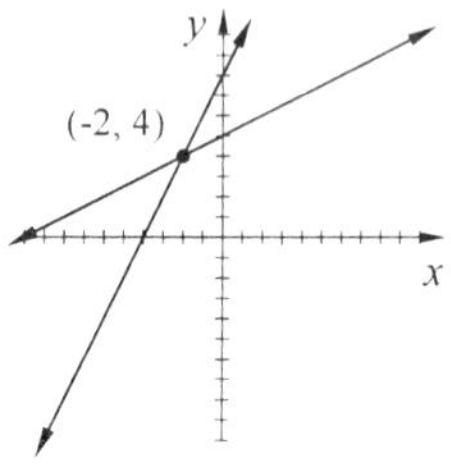

$(-2,4)$

2. $-2x+y=4$

$x-3y=3$

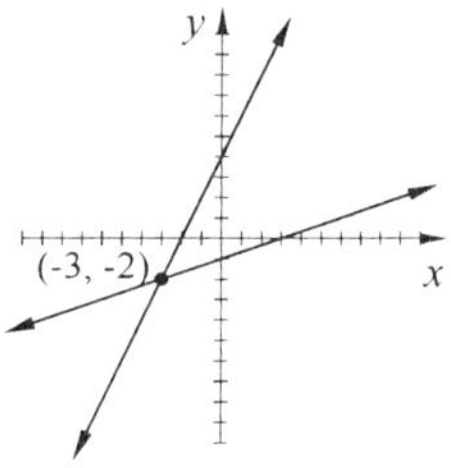

$(-3,-2)$

3. $5x + y = 26$ (1)

   $3x + 2y = 10$ (2)

   Solve (1) for y

   $y = -5x + 26$ (3)

   Substitute $-5x + 26$ for y in (2)

   $3x + 2(-5x + 26) = 10$

   $3x - 10x + 52 = 10$

   $-7x + 52 = 10$

   $-7x = -42$

   $x = 6$

   Substitute 6 for x in (3)

   $y = -5(6) + 26 = -4$

   $(6, -4)$

4. $4x - 3y = 18$ (1)

   $5x - y = 17$ (2)

   Solve (2) for y

   $y = 5x - 17$ (3)

   Substitute $5x - 17$ for y in (1)

   $4x - 3(5x - 17) = 18$

   $4x - 15x + 51 = 18$

   $-11x + 51 = 18$

   $-11x = -33$

   $x = 3$

   Substitute 3 for x in (3)

   $y = 5(3) - 17$

   $y = -2$

   $(3, -2)$

5. $2x + 3y = 5$ (1)

   $\frac{1}{2}x + \frac{1}{6}y = -\frac{1}{2}$ (2)

   Multiply (2) by 6

   $3x + y = -3$

   Solve for y

   $y = -3x - 3$ (3)

   Substitute $-3x - 3$ for y in (1)

   $2x + 3(-3x - 3) = 5$

   $2x - 9x - 9 = 5$

   $-7x - 9 = 5$

   $-7x = 14$

   $x = -2$

   Substitute -2 for x in (3)

   $y = -3(-2) - 3$

   $y = 3$

   $(-2, 3)$

6. $7x - 5y - 17 = 0$ (1)

   $x + 8y + 15 = 0$ (2)

   Solve (2) for x

   $x = -8y - 15$ (3)

   Substitute $-8y - 15$ for x in (1)

   $7(-8y - 15) - 5y - 17 = 0$

   $-56y - 105 - 5y - 17 = 0$

   $-61y - 122 = 0$

   $-61y = 122$

   $y = -2$

   Substitute -2 for y in (3)

   $x = -8(-2) - 15 = 1$

   $(1, -2)$

7. $3x - 2y = 19 \quad (1)$

$x + y = 8 \quad (2)$

Multiply (2) by 2

$3x - 2y = 19$

$\underline{2x + 2y = 16}$

$5x \quad = 35$

$x = -7$

Substitute 7 for x in (2)

$7 + y = 8$

$y = 1$

$(7, 1)$

8. $9x - 8y = 3 \quad (1)$

$6x + 3y = 27 \quad (2)$

Multiply (1) by 2 and (2) by -3

$18x - 16y = 6 \quad (3)$

$\underline{-18x - 9y = -81} \quad (4)$

$-25y = -75$

$y = 3$

Substitute 3 for y in (2)

$6x + 3(3) = 27$

$6x + 9 = 27$

$6x = 18$

$x = 3$

$(3, 3)$

9. $\frac{2}{3}x - \frac{3}{4}y = 3 \quad (1)$

$\frac{1}{9}x + \frac{1}{12}y = -2 \quad (2)$

Multiply (1) by 12 and (2) by 36

$8x - 9y = 36 \quad (3)$

$4x + 3y = -72 \quad (4)$

Multiply (4) by 3

$8x - 9y = \quad 36$

$\underline{12x + 9y = -216}$

$20x \quad = -180$

$x = -9$

Substitute -9 for x in (1)

$\frac{2}{3}(-9) - \frac{3}{4}y = 3$

$-6 - \frac{3}{4}y = 3$

$-\frac{3}{4}y = 9$

$y = -12$

$(-9, -12)$

10. $0.6x - 0.5y = 2.1 \quad (1)$

$0.4x - 0.3y = 1.3 \quad (2)$

Multiply (1) and (2) by 10

$6x - 5y = 21 \quad (3)$

$4x - 3y = 13 \quad (4)$

Multiply (3) by 2 and (4) by -3

$12x - 10y = \quad 42$

$\underline{-12x + 9y = -39}$

$-y = 3$

$y = -3$

Substitute -3 for y in (1)

$0.5x - 0.5(-3) = 2.1$

$0.6x + 1.5 = 2.1$

$0.6x = 0.6$

$x = 1$

$(1, -3)$

8.4 Exercises

1. If there is no solution to a system of linear equations, the graphs of the equations are parallel lines. Solving the system algebraically, you will obtain an equation that is inconsistent with known facts.

3. If there is exactly one solution, the graphs of the equations intersect. This system is said to be independent and consistent.

5. $-2x - 3y = 15$ (1)

$5x + 2y = 1$ (2)

Multiply (1) by 2 and (2) by 3

$-4x - 6y = 30$

$15x + 6y = 3$

$11x = 33$

$x = 3$

Substitute 3 for x in (2)

$5(3) + 2y = 1$

$2y = -14$

$y = -7$

$(3, -7)$

7. $4x - 5y = 1$ (1)

$-3x + 4y = 4$ (2)

Multiply (1) by 3 and (2) by 4

$12x - 15y = 3$

$-12x + 16y = 16$

$y = 19$

Substitute 19 for y in (2)

$-3x + 4(19) = 4$

$-3x + 76 = 4$

$x = 24$

$(24, 19)$

9. $2x - 4y = 5$ (1)

$-4x + 8y = 9$ (2)

Multiply (1) by 2

$4x - 8y = 10$

$-4x + 8y = 9$

$0 = 19$

There is no solution.

11. $-5x + 2y = 2$ (1)

$15x - 6y = -6$ (2)

Multiply (1) by 3

$-15x + 6y = 6$

$15x - 6y = -6$

$0 = 0$

There is an infinite number of solutions.

13. $5x - 3y = 13$ (1)

$7x + 2y = 43$ (2)

Multiply (1) by 2 and (2) by 3

$$\begin{array}{r} 10x - 6y = 26 \\ \underline{21x + 6y = 129} \\ 31x \quad = 155 \end{array}$$

$$x = 5$$

Substitute 5 for x in (1)

$$5(5) - 3y = 13$$
$$25 - 3y = 13$$
$$-3y = -12$$
$$y = 4$$

$(5, 4)$

15. $3x - 2y = 70$ (1)

$0.6x + 0.5y = 50$ (2)

Multiply (1) by -2 and (2) by 10

$$\begin{array}{r} -6x + 4y = -140 \\ \underline{6x + 5y = 500} \\ 9y = 360 \end{array}$$

$$y = 40$$

Substitute 40 for y in (1)

$$3x - 2(40) = 70$$
$$3x = 150$$
$$x = 50$$

$(50, 40)$

17. $0.3x - 0.5y = 0.4$ (1)

$0.6x + 1.0y = 2.8$ (2)

Multiply (1) by 20 and (2) by -10

$$\begin{array}{r} 6x - 10y = 8 \\ \underline{-6x - 10y = -28} \\ -20y = -20 \end{array}$$

$$y = 1$$

Substitute 1 for y in (1)

$$0.3x - 0.5(1) = 0.4$$
$$0.3x = 0.9$$
$$x = 3$$

$(3, 1)$

19. $\frac{4}{3}x + \frac{1}{2}y = 1$ (1)

$\frac{1}{3}x - y = -\frac{1}{2}$ (2)

Multiply (1) by 6 and (2) by -24

$$\begin{array}{r} 8x + 3y = 6 \\ \underline{-8x + 24y = 12} \\ 27y = 18 \end{array}$$

$$y = \frac{2}{3}$$

Substitute $\frac{2}{3}$ for y in (2)

$$\frac{1}{3}x - \frac{2}{3} = -\frac{1}{2}$$
$$\frac{1}{3}x = \frac{1}{6}$$
$$x = \frac{1}{2}$$

$\left(\frac{1}{2}, \frac{2}{3}\right)$

21. $\frac{2}{3}x+\frac{1}{6}y=2 \quad (1)$

$\frac{1}{4}x-\frac{1}{2}y=-\frac{3}{4} \quad (2)$

Multiply (1) by 12 and (2) by 4

$$\begin{array}{r} 8x+\ 2y=24 \\ x-\ 2y=-3 \\ \hline 9x \qquad =21 \end{array}$$

$$x=\frac{7}{3}$$

Substitute $\frac{7}{3}$ for x in (1)

$$\frac{2}{3}\left(\frac{7}{3}\right)+\frac{1}{6}y=2$$

$$\frac{1}{6}y=\frac{4}{9}$$

$$y=\frac{8}{3}$$

$\left(\frac{7}{3}, \frac{8}{3}\right)$

23. $4x+3y=-2 \quad (1)$

$5x-2y=49 \quad (2)$

Multiply (1) by 2 and (2) by 3

$$\begin{array}{r} 8x+6y=-4 \\ 15x-6y=27 \\ \hline 23x \qquad =23 \end{array}$$

$$x=1$$

Substitute 1 for x in (1)

$$4(1)+3y=-2$$

$$3y=-6$$

$$y=-2$$

$(1, -2)$

25. $\frac{2}{3}x+y=1 \quad (1)$

$\frac{3}{4}x+\frac{1}{2}y=\frac{3}{4} \quad (2)$

Multiply (1) by 3 and (2) by 4

$2x+3y=3 \quad (3)$

$3x+2y=3 \quad (4)$

Multiply (3) by 3 and (2) by -2

$$\begin{array}{r} 6x+9y=9 \\ -6x-4y=-6 \\ \hline 5y=3 \end{array}$$

$$y=\frac{3}{5}$$

Substitute $\frac{3}{5}$ for y in (1)

$$\frac{2}{3}x+\frac{3}{5}=1$$

$$10x+9=15$$

$$10x=6$$

$$x=\frac{3}{5}$$

$\left(\frac{3}{5}, \frac{3}{5}\right)$

27. $x-2y+2=0 \quad (1)$

$3(x-2y+1)=15$

$3x-6y+3=15 \quad (2)$

Multiply (1) by -3

$$\begin{array}{r} -3x+6y-6=0 \\ 3x-6y+3=15 \\ \hline -3=15 \end{array}$$

No solution. The system is inconsistent.

29. $2(a+3)=b+1$

$3(a-b)=a+1$

Expand

$2a+6=b+1$

$3a-3b=a+1$

Rearrange

$$\begin{array}{rl} 2a-\ b=-5 & (1) \\ \underline{-2a+3b=-1} & (2) \\ 2b=-6 & \\ b=-3 & \end{array}$$

Substitute -3 for b in (1)

$2a-(-3)=-5$

$2a=-8$

$a=-4$

$(-4,-3)$

Cumulative Review

31. $20x+12\left(\frac{9}{2}-x\right)=70$

$20x+54-12x=70$

$8x=16$

$x=2$

33. x = Robert's taxable income

$0.15x=3225$

$x=21,500$

Robert's taxable income was $21,500

35. $100\%-40\%=60\%$

$\frac{0.60(92,500)}{44}\approx 1261$ tickets

8.5 Exercises

1. x = number of first class tickets

y = number of coach tickets

$x+y=14$ (1)

$1150x+250y=10,700$ (2)

Solve (1) for x

$x=14-y$ (3)

Substitute $14-y$ for x in (2)

$1150(14-y)+250y=10,700$

$16,100-1150y+250y=10,700$

$16,100-900y=10,700$

$-900y=-5400$

$y=6$

Substitute 6 for y in (3)

$x=14-6$

$x=8$

8 first class tickets

6 coach tickets

3. x = number of etchings sold at \$35.

   y = number of etchings sold at \$40

   $35x + 40y = 455$ (1)

   $x + y = 12$ (2)

   Solve (2) for y

   $y = -x + 12$ (3)

   Substitute $-x + 12$ for y in (1)

   $35x + 40(-x + 12) = 455$

   $35x - 40x + 480 = 455$

   $-5x = -25$

   $x = 5$

   Substitute 5 for x in (3)

   $y = -5 + 12 = 7$

   Sold 5 etchings at \$35.

   Sold 7 etchings at \$40.

5. x = hours on day shift

   y = hours on night shift

   $x + y = 23$ (1)

   $13.50x + 16.50y = 352.50(2)$ (2)

   Solve for x

   $x = 23 - y$ (3)

   Substitute $23 - y$ for x in (2)

   $13.50(23 - y) + 16.50y = 352.50$

   $310.50 - 13.50y + 16.50y = 352.50$

   $310.50 + 3y = 352.50$

   $3y = 42$

   $y = 14$

   Substitute 14 for y in (3)

   $x = 23 - 14$

   $x = 9$

   9 hours on day shift

   14 hours on night shift

7. L = original length

   W = original width

   $2L + 2W = 38$ (1)

   $2(2L - 6) + 2(W + 5) = 56$

   $4L - 12 + 2W + 10 = 56$

   $4L + 2W = 58$ (2)

   Multiply (1) by -1

   $-2L - 2W = -38$

   $4L + 2W = 58$

   $2L = 20$

   $L = 10$

   Substitute 10 for L in (1)

   $2(10) + 2W = 38$

   $2W = 18$

   $W = 9$

   original length = 10 feet

   original width = 9 feet

   New length = $2(10) - 6 = 14$ feet

   New width = $9 + 5 = 14$ feet

9. x = quarts before

   y = quarts added

   $x + y = 16$ (1)

   $0.50x + 0.80y = 0.65(16)$ (2)

   Multiply (1) by -5 and (2) by 10

   $-5x - 5y = -80$

   $5x + 8y = 104$

   $3y = 24$

   $y = 8$

   Substitute 8 for y in (1)

   $x + 8 = 16$

   $x = 8$

   8 quarts before

   8 quarts added

11. x = weight of nuts

    y = weight of raisins

    $x + y = 50$ (1)

    $2.00x + 1.50y = 50(1.80)$ (2)

    Multiply (1) by -2

    $-2x - 2y = -100$

    $2x + 1.5y = 90$

    $-0.5y = -10$

    $y = 20$

    Substitute 20 for y in (1)

    $x + 20 = 50$

    $x = 30$

    30 lbs of nuts

    20 lbs of raisins

13. w = speed of wind

    r = airspeed of airplane

    $$\text{speed} = \frac{\text{distance}}{\text{time}}$$

    $$r + w = \frac{3000}{5} = 600 \quad (1)$$

    $$r - w = \frac{3000}{6} = 500 \quad (2)$$

    $2r = 1100$

    $r = 550$

    Substitute 550 for r in (1)

    $550 + w = 600$

    $w = 50$

    Airspeed = 550 kilometers per hour

    Windspeed = 50 kilometers per hour

13. $x - y = -65$ (1)

    $3x + 5y = 725$ (2)

    Multiply (1) by -3

    $-3x + 3y = 195$

    $3x + 5y = 725$

    $8y = 920$

    $y = 115$

    Substitute 115 for y in (1)

    $x - 115 = -65$

    $x = 50$

    There will be 115,000 employees in 2040.

15. x = number of \$10 bills for Jewelry
y = number of \$20 bills for Jewelry
Total amount in \$10 and \$20 bills
$= 320 - 25 - 55 = 240$

$$10x + 20y = 240 \quad (1)$$
$$10(2x-2) + 20(y-4) = 240$$
$$20x - 20 + 20y - 80 = 240$$
$$20x + 20y = 340 \quad (2)$$

Multiply (1) by -1

$$\begin{aligned} -10x - 20y &= -240 \\ 20x + 20y &= 340 \\ \hline 10x \quad &= 100 \\ x &= 10 \end{aligned}$$

Substitute 10 for x in (1)

$$10(10) + 20y = 240$$
$$100 + 20y = 240$$
$$20y = 140$$
$$y = 7$$

Jewelry: 10 - \$10 bills and 7 - \$20 bills
Cosmetics: $2(10) - 2 = 18$ \$10 bills
$7 - 4 = 3$ \$20 bills

17. x = price of oranges
y = price of apples

$$7x + 5y = 8.78 \quad (1)$$
$$3x + 8y = 8.39 \quad (2)$$

Multiply (1) by 3 and (2) by -7

$$\begin{aligned} 21x + 15y &= 26.34 \\ -21x - 56y &= -58.73 \\ \hline -41y &= -32.39 \\ y &= 0.79 \end{aligned}$$

Substitute 0.79 for y in (1)

$$7x + 5(0.79) = 8.78$$
$$7x + 3.95 = 8.78$$
$$7x = 4.83$$
$$x = 0.69$$

Oranges are \$0.69 per pound.
Apples are \$0.79 per pound.

19. x = number of general admission tickets
y = number of student admission tickets

$$4.5x + 3y = 1110 \quad (1)$$
$$x + y = 290 \quad (2)$$

Solve (2) for x
$x = 290 - y$
Substitute $290 - y$ for x in (1)

$$4.5(290 - y) + 3y = 1110$$
$$1305 - 4.5y + 3y = 1110$$
$$1305 - 1.5y = 1110$$
$$-1.5y = -195$$
$$y = 130$$

Substitute 130 for y in (3)
$x = 290 - 130$
$x = 160$
160 general admission tickets
130 student admission tickets

21. $40x - y = -5275 \quad (1)$

$28x - 2y = -11{,}078 \quad (2)$

Multiply (1) by -2

$$\begin{aligned} -80x + 2y &= 10{,}550 \\ 28x - 2y &= -11{,}078 \\ \hline -52x &= -528 \\ x &\approx 10.154 \end{aligned}$$

Substitute 10.154 for x in (2)

$$\begin{aligned} 28(10.154) - 2y &= -11{,}078 \\ 284.312 - 2y &= -11{,}078 \\ -2y &= -11362.312 \\ y &\approx 5680 \end{aligned}$$

The population will be 5,680,000 employees in 2013.

23. x = cost of a kidney kit

y = cost of a heart kit

$0.06x + 0.08y = 4480 \quad (1)$

$0.08x + 0.10y = 5780 \quad (2)$

Multiply (1) by 400 and (2) by -300

$$\begin{aligned} 24x + 32y &= 1{,}792{,}000 \\ -24x - 30y &= -1{,}734{,}000 \\ \hline 2y &= 58{,}000 \\ y &= 29{,}000 \end{aligned}$$

Substitute 29,000 for y in (2)

$$\begin{aligned} 0.08x + 0.10(29{,}000) &= 5780 \\ 0.08x &= 2880 \\ x &= 36{,}000 \end{aligned}$$

\$36,000 for a kidney kit

\$29,000 for a heart kit

Cumulative Review

25. $\dfrac{1}{x-3} - \dfrac{1}{x+3} + 2$

$$\begin{aligned} &= \frac{1(x+3)}{(x-3)(x+3)} - \frac{1(x-3)}{(x+3)(x-3)} + \frac{2(x+3)(x-3)}{(x+3)(x-3)} \\ &= \frac{(x+3) - (x-3) + 2(x^2 - 9)}{(x+3)(x-3)} \\ &= \frac{x + 3 - x + 3 + 2x^2 - 18}{(x+3)(x-3)} \\ &= \frac{2x^2 - 12}{x^2 - 9} \end{aligned}$$

27. $$\begin{aligned} \left(-4x^3y^2\right)^3 &= (-4)^3\left(x^3\right)^3\left(y^2\right)^3 \\ &= -64x^9y^6 \end{aligned}$$

29. $$\begin{aligned} \text{Thickness} &= \frac{1}{8} + \frac{1}{8} + 1\frac{3}{4} \\ &= \frac{1}{8} + \frac{1}{8} + 1\frac{6}{8} \\ &= 1\frac{8}{8} \\ &= 2 \end{aligned}$$

Total $= 3(2) = 6$ inches

Putting Your Skills to Work

1. $-5.4x + y = 372$

   $y = 372$ million in 1980

   $y = 5.4(20) + 372$

   $= 480$ million in 2000

   $y = 5.4(0) + 372$

   $y = 534$ million in 2010

2. $-5.25x + y = 242$

   $y = 242$ million in 1980

   $y = 5.25(20) + 242$

   $y = 347$ million in 2000

   $y = 5.25(30) + 242$

   $y = 399.5$ million in 2010

3. North America $480 - 372 = 108$ million

   South America $347 - 242 = 105$ million

   North America experienced greater growth. It grew by 3 million more peole in the period from 1980-2000.

4. North America $y = 5.4(40) + 372$

   $y = 588$ million

   South America $y = 5.25(40) + 242$

   $y = 452$ million

   Difference: $588 - 452 = 136$ million

   The difference will be 136 million people in 2020. Yes

5. 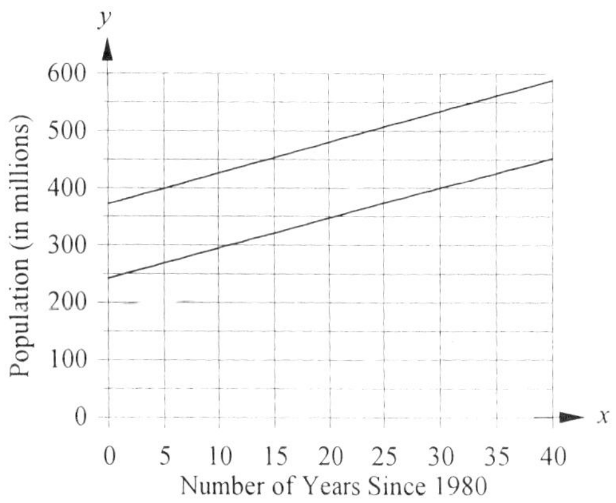

6. They do not intersect on the graph paper. For the time period shown the population of North America is never the same as the population of South America. If they intersect it would have to be for values of x less than 0.

7. $-5.4x + y = 372 \quad (1)$

   $-5.25x + y = 242 \quad (2)$

   Multiply (2) by -1

   $-5.4x + y = 372$

   $\underline{5.25x - y = -242}$

   $-0.15x \quad = 130$

   $x = -866.7$

   Substitute -866.67 for x in (1)

   $-5.4(-866.67) + y = 372$

   $y = -4308.0$

   $(-866.7, -4308.0)$

8. This would have no meaning since y is negative and you cannot have a negative population.

Chapter 8 - Review Problems

1. $2x+3y=0$

   $-x+3y=9$

   $(-3, 2)$

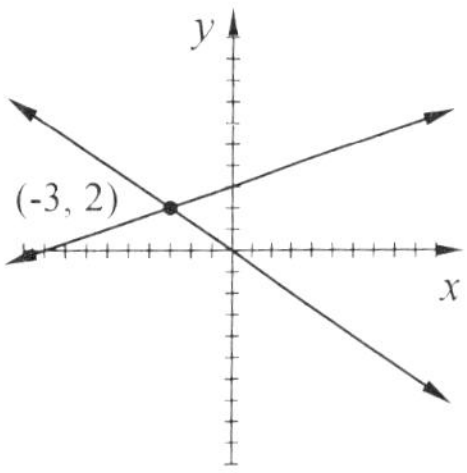

2. $-3x+y=-2$

   $-2x-y=-8$

   $(2, 4)$

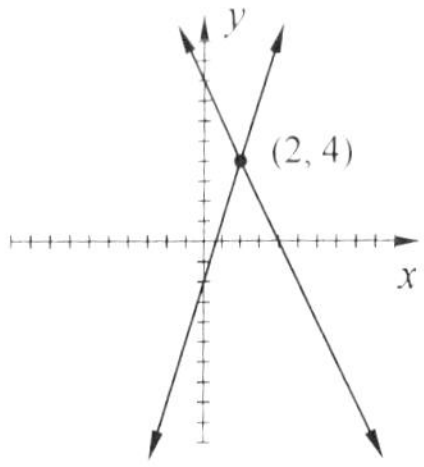

3. $2x-y=6$

   $6x+3y=6$

   $(-3, 1)$

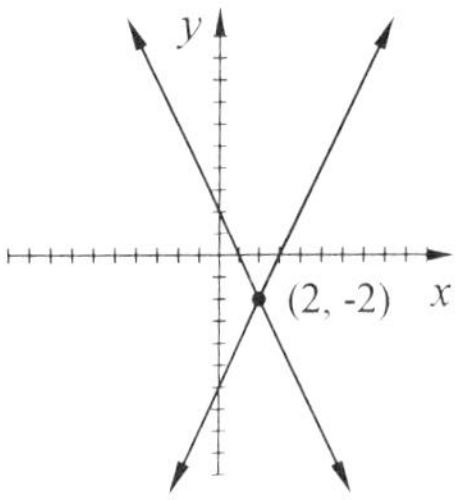

4. $2x-y=1$

   $3x+y=-6$

   $(-1, -3)$

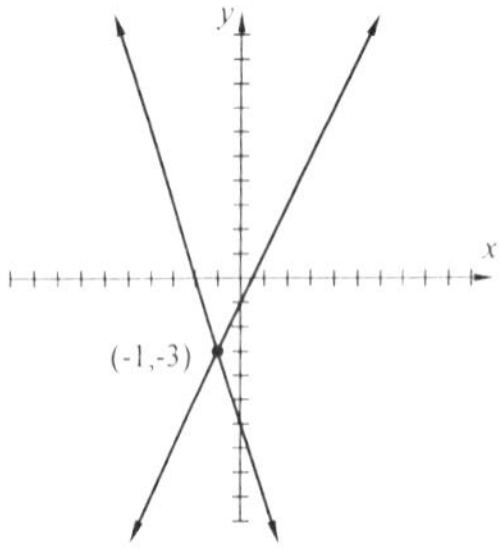

5. $x+y=6 \quad (1)$

   $-2x+y=-3 \quad (2)$

   Solve (2) for y

   $y=2x-3 \quad (3)$

   Substitute $2x-3$ for y in (1)

   $x+(2x-3)=6$

   $x+2x-3=6$

   $3x=9$

   $x=3$

   Substitute 3 for x in (3)

   $y=2(3)-3=3$

   $(3, 3)$

   Check: $3+3\stackrel{?}{=}6$

   $6=6$

6. $x+3y=18 \quad (1)$

   $2x+y=11 \quad (2)$

   Solve (2) for y

   $y=-2x+11 \quad (3)$

   Substitute $-2x+11$ for y in (1)

   $x+3(-2x+11)=18$

   $x-6x+33=18$

   $-5x=-15$

   $x=3$

   Substitute 3 for x in (3)

   $y=-2(3)+11=5$

   $(3, 5)$

   Check: $3+3(5)\stackrel{?}{=}18$

   $18=18$

7. $3x - 2y = 3 \quad (1)$

$x - \frac{1}{3}y = 8$

$3x - y = 24 \quad (2)$

Solve (2) for y

$y = 3x - 24 \quad (3)$

Substitute $3x - 24$ for y in (1)

$3x - 2(3x - 24) = 3$

$3x - 6x + 48 = 3$

$-3x = -45$

$x = 15$

Substitute 15 for x in (3)

$y = 3(15) - 24 = 21$

$(15, 21)$

Check: $15 - \frac{1}{3}(21) \stackrel{?}{=} 8$

$15 - 7 \stackrel{?}{=} 6$

$8 = 8$

8. $0.5x + y = 16 \quad (1)$

$4x - 2y = 8 \quad (2)$

Solve (1) for y

$y = -0.5x + 16 \quad (3)$

Substitute $-0.5x + 16$ for y in (2)

$4x - 2(-0.5x + 16) = 8$

$4x + x - 32 = 8$

$5x - 32 = 8$

$5x = 40$

$x = 8$

Substitute 8 for x in (3)

$y = -0.5(8) + 16 \quad y = 12$

Check: $4(8) - 2(12) \stackrel{?}{=} 8$

$8 = 8$

$(8,12)$

9. $6x - 2y = 10 \quad (1)$

$2x + 3y = 7 \quad (2)$

Multiply (1) by 3 and (2) by 2

$18x - 6y = 30$

$4x + 6y = 14$

$22x = 44$

$x = 2$

Substitute 2 for x in (2)

$2(2) + 3y = 7$

$3y = 3$

$y = 1$

$(2, 1)$

Check: $6(2) - 2(1) \stackrel{?}{=} 10$

$10 = 10$

10. $4x - 5y = -4 \quad (1)$

$-3x + 2y = 3 \quad (2)$

Multiply (1) by 2 and (2) by 5

$8x - 10y = -8$

$-15x + 10y = 15$

$-7x = 7$

$x = -1$

Substitute -1 for x in (2)

$-3(-1) + 2y = 3$

$2y = 0$

$y = 0$

$(-1, 0)$

Check: $4(-1) - 5(0) \stackrel{?}{=} -4$

$-4 = -4$

11. $5x-4y=12$ (1)

$-2x+3y=-9$ (2)

Multiply (1) by 3 and (2) by 4

$$\begin{aligned} 15x-12y&=36\\ -8x+12y&=-36\\ \hline 7x\quad&=0\\ x&=0 \end{aligned}$$

Substitute 0 for x in (2)

$$\begin{aligned} -2(0)+3y&=-9\\ 3y&=-9\\ y&=-3 \end{aligned}$$

$(0, -3)$

Check: $5(0)-4(-3)\stackrel{?}{=}12$

$12=12$

12. $7x+2y=4$ (1)

$2x-3y=19$ (2)

Multiply (1) by 3 and (2) by 2

$$\begin{aligned} 21x+6y&=12\\ 4x-6y&=38\\ \hline 25x\quad&=50\\ x&=2 \end{aligned}$$

Substitute 2 for x in (1)

$$\begin{aligned} 7(2)+2y&=4\\ 14+2y&=4\\ 2y&=-10\\ y&=-5 \end{aligned}$$

$(2, -5)$

Check: $2(2)-3(-5)\stackrel{?}{=}19$

$19=19$

13. $5x-3y=32$ (1)

$5x+7y=-8$ (2)

Multiply (2) by -1

$$\begin{aligned} 5x-3y&=32\\ -5x-7y&=8\\ \hline -10y&=40\\ y&=-4 \end{aligned}$$

Substitute -4 for y in (1)

$$\begin{aligned} 5x-3(-4)&=32\\ 5x&=20\\ x&=4 \end{aligned}$$

$(4, -4)$

14. $4x+3y=46$ (1)

$-2x+3y=-14$ (2)

Multiply (2) by -1

$$\begin{aligned} 4x+3y&=46\\ 2x-3y&=14\\ \hline 6x\quad&=60\\ x&=10 \end{aligned}$$

Substitute 10 for x in (1)

$$\begin{aligned} 4(10)+3y&=46\\ 3y&=6\\ y&=2 \end{aligned}$$

$(10, 2)$

15. $7x + 3y = 2 \quad (1)$

$-8x - 7y = 2 \quad (2)$

Multiply (1) by 7 and (2) by 3

$49x + 21y = 14$

$\underline{-24x - 21y = 6}$

$25x \qquad = 20$

$x = \frac{4}{5}$

Substitute $\frac{4}{5}$ for x in (1)

$7\left(\frac{4}{5}\right) + 3y = 2$

$3y = -\frac{18}{5}$

$y = -\frac{6}{5}$

$\left(\frac{4}{5}, -\frac{6}{5}\right)$

16. $7x + 5y = 7 \quad (1)$

$-3x - 4y = 1 \quad (2)$

Multiply (1) by 4 and (2) by 5

$28x + 20y = 28$

$\underline{-15x - 20y = 5}$

$13x \qquad = 33$

$x = \frac{33}{13}$

Substitute $\frac{33}{13}$ for x in (1)

$7\left(\frac{33}{13}\right) + 5y = 7$

$\frac{231}{13} + 5y = 7$

$5y = -\frac{140}{13}$

$y = -\frac{28}{13}$

$\left(\frac{33}{13}, -\frac{28}{13}\right)$

17. $4x + 5y = 4 \quad (1)$

$-5x - 8y = 2 \quad (2)$

Multiply (1) by 5 and (2) by 4

$20x + 25y = 20$

$\underline{-20x - 32y = 8}$

$-7y = 28$

$y = -4$

Substitute -4 for y in (1)

$4x + 5(-4) = 4$

$4x - 20 = 4$

$4x = 24$

$x = 6$

$(6, -4)$

18. $7x - 2y = 1 \quad (1)$

$-5x + 3y = -7 \quad (2)$

Multiply (1) by 3 and (2) by 2

$21x - 6y = 3$

$\underline{-10x + 6y = -14}$

$11x \qquad = -11$

$x = -1$

Substitute -1 for x in (1)

$7(-1) - 2y = 1$

$-7 - 2y = 1$

$-2y = 8$

$y = -4$

$(-1, -4)$

19. $2x-4y=6$ (1)

$-3x+6y=7$ (2)

Multiply (1) by 3 and (2) by 2

$6x-12y=18$

$-6x+12y=14$

$0=32$

No solution, inconsistent system.

20. $4x-7y=8$ (1)

$5x+9y=81$ (2)

Multiply (1) by 9 and (2) by 7

$36x-63y=72$

$35x+63y=567$

$71x=639$

$x=9$

Substitute 9 for x in (2)

$5(9)+9y=81$

$9y=36$

$y=4$

$(9, 4)$

21. $2x-9y=0$ (1)

$3x+5=6y$ (2)

Rearrange (2)

$3x-6y=-5$ (3)

Multiply (1) by 3 and (3) by -2

$6x-27y=0$

$-6x+12y=10$

$-15y=10$

$y=-\frac{2}{3}$

Substitute $-\frac{2}{3}$ for y in (1)

$2x-9\left(-\frac{2}{3}\right)=0$

$2x=-6$

$x=-3$

$\left(-3, -\frac{2}{3}\right)$

22. $2x+10y=1$ (1)

$-4x-20y=-2$ (2)

Multiply (1) by 2

$4x+20y=2$

$-4x-20y=-2$

$0=0$

Infinite number of solutions, dependent equations

23. $5x-8y=3x+12$ (1)

$7x+y=6y-4$ (2)

Rearrange and simplify

$2x-8y=12$ (3)

$7x-5y=-4$ (4)

Multiply (3) by 7 and (4) by -2

$14x-56y=84$

$-14x+10y=8$

$-46y=92$

$y=-2$

Substitute -2 for y in (3)

$2x-8(-2)=12$

$2x=-4$

$x=-2$

$(-2, -2)$

24. $1 + x - y = y + 4$

$4(x - y) = 3 - x$

Rearrange and simplify

$x - 2y = 3 \quad (1)$

$5x - 4y = 3 \quad (2)$

Multiply (1) by -5

$-5x + 10y = -15$

$5x - 4y = 3$

$6y = -12$

$y = -2$

Substitute -2 for y in (1)

$x - 2(-2) = 3$

$x = -1$

$(-1, -2)$

25. $3x + y = 9 \quad (1)$

$x - 2y = 10 \quad (2)$

Multiply (1) by 2

$6x + 2y = 18$

$x - 2y = 10$

$7x = 28$

$x = 4$

Substitute 4 for x in (1)

$3(4) + y = 9$

$y = -3$

$(4, -3)$

26. $x + 12y = 0 \quad (1)$

$3x - 5y = -2 \quad (2)$

Multiply (1) by -3

$-3x - 36y = 0$

$3x - 5y = -2$

$-41y = -2$

$y = \frac{2}{41}$

Substitute $\frac{2}{41}$ for y in (1)

$x + 12\left(\frac{2}{41}\right) = 0$

$x = -\frac{24}{41}$

$\left(-\frac{24}{41}, \frac{2}{41}\right)$

27. $2(x + 3) = y + 4$

$4x - 2y = -4$

Expand, rearrange and simplify

$2x - y = -2 \quad (1)$

$4x - 2y = -4 \quad (2)$

Multiply (1) by -2

$-4x + 2y = 4$

$4x - 2y = -4$

$0 = 0$

Infinite number of solutions, dependent equations.

28. $x + y = 3000$ (1)

$x - 2y = -120$ (2)

Multiply (2) by -1

$$\begin{array}{r} x + y = 3000 \\ -x + 2y = 120 \\ \hline 3y = 3120 \\ y = 1040 \end{array}$$

Substitute 1040 for y in (1)

$x + 1040 = 3000$

$x = 1960$

$(1960, 1040)$

29. $5x + 4y + 3 = 23$

$8x - 3y - 4 = 75$

Rearrange and simplify

$5x + 4y = 20$ (1)

$8x - 3y = 79$ (2)

Multiply (1) by 3 and (2) by 4

$$\begin{array}{r} 15x + 12y = 60 \\ 32x - 12y = 316 \\ \hline 47x \qquad = 376 \end{array}$$

$x = 8$

Substitute 8 for x in (1)

$5(8) + 4y = 20$

$4y = -20$

$y = -5$

$(8, -5)$

30. $4x - 3y + 1 = 6$ (1)

$5x + 8y + 2 = -74$ (2)

Rearrange and simplify

$4x - 3y = 5$ (3)

$5x + 8y = -76$ (4)

Multiply (3) by 8 and (4) by 3

$$\begin{array}{r} 32x - 24y = 40 \\ 15x + 24y = -228 \\ \hline 47x = -188 \\ x = -4 \end{array}$$

Substitute -4 for x in (2)

$5(-4) + 8y + 2 = -74$

$8y = -56$

$y = -7$

$(-4, -7)$

31. $\frac{2x}{3} - \frac{3y}{4} = \frac{7}{12}$ (1)

$8x + 5y = 9$ (2)

Multiply (1) by -12

$$\begin{array}{r} -8x + 9y = -7 \\ 8x + 5y = 9 \\ \hline 14y = 2 \end{array}$$

$y = \frac{1}{7}$

Substitute $\frac{1}{7}$ for y in (2)

$$8x + 5\left(\frac{1}{7}\right) = 9$$
$$8x = \frac{58}{7}$$
$$x = \frac{29}{28}$$
$$\left(\frac{29}{28}, \frac{1}{7}\right)$$

32. $\frac{x}{2} + \frac{y}{5} = 4 \quad (1)$

$\frac{x}{3} + \frac{y}{5} = \frac{10}{3} \quad (2)$

Multiply (1) by 10 and (2) by -15

$$5x + 2y = 40$$
$$\underline{-5x - 3y = -50}$$
$$-y = -10$$
$$y = 10$$

Substitute 10 for y in (1)

$$\frac{x}{2} + \frac{10}{5} = 4$$
$$\frac{x}{2} = 2$$
$$x = 4$$
$$(4, 10)$$

33. $\frac{1}{5}a + \frac{1}{2}b = 6 \quad (1)$

$\underline{\frac{3}{5}a - \frac{1}{2}b = 2} \quad (2)$

$$\frac{4}{5}a \qquad = 8$$
$$a = 10$$

Substitute 10 for a in (1)

$$\frac{1}{5}(10) + \frac{1}{2}b = 6$$
$$\frac{1}{2}b = 4$$
$$b = 8$$
$$(10, 8)$$

34. $\frac{2}{3}a + \frac{3}{5}b = -17 \quad (1)$

$\frac{1}{2}a - \frac{1}{3}b = -1 \quad (2)$

Multiply (1) by 30 and (2) by 54

$$20a + 18b = -510$$
$$\underline{27a - 18b = -54}$$
$$47a \qquad = -564$$
$$a = -12$$

Substitute -12 for a in (2)

$$\frac{1}{2}(-12) - \frac{1}{3}b = -1$$
$$-\frac{1}{3}b = 5$$
$$b = -15$$
$$(-12, -15)$$

35. $4.8 + 0.6m = 0.9n$

$0.6m + 0.9n = -4.8$

$6m - 9n = -48 \quad (1)$

$0.2m - 0.3n = 1.6$

$2m - 3n = 16 \quad (2)$

Multiply (2) by -3

$$6m - 9n = -48$$
$$\underline{-6m + 9n = -48}$$
$$0 + 0 = -9$$

No solution. Inconsistent system.

Substitute $\frac{1}{2}$ for s in (4)

$$4\left(\frac{1}{2}\right)+6t=-2$$
$$6t=-4$$
$$t=-\frac{2}{3}$$
$$\left(\frac{1}{2}, -\frac{2}{3}\right)$$

36. $8.4-0.8m=0.4n$

$$-0.8m-0.4n=-8.4$$
$$-8m-4n=-84 \quad (1)$$
$$0.2m+0.1n=2.1$$
$$2m+n=21 \quad (2)$$

Multiply (2) by 4

$$-8m-4n=-84$$
$$8m+4n=84$$
$$0=0$$

Infinite number of solutions.

Dependent equations.

37. $3s-8t=5 \quad (1)$

$4s+3t=2 \quad (2)$

Multiply (1) by 4 and (2) by -3

$$12s-32t=20$$
$$-12s-9t=-6$$
$$-41t=14$$
$$t=-\frac{14}{41}$$

Substitute $-\frac{14}{41}$ for t in (2)

$$4s+3\left(-\frac{14}{41}\right)=2$$
$$4s-\frac{42}{41}=2$$
$$4s=\frac{42}{41}+\frac{82}{41}$$
$$4s=\frac{124}{41}$$
$$s=\frac{31}{41}$$
$$\left(\frac{31}{41}, -\frac{14}{41}\right)$$

38. $5s+6t=2 \quad (1)$

$3s-5t=4 \quad (2)$

Multiply (1) by 5 and (2) by 6

$$25s+30t=10$$
$$18s-30t=24$$
$$43s=34$$
$$s=\frac{34}{43}$$

Substitute $\frac{34}{43}$ for s in (2)

$$3\left(\frac{34}{43}\right)-5t=4$$
$$102-215t=172$$
$$-215t=70$$
$$t=-\frac{70}{215}$$
$$t=-\frac{14}{43}$$
$$\left(\frac{34}{43}, -\frac{14}{43}\right)$$

39. $3(x+2)=-2-(x+3y)$

$3(x+y)=3-2(y-1)$

Expand, rearrange and simplify

$4x+3y=-8 \quad (1)$

$3x+5y=5 \quad (2)$

Multiply (1) by 3 and (2) by -4

$$\begin{array}{r} 12x+9y=-24 \\ -12x-20y=-20 \\ \hline -11y=-44 \\ y=4 \end{array}$$

Substitute 4 for y in (1)

$$4x+3(4)=-8$$
$$4x=-20$$
$$x=-5$$

$(-5, 4)$

40. $13-x=3(x+y)+1$

$14+2x=5(x+y)+3x$

Expand, rearrange and simplify

$4x+3y=12 \quad (1)$

$6x+5y=14 \quad (2)$

Multiply (1) by -3 and (2) by 2

$$\begin{array}{r} -12x-9y=-36 \\ 12x+10y=28 \\ \hline y=-8 \end{array}$$

Substitute -8 for y in (1)

$$4x+3(-8)=12$$
$$4x=36$$
$$x=9$$

$(9, -8)$

41. $0.2b-1.4-0.3a$

$0.1b+0.6=0.5a$

Rearrange

$0.3a+0.2b=1.4 \quad (1)$

$0.5a-0.1b=0.6 \quad (2)$

Multiply (1) by 10 and (2) by 20

$$\begin{array}{l} 3a+2b=14 \\ 10a-2b=12 \\ \hline 13a \quad =26 \\ a=2 \end{array}$$

Substitute 2 for a in (1)

$$0.3(2)+0.2b=1.4$$
$$0.2b=0.8$$
$$b=4$$

$(2, 4)$

42. $0.3a=1.1-0.2b$

$0.3b=0.4a-0.9$

Rearrange and simplify

$0.3a+0.2b=1.1 \quad (1)$

$0.4a-0.3b=0.9 \quad (2)$

Multiply (1) by 30 and (2) by 20

$$\begin{array}{l} 9a+6b=33 \\ 8a-6b=18 \\ \hline 17a \quad =51 \\ a=3 \end{array}$$

Substitute 3 for a in (1)

$0.3(3)+0.2b=1.1$

$0.2b=-0.2$

$b=1$

$(3, 1)$

43. $\frac{b}{5}=\frac{2}{5}-\frac{a-3}{2}$

$4(a-b)=3b-2(a-2)$

Multiply the first equation by 10, then Expand, rearrange, and simplify.

$5a+2b=19$ (1)

$6a-7b=4$ (2)

Multiply (1) by 7 and (2) by 2

$35a+14b=133$

$\underline{12a-14b=8}$

$47a=141$

$a=3$

Substitute 3 for a in (1)

$5(3)+2b=19$

$2b=4$

$b=2$

$(3, 2)$

44. $9(b+4)=2(2a+5b)$

Expand and simplify

$4a+b=36$ (1)

$\frac{b}{5}+\frac{a}{2}=\frac{18}{5}$

Rearrange and multiply by 10

$5a+2b=36$ (2)

Multiply (1) by -2

$-8a-2b=-72$

$\underline{5a+2b=36}$

$-3a=-36$

$a=12$

Substitute 12 for a in (1)

$4(12)+b=36$

$b=-12$

$(12, -12)$

45. x = number of reserved seat tickets

y = number of general admission tickets

$x+y=760$ (1)

$6x+4y=3280$ (2)

Multiply (1) by -4

$-4x-4y=-3040$

$\underline{6x+4y=3280}$

$2x=240$

$x=120$

Substitute 120 for x in (1)

$120+y=760$

$y=640$

120 reserved seats

640 general admission seats

46. x = number of 2-pound bags
y = number of 5-pound bags
$2x + 5y = 252$ (1)
$y = 2x$ (2)
Substitute $2x$ for y in (1)
$2x + 5(2x) = 252$
$2x + 10x = 252$
$12x = 252$
$x = 21$
Substitute 21 for x in (2)
$y = 2(21) = 42$
He prepared 21 2-lb bags
and 42 5-lb bags

47. r = airspeed
w = wind speed
$\text{speed} = \dfrac{\text{distance}}{\text{time}}$
$r + w = \dfrac{1500}{5}$ (1)
$r - w = \dfrac{1500}{6}$ (2)
$2r = 550$

$r = 275$
Substitute 275 for r in (1)
$275 + w = 300$
$w = 25$
Speed in still air = 275 mph
Wind speed = 25 mph

48. x = amount of 30% acid solution
y = amount of 20% acid solution
$y = x + 40$ (1)
$0.30x + 0.12(40) = 0.20y$
Multiply by 10 and rearrange
$3x - 2y = -48$ (2)
Substitute $x + 40$ for y in (2)
$3x - 2(x + 40) = -48$
$3x - 2x - 80 = -48$
$x = 32$
32 liters of 30% acid solution
should be added.

49. x = number of regular-sized cars
y = number of compact cars
$x + y = 86$ (1)
$4x + 3.50y = 323$ (2)
Multiply (1) by -4
$-4x - 4y = -344$
$4x + 3.50y = 323$
$-0.50y = -21$
$y = 42$
Substitute 42 for y in (1)
$x + 42 = 86$
$x = 44$

44 regular-sized cars
42 compact cars

50. $x =$ tons of 15% salt mixture

$y =$ tons of 30% salt mixture

$x + y = 24$ (1)

$0.15x + 0.30y = 0.25(24)$

Multiply by 100 and simplify

$15x + 30y = 600$ (2)

Solve (1) for x

$x = 24 - y$ (3)

Substitute $24 - y$ for x in (2)

$15(24 - y) + 30y = 600$

$360 - 15y + 30y = 600$

$15y = 240$

$y = 16$

Substitute 16 for y in (3)

$x = 24 - 16 = 8$

8 tons of 15% salt mixture

16 tons of 30% salt mixture

51. $r =$ speed of boat in still water

$c =$ speed of current

$r + c = 23$ (1)

$r - c = 15$ (2)

$2r = 38$

$r = 19$

Substitute 19 for r in (1)

$19 + c = 23$

$c = 4$

Speed of boat = 19 kph

Speed of current = 4 kph

52. $x =$ cost of copper per gram

$y =$ hourly labor rate

$5x + 3y = 27$ (1)

$4x + 5y = 32$ (2)

Multiply (1) by 5 and (2) by -3

$25x + 15y = 135$

$-12x - 15y = -96$

$13x = 39$

$x = 3$

Substitute 3 for x in (1)

$5(3) + 3y = 27$

$3y = 12$

$y = 4$

It costs \$3 per gram of copper.

The hourly labor rate is \$4.

53. $x =$ number of lower seats

$y =$ number of upper seats

$x + y = 23$ (1)

$22x + 6.5y = 320$ (2)

Solve (1) for x

$x = 23 - y$ (3)

Substitute $23 - y$ for x in (2)

$22(23 - y) + 6.5y = 320$

$506 - 22y + 6.5y = 320$

$506 - 15.5y = 320$

$-15.5y = -186$

$y = 12$

Substitute 12 for y in (3)

$x = 23 - 12$

$x = 11$

11 lower seats

9 upper seats

54. x = number of infield seats

y = number of bases seats

$x + y = 27$ (1)

$42x + 34y = 1070$ (2)

Solve (1) for x

$x = 27 - y$ (3)

Substitute $27 - y$ for x in (2)

$42(27-y)+34y = 1070$

$1134 - 42y + 34y = 1070$

$1134 - 8y = 1070$

$-8y = -64$

$y = 8$

19 upper infield seats

8 lower bases seats

How Am I Doing? Chapter 8 Test

1. $3x - y = -5$ (1)

$-2x + 5y = -14$ (2)

Solve (1) for y

$y = 3x + 5$ (3)

Substitute $3x + 5$ for y in (2)

$-2x + 5(3x+5) = -14$

$-2x + 15x + 25 = -14$

$13x = -39$

$x = -3$

Substitute -3 for x in (1)

$3(-3) - y = -5$

$-y = 4$

$y = -4$

$(-3, -4)$

2. $3x + 4y = 7$ (1)

$2x + 3y = 6$ (2)

Multiply (1) by 2 and (2) by -3

$6x + 8y = 14$

$-6x - 9y = -18$

$-y = -4$

$y = 4$

Substitute 4 for y in (2)

$2x + 3(4) = 6$

$2x = -6$

$x = -3$

$(-3, 4)$

3. $2x - y = 4$

$4x + y = 2$

$(1, -2)$

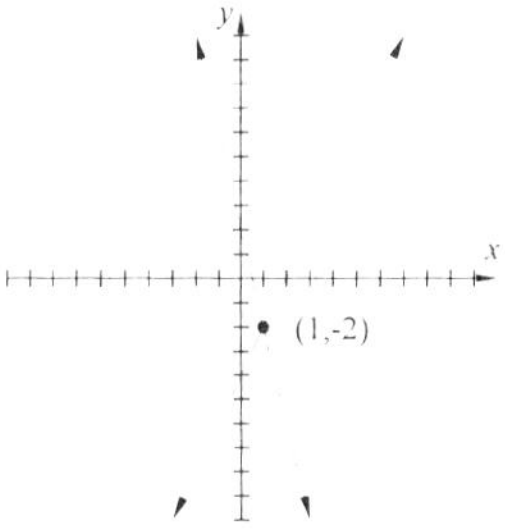

4. $x + 3y = 12$ (1)

$2x - 4y = 4$ (2)

Multiply (1) by -2

$-2x - 6y = -24$

$2x - 4y = 4$

$-10y = -20$

$y = 2$

Substitute 2 for y in (1)

$x + 3(2) = 12$

$x = 6$

$(6, 2)$

5. $2x - y = 5$ (1)

$-x + 3y = 5$ (2)

Multiply (1) by 3

$6x - 3y = 15$

$\underline{-x + 3y = 5}$

$5x = 20$

$x = 4$

Substitute 4 for x in (1)

$2(4) - y = 5$

$-y = -3$

$y = 3$

$(4, 3)$

6. $2x + 3y = 13$ (1)

$3x - 5y = 10$ (2)

Multiply (1) by 5 and (2) by 3

$10x + 15y = 65$

$\underline{9x - 15y = 30}$

$19x = 95$

$x = 5$

Substitute 5 for x in (1)

$2(5) + 3y = 13$

$3y = 3$

$y = 1$

$(5, 1)$

7. $\frac{2}{3}x - \frac{1}{5}y = 2$ (1)

$\frac{4}{3}x + 4y = 4$ (2)

Multiply (1) by 15 and (2) by 3

$10x - 3y = 30$ (3)

$4x + 12y = 12$ (4)

Multiply (3) by 4

$40x - 12y = 120$

$\underline{4x + 12y = 12}$

$44x = 132$

$x = 3$

Substitute 3 for x in (2)

$\frac{4}{3}(3) + 4y = 4$

$4y = 0$

$y = 0$

$(3, 0)$

8. $3x - 6y = 5$ (1)

$-\frac{1}{2}x + y = \frac{7}{2}$ (2)

Multiply (2) by 6

$3x - 6y = 5$ (3)

$\underline{-3x + 6y = 21}$ (4)

$0 = 26$

No solution, inconsistent system

9. $5x - 2 = y \quad (1)$

$10x = 4 + 2y \quad (2)$

Substitute $5x - 2$ for y in (2)

$10x = 4 + 2(5x - 2)$

$10x = 4 + 10x - 4$

$0 = 0$

Infinite number of solutions, dependent equations.

10. $0.3x + 0.2y = 0 \quad (1)$

$1.0x + 0.5y = -0.5 \quad (2)$

Multiply (1) by -50 and (2) by 20

$-15x - 10y = 0 \quad (3)$

$20x + 10y = -10 \quad (4)$

$5x \quad = -10$

$x = -2$

Substitute -2 for x in (4)

$20(-2) + 10y = -10$

$10y = 30$

$y = 3$

$(-2, 3)$

11. $2(x + y) = 2(1 - y) \quad (1)$

$5(-x + y) = 2(23 - y) \quad (2)$

Expand, rearrange, and simplify

$2x + 4y = 2 \quad (3)$

$-5x + 7y = 46 \quad (4)$

Multiply (3) by 5 and (4) by 2

$10x + 20y = 10$

$-10x + 14y = 92$

$34y = 102$

$y = 3$

Substitute 3 for y in (3)

$2x + 4(3) = 2$

$2x = -10$

$x = -5$

$(-5, 3)$

12. $3(x - y) = 12 + 2y \quad (1)$

$8(y + 1) = 6x - 7 \quad (2)$

Expand, rearrange, and simplify

$3x - 5y = 12 \quad (3)$

$-6x + 8y = -15 \quad (4)$

Multiply (3) by 2

$6x - 10y = 24$

$-6x + 8y = -15$

$-2y = 9$

$y = -4.5$

Substitute -4.5 for y in (3)

$3x - 5(-4.5) = 12$

$3x + 22.5 = 12$

$3x = -10.5$

$x = -3.5$

$(-3.5, -4.5)$

13. $x =$ first number

$y =$ second number

$2x+3y=1 \qquad (1)$

$3x+2y=9 \qquad (2)$

Multiply (1) by 3 and (2) by -2

$$\begin{aligned} 6x+9y&=3 \\ \underline{-6x-4y}&\underline{=-18} \\ 5y&=-15 \\ y&=-3 \end{aligned}$$

Substitute -3 for y in (1)

$$\begin{aligned} 2x+3(-3)&=1 \\ 2x&=10 \\ x&=5 \end{aligned}$$

The first number is 5 and the second number is -3

14. $x =$ the number of \$8 tickets

$y =$ the number of \$12 tickets

$x + \quad y = 30,500 \qquad (1)$

$8x+12y=308,000 \qquad (2)$

Solve (1) for y

$y=30,500-x \qquad (3)$

Substitute $30,500-x$ for y in (2)

$$\begin{aligned} 8x+12(30,500-x)&=308,000 \\ 8x+366,000-12x&=308,000 \\ -4x&=-58,000 \\ x&=14,500 \end{aligned}$$

Substitute 14,500 for x in (3)

$y=30,500-14,500=16,000$

14,500 of the \$8 tickets

16,000 of the \$12 tickets

15. $x =$ cost of a shirt

$y =$ cost of a pair of slacks

$5x+3y=172 \qquad (1)$

$3x+4y=156 \qquad (2)$

Multiply (1) by 4 and (2) by -3

$$\begin{aligned} 20x+12y&=688 \\ \underline{-9x-12y}&\underline{=-468} \\ 11x \qquad &=220 \\ x&=20 \end{aligned}$$

Substitute 20 for x in (1)

$$\begin{aligned} 5(20)+3y&=172 \\ 3y&=72 \\ y&=24 \end{aligned}$$

A shirt costs \$20

A pair of slacks costs \$24

16. $x =$ number of booklets

$y =$ cost

$y=0.25x+200 \qquad (1)$

$y=0.2x+250 \qquad (2)$

Substitute $0.25x+200$ for y in (2)

$$\begin{aligned} 0.25x+200&=0.2x+250 \\ 0.05x+200&=250 \\ 0.05x&=50 \\ x&=1000 \end{aligned}$$

Substitute 1000 for x in (1)

$y=0.25(1000)+200$

$y=450$

Order 1000 booklets at a cost of \$450.

17. r = airspeed

w = wind speed

$$r + w = \frac{2000}{4} \quad (1)$$

$$r - w = \frac{2000}{5} \quad (2)$$

$$2r = 900$$

$$r = 450$$

Substitute 450 for r in (1)

$$450 + w = 500$$

$$w = 50$$

Speed of plane in still air = 450 kmph

Wind speed = 50 kmph

Cumulative Test for Chapters 0-8

1. 11% of $37.20 = 0.11(37.20) = \$4.09$

2. $\dfrac{-2^4(2^2)}{2^3} = \dfrac{-16(4)}{8} = -8$

3. $(3x^2 - 2x + 1)(5x - 3)$

$= (3x^2 - 2x + 1)(5x) - (3x^2 - 2x + 1)(3)$

$= 15x^3 - 10x^2 + 5x - 9x^2 + 6x - 3$

$= 15x^3 - 19x^2 + 11x - 3$

4. $\dfrac{x^2 - 7x + 12}{x^2 - 9} \div \dfrac{x^2 - 8x + 16}{1}$

$= \dfrac{x^2 - 7x + 12}{x^2 - 9} \cdot \dfrac{1}{x^2 - 8x + 16}$

$= \dfrac{(x-4)(x-3)}{(x+3)(x-3)} \cdot \dfrac{1}{(x-4)(x-4)}$

$= \dfrac{1}{(x+3)(x-4)}$

5. $2(x+3) < 3(x-5)$

$2x + 6 < 3x - 15$

$-x + 6 < -15$

$-x < -21$

$x > 21$

6. $6x^2 + x - 2 = 0$

$(2x-1)(3x+2) = 0$

$2x - 1 = 0 \qquad 3x + 2 = 0$

$2x = 1 \qquad 3x = -2$

$x = \dfrac{1}{2} \qquad x = -\dfrac{2}{3}$

7. 12% of x is 360

$0.12x = 360$

$x = 3000$

8. $\dfrac{1}{10}x + \dfrac{1}{3} = \dfrac{2}{5}x - \dfrac{1}{6}$

$$30\left(\frac{1}{10}x\right) + 30\left(\frac{1}{3}\right) = 30\left(\frac{2}{5}\right)x - 30\left(\frac{1}{6}\right)$$

$$3x + 10 = 12x - 5$$

$$10 = 9x - 5$$
$$15 = 9x$$
$$\frac{5}{3} = x$$

9. $10t^2 - 25t^3 + 15t^4$
$= 5t^2(3t^2 - 5t + 2)$
$= 5t^2(3t - 2)(t - 1)$

10. $4x = 2(12 - 2x)$
$4x = 24 - 4x$
$8x = 24$
$x = 3$

11. $2x + y = 8$ (1)
$3x + 4y = -8$ (2)
Multiply (1) by -4
$-8x - 4y = -32$
$3x + 4y = -8$
$-5x = -40$
$x = 8$
Substitute 8 for x in (1)
$2(8) + y = 8$
$y = -8$
$(8, -8)$

12. $\frac{1}{3}x - \frac{1}{2}y = 4$
$\frac{3}{8}x - \frac{1}{4}y = 7$
Clear the fractions
$2x - 3y = 24$ (1)
$3x - 2y = 56$ (2)
Multiply (1) by -3 and (2) by 2
$-6x + 9y = -72$
$6x - 4y = 112$
$5y = 40$
$y = 8$
Substitute 8 for y in (1)
$2x - 3(8) = 24$
$2x = 48$
$x = 24$
$(24, 8)$

13. $1.3x - 0.7y = 0.4$ (1)
$-3.9x + 2.1y = -1.2$ (2)
Multiply (1) by 30 and (2) by 10
$39x - 21y = 12$
$-39x + 21y = -12$
$0 = 0$
Infinite number of solutions, dependent system.

14. $10x - 5y = 45$ (1)
$3x - 8y = 7$ (2)
Multiply (1) by -8 and (2) by 5
$-80x + 40y = -360$
$15x - 40y = 35$
$-65x = -325$
$x = 5$
Substitute 5 for x in (2)
$3(5) - 8y = 7$
$-8y = -8$
$y = 1$
$(5, 1)$

15. Parallel lines represent an inconsistent system for which there is no solution.

16. x = amount of 8% solution
y = amount of 20% solution
$x + y = 12{,}000$ (1)
$0.08x + 0.20y = 0.16(12{,}000)$
Multiply by 100 and simplify
$8x + 20y = 192{,}000$ (2)
Solve (1) for x
$x = 12{,}000 - y$ (3)
Substitute $24 - y$ for x in (2)
$8(12{,}000 - y) + 20y = 192{,}000$
$96{,}000 - 8y + 20y = 192{,}000$
$12y = 96{,}000$
$y = 8000$
Substitute 8000 for y in (3)
$x = 12{,}000 - 8000 = 4000$
4000 liters of 8% solution
8000 liters of 20% solution

17. r = speed of boat in still water
s = speed of the stream
$r - s = \frac{6}{3}$ (1)
$r + s = \frac{6}{1.5}$ (2)
Multiply (1) by -1
$-r + s = -2$
$r + s = 4$
$2s = 2$
$s = 1$
Speed of the stream is 1 mph.

18. x = first printer's rate
y = second printer's rate
$5x + 5y = 15{,}000$ (1)
$7x + 2y = 15{,}000$ (2)
Multiply (1) by 2 and (2) by -5
$10x + 10y = 30{,}000$
$-35x - 10y = -75{,}000$
$-25x = -45{,}000$
$x = 1800$
Substitute 1800 for x in (1)
$5(1800) + 5y = 15{,}000$
$9000 + 5y = 15{,}000$
$5y = 6000$
$y = 1200$
The printer that broke prints 1200 labels/hr. The other printer prints 1800 labels/hr.

Chapter 9

9.1 Exercises

1. Answers will vary.

3. No, because $(0.3)(0.3) = 0.09$.

5. $\pm\sqrt{9} = \pm 3$

7. $\pm\sqrt{49} = \pm 7$

9. $\sqrt{16} = 4$

11. $\sqrt{81} = 9$

13. $-\sqrt{36} = -6$

15. $\sqrt{0.81} = 0.9$

17. $\sqrt{\frac{36}{121}} = \frac{6}{11}$

19. $\sqrt{\frac{49}{64}} = \frac{7}{8}$

21. $\sqrt{400} = 20$

23. $-\sqrt{10{,}000} = -100$

25. $\sqrt{169} = 13$

27. $-\sqrt{\frac{1}{64}} = -\frac{1}{8}$

29. $\sqrt{\frac{9}{16}} = \frac{3}{4}$

31. $\sqrt{0.0036} = 0.06$

33. $\sqrt{16{,}900} = 130$

35. $-\sqrt{289} = -17$

37. $\sqrt{27} = 5.176$

39. $\sqrt{74} \approx 8.602$

41. $-\sqrt{133} = -11.533$

43. $-\sqrt{195} = -13.964$

45. $\sqrt{169} = 13$ feet

47. If $a = 900$,

$$t = \frac{1}{4}\sqrt{a} = \frac{1}{4}\sqrt{900} = 7.5 \text{ seconds}$$

49. If $a = 1444$,

$$t = \frac{1}{4}\sqrt{a} = \frac{1}{4}\sqrt{1444} = 9.5 \text{ seconds}$$

51. $\sqrt[3]{27} = 3$

53. $\sqrt[3]{-216} = -6$

55. $\sqrt[4]{81} = 3$

57. $\sqrt[5]{243} = 3$

59. No; a fourth power of a real number must be nonnegative.

Cumulative Review

61. $3x + 2y = 8 \quad (1)$

$7x - 3y = 11 \quad (2)$

Multiply (1) by 3 and (2) by 2

$$\begin{array}{r} 9x + 6y = 24 \\ 14x - 6y = 22 \\ \hline 23x \quad = 46 \\ x = 2 \end{array}$$

Substitute 2 for x in (1)

$$3(2) + 2y = 8$$
$$2y = 2$$
$$y = 1$$

$(2, 1)$

63. s = cost per snowboard

g = cost per pair of goggles

$3s + 2g = 850 \quad (1)$

$4s + 3g = 1150 \quad (2)$

Multiply (1) by -3 and (2) by 2.

$$\begin{array}{r} -9s - 6g = -2550 \\ 8s + 6g = \ \ 2300 \\ \hline -s \ = -250 \\ s = 250 \end{array}$$

Substitute 250 for s in (1).

$$3(250) + 2g = 850$$
$$750 + 2g = 850$$
$$2g = 100$$
$$g = 50$$

\$250 per snowboard

\$50 per pair of goggles

65. $2000-2010:$

$$\text{Decrease} = 10.6 - 9.4 = 1.2$$

$$\text{Percent decrease} = \frac{1.2}{10.6} = 0.113 = 11.3\%$$

$2010-2020:$

$$100\% - 11.3\% = 88.7\%$$

$$\text{Length of stay} = 88.7\% \text{ of } 9.4$$
$$= 0.887(9.4)$$
$$= 8.3 \text{ days}$$

9.2 Exercises

1. Yes. $\sqrt{6^2} = \sqrt{6 \cdot 6} = \sqrt{6} \cdot \sqrt{6} = \left(\sqrt{6}\right)^2$

3. No. $\sqrt{(-9)^2} = \sqrt{81} = \sqrt{9^2}$

5. $\sqrt{8^2} = 8$

7. $\sqrt{18^4} = \sqrt{\left(18^2\right)^2} = 18^2$

9. $\sqrt{9^6} = \sqrt{\left(9^3\right)^2} = 9^3$

11. $\sqrt{33^8} = \sqrt{\left(33^4\right)^2} = 33^4$

13. $\sqrt{5^{140}} = \sqrt{\left(5^{70}\right)^2} = 5^{70}$

15. $\sqrt{x^{12}} = \sqrt{\left(x^6\right)^2} = x^6$

17. $\sqrt{t^{18}} = \sqrt{\left(t^9\right)^2} = t^9$

19. $\sqrt{y^{26}} = \sqrt{\left(y^{13}\right)^2} = y^{13}$

21. $\sqrt{36x^8} = \sqrt{6^2\left(x^4\right)^2} = 6x^4$

23. $\sqrt{144x^2} = \sqrt{144}\sqrt{x^2} = 12x$

25. $\sqrt{x^6y^4} = \sqrt{\left(x^3\right)^2\left(y^2\right)^2} = x^3y^2$

27. $\sqrt{16x^2y^{20}} = \sqrt{4^2x^2\left(y^{10}\right)^2} = 4xy^{10}$

29. $\sqrt{100x^{12}y^8} = \sqrt{10^2\left(x^6\right)^2\left(y^4\right)^2} = 10x^6y^4$

31. $\sqrt{24} = \sqrt{4\cdot 6} = 2\sqrt{6}$

33. $\sqrt{45} = \sqrt{9\cdot 5} = 3\sqrt{5}$

35. $\sqrt{18} = \sqrt{9\cdot 2} = 3\sqrt{2}$

37. $\sqrt{72} = \sqrt{36\cdot 2} = 6\sqrt{2}$

39. $\sqrt{90} = \sqrt{9\cdot 10} = 3\sqrt{10}$

41. $\sqrt{128} = \sqrt{64\cdot 2} = 8\sqrt{2}$

43. $\sqrt{8x^3} = \sqrt{4\cdot 2x^2x} = 2x\sqrt{2x}$

45. $\sqrt{27w^5} = \sqrt{9\cdot 3w^4w}$
$= \sqrt{3^2\cdot 3\left(w^2\right)^2w}$
$= 3w^2\sqrt{3w}$

47. $\sqrt{32z^9} = \sqrt{16\cdot 2z^8z} = 4z^4\sqrt{2z}$

49. $\sqrt{28x^5y^7} = \sqrt{4\cdot 7x^4xy^6y} = 2x^2y^3\sqrt{7xy}$

51. $\sqrt{48y^3w} = \sqrt{16\cdot 3y^2yw} = 4y\sqrt{3yw}$

53. $\sqrt{75x^2y^3} = \sqrt{25\cdot 3x^2y^2y} = 5xy\sqrt{3y}$

55. $\sqrt{81x^8} = 9x^4$

57. $\sqrt{y^7} = \sqrt{y^6 y} = y^3\sqrt{y}$

59. $\sqrt{x^5 y^{11}} = \sqrt{x^4 x y^{10} y} = x^2 y^5\sqrt{xy}$

61. $\sqrt{135x^5 y^7} = \sqrt{9\cdot 15x^4 xy^6 y} = 3x^2 y^3\sqrt{15xy}$

63. $\sqrt{80a^3 b^3 c^6} = \sqrt{16\cdot 5a^2 ab^2 bc^6} = 4abc^3\sqrt{5ab}$

65. $\sqrt{169a^3 b^8 c^5} = \sqrt{169a^2 ab^8 c^4 c}$
$= 13ab^4 c^2\sqrt{ac}$

67. $\sqrt{(x+5)^2} = x+5$

69. $\sqrt{16x^2 + 8x + 1} = \sqrt{(4x+1)^2} = 4x+1$

71. $\sqrt{x^2 y^2 + 14xy^2 + 49y^2} = \sqrt{y^2(x^2 + 14x + 49)}$
$= \sqrt{y^2(x+7)^2}$
$= y(x+7)$

73. $\sqrt{75} = \sqrt{25\cdot 3} = 5\sqrt{3}$ mm^2

75. (a) $\sqrt{1200} = \sqrt{400\cdot 3} = 20\sqrt{3}$ mm^2
(b) $\frac{20\sqrt{3}}{5\sqrt{3}} = 4$ times larger

Cumulative Review

77. $3x - 2y = 14$ (1)
$y = x - 5$ (2)
$3x - 2(x-5) = 14$ Substitute $x - 5$ into (1)
$3x - 2x + 10 = 14$
$x = 4$
$y = 4 - 5 = -1$ Substitute 4 into (2)
$(4, -1)$

79. 1984 – 1994
Rate of increase $= \frac{560-310}{10} = 25$
Use in 2007 $= 560 + 13(25)$
$= 885$ million tons

81. Rate of increase $= \frac{990-560}{560} \approx 0.768 = 76.8\%$

9.3 Exercises

1. (1) Simplify each radical term.
(2) Combine like radicals.

3. $3\sqrt{5} - \sqrt{5} + 4\sqrt{5} = 6\sqrt{5}$

5. $\sqrt{2} + 8\sqrt{3} - 5\sqrt{3} + 4\sqrt{2}$
$= 5\sqrt{2} + 3\sqrt{3}$

7. $\sqrt{5} + \sqrt{45} = \sqrt{5} + \sqrt{9\cdot 5}$
$= \sqrt{5} + 3\sqrt{5}$
$= 4\sqrt{5}$

9. $\sqrt{80}+4\sqrt{20}=\sqrt{16\cdot5}+4\sqrt{4\cdot5}$
$=4\sqrt{5}+8\sqrt{5}$
$=12\sqrt{5}$

11. $2\sqrt{8}-3\sqrt{2}=2\sqrt{4\cdot2}-3\sqrt{2}$
$=4\sqrt{2}-3\sqrt{2}$
$=\sqrt{2}$

13. $-\sqrt{2}+\sqrt{18}+\sqrt{98}$
$=-\sqrt{2}+\sqrt{9\cdot2}+\sqrt{49\cdot2}$
$=-\sqrt{2}+3\sqrt{2}+7\sqrt{2}$
$=9\sqrt{2}$

15. $\sqrt{25}+\sqrt{72}+3\sqrt{12}=\sqrt{25}+\sqrt{36\cdot2}+3\sqrt{4\cdot3}$
$=5+6\sqrt{2}+6\sqrt{3}$

17. $2\sqrt{48}-3\sqrt{8}+\sqrt{50}$
$=2\sqrt{16\cdot3}-3\sqrt{4\cdot2}+\sqrt{25\cdot2}$
$=8\sqrt{3}-\sqrt{2}$

19. $8\sqrt{2x}+\sqrt{50x}=8\sqrt{2x}+\sqrt{25\cdot2x}$
$=8\sqrt{2x}+5\sqrt{2x}$
$=13\sqrt{2x}$

21. $1.2\sqrt{3x}-0.5\sqrt{12x}=1.2\sqrt{3x}-0.5\sqrt{4\cdot3x}$
$=1.2\sqrt{3x}-\sqrt{3x}$
$=0.2\sqrt{3x}$

23. $\sqrt{20y}+2\sqrt{45y}-\sqrt{5y}$
$=\sqrt{4\cdot5y}+2\sqrt{9\cdot5y}-\sqrt{5y}$
$=2\sqrt{5y}+6\sqrt{5y}-\sqrt{5y}$
$=7\sqrt{5y}$

25. $\sqrt{50}-3\sqrt{8}=\sqrt{25\cdot2}-3\sqrt{4\cdot2}$
$=5\sqrt{2}-6\sqrt{2}$
$=-\sqrt{2}$

27. $3\sqrt{28x}-5x\sqrt{63x}$
$=3\sqrt{4\cdot7x}-5x\sqrt{9\cdot7x}$
$=6\sqrt{7x}-15x\sqrt{7x}$

29. $5\sqrt{8x^3}-3x\sqrt{50x}$
$=5\sqrt{4\cdot2xx^2}-3x\sqrt{25\cdot2x}$
$=10x\sqrt{2x}-15x\sqrt{2x}$
$=-5x\sqrt{2x}$

31. $5x\sqrt{48}-2x\sqrt{75}$
$=5x\sqrt{16\cdot3}-2x\sqrt{25\cdot3}$
$=5x(4)\sqrt{3}-2x(5)\sqrt{3}$
$=20x\sqrt{3}-10x\sqrt{3}$
$=10x\sqrt{3}$

33. $2\sqrt{6y^3}-2y\sqrt{54}=2\sqrt{6yy^2}-2y\sqrt{9\cdot6}$
$=2y\sqrt{6y}-6y\sqrt{6}$

35. $5x\sqrt{8x} - 24\sqrt{50x^3}$
$= 5x\sqrt{4 \cdot 2x} - 24\sqrt{25 \cdot 2xx^2}$
$= 10x\sqrt{2x} - 120x\sqrt{2x}$
$= -110x\sqrt{2x}$

37. $P = 2\left(\sqrt{5} + 2\sqrt{3}\right) + 2\left(3\sqrt{5} + \sqrt{3}\right)$
$= 2\sqrt{5} + 4\sqrt{3} + 6\sqrt{5} + 2\sqrt{3}$
$= 8\sqrt{5} + 6\sqrt{3}$ miles

39. $\left(50\sqrt{3} + 70\sqrt{2}\right) - \left(42\sqrt{3} + 16\sqrt{2}\right)$
$= 50\sqrt{3} + 70\sqrt{2} - 42\sqrt{3} - 16\sqrt{2}$
$= 8\sqrt{3} + 54\sqrt{2}$ meters above

41. $7 - 3x \le 11$
$-3x \le 4$
$x \ge -\frac{4}{3}$

43. x = number of levels above the lobby
$\frac{1}{6}(5280) = 45 + 31 + 17x$
$880 = 76 + 17x$
$804 = 17x$
$x \approx 47$ levels above the lobby
Total number of levels is 48.

9.4 Exercises

1. $\sqrt{3}\sqrt{10} = \sqrt{30}$

3. $\sqrt{2}\sqrt{22} = 2\sqrt{11}$

5. $\sqrt{18}\sqrt{3} = \sqrt{18 \cdot 3} = \sqrt{54} = \sqrt{9 \cdot 6} = 3\sqrt{6}$

7. $\left(3\sqrt{5}\right)\left(2\sqrt{6}\right) = 6\sqrt{30}$

9. $\sqrt{5}\sqrt{10a} = \sqrt{50a} = \sqrt{25 \cdot 2a} = 5\sqrt{2a}$

11. $\left(2\sqrt{14x}\right)\left(3x\sqrt{7x}\right) = 6x\sqrt{98x^2}$
$= 6x\sqrt{49 \cdot 2x^2}$
$= 42x^2\sqrt{2}$

13. $\left(-4\sqrt{8a}\right)\left(-2\sqrt{3a}\right) = 8\sqrt{24a^2}$
$= 8\sqrt{4 \cdot 6a^2}$
$= 16a\sqrt{6}$

15. $\left(-3\sqrt{ab}\right)\left(2\sqrt{b}\right) = -6\sqrt{ab^2}$
$= -6b\sqrt{a}$

17. $\left(3\sqrt{5}\right)\left(2\sqrt{8}\right) = 6\sqrt{40}$
$= 6\sqrt{4 \cdot 10}$
$= 12\sqrt{10}$

19. $\sqrt{3}\left(2\sqrt{6}+5\sqrt{15}\right)=2\sqrt{18}+5\sqrt{45}$
$=2\sqrt{9\cdot 2}+5\sqrt{9\cdot 5}$
$=6\sqrt{2}+15\sqrt{5}$

21. $2\sqrt{x}\left(\sqrt{x}-8\sqrt{5}\right)=2\sqrt{x^2}-16\sqrt{5x}$
$=2x-16\sqrt{5x}$

23. $\sqrt{6}\left(\sqrt{2}-3\sqrt{6}+2\sqrt{10}\right)$
$=\sqrt{12}-3\sqrt{36}+2\sqrt{60}$
$=\sqrt{4\cdot 3}-18+2\sqrt{4\cdot 15}$
$=2\sqrt{3}-18+4\sqrt{15}$

25. $\left(2\sqrt{3}+\sqrt{6}\right)\left(\sqrt{3}-2\sqrt{6}\right)$
$=2\sqrt{9}-4\sqrt{18}+\sqrt{18}-2\sqrt{36}$
$=6-3\sqrt{9\cdot 2}-12$
$=-6-9\sqrt{2}$

27. $\left(5+3\sqrt{2}\right)\left(3+\sqrt{2}\right)$
$=15+5\sqrt{2}+9\sqrt{2}+3\sqrt{4}$
$=15+14\sqrt{2}+6$
$=21+14\sqrt{2}$

29. $\left(2\sqrt{7}-3\sqrt{3}\right)\left(\sqrt{7}+\sqrt{3}\right)$
$=2\sqrt{49}+2\sqrt{21}-3\sqrt{21}-3\sqrt{9}$
$=14-\sqrt{21}-9$
$=5-\sqrt{21}$

31. $\left(\sqrt{3}+2\sqrt{6}\right)\left(2\sqrt{3}-\sqrt{6}\right)$
$=2\sqrt{9}-\sqrt{18}+4\sqrt{18}-2\sqrt{36}$
$=6+3\sqrt{18}-12$
$=-6+3\sqrt{9\cdot 2}$
$=-6+9\sqrt{2}$

33. $\left(3\sqrt{7}-\sqrt{8}\right)\left(\sqrt{8}+2\sqrt{7}\right)$
$=3\sqrt{56}+6\sqrt{49}-\sqrt{64}-2\sqrt{56}$
$=\sqrt{56}+42-8$
$=\sqrt{4\cdot 14}+34$
$=2\sqrt{14}+34$

35. $\left(2\sqrt{5}-3\right)^2=4\sqrt{25}+2\left(-6\sqrt{5}\right)+9$
$=20-12\sqrt{5}+9$
$=29-12\sqrt{5}$

37. $\left(\sqrt{3}+5\sqrt{2}\right)^2=\sqrt{9}+2\left(5\sqrt{6}\right)+25\sqrt{4}$
$=3+10\sqrt{6}+50$
$=53+10\sqrt{6}$

39. $\left(3\sqrt{5}-\sqrt{3}\right)^2$
$=9\sqrt{25}-6\sqrt{15}+\sqrt{9}$
$=45-6\sqrt{15}+3$
$=48-6\sqrt{15}$

41. $\left(3\sqrt{5}\right)\left(2\sqrt{8}\right)=6\sqrt{40}$
$=6\sqrt{4\cdot 10}$
$=12\sqrt{10}$

43. $\sqrt{7}\left(\sqrt{2}+3\sqrt{14}-\sqrt{6}\right)$
$=\sqrt{14}+3\sqrt{98}-\sqrt{42}$
$=\sqrt{14}+3\sqrt{49\cdot 2}-\sqrt{42}$
$=\sqrt{14}+2\sqrt{2}-\sqrt{42}$

45. $\left(5\sqrt{2}-3\sqrt{7}\right)\left(3\sqrt{2}+\sqrt{7}\right)$
$=15\sqrt{4}+5\sqrt{14}-9\sqrt{14}-3\sqrt{49}$
$=30-4\sqrt{14}-21$
$=9-4\sqrt{14}$

47. $2\sqrt{a}\left(3\sqrt{b}+\sqrt{ab}-2\sqrt{a}\right)$
$=6\sqrt{ab}+2\sqrt{a^2b}-4\sqrt{a^2}$
$=6\sqrt{ab}+2a\sqrt{b}-4a$

49. $\left(3x\sqrt{y}+\sqrt{5}\right)\left(3x\sqrt{y}-\sqrt{5}\right)$
$=\left(3x\sqrt{y}\right)^2-\left(\sqrt{5}\right)^2$
$=9x^2y-5$

51. (*a*) $\sqrt{a}\cdot\sqrt{a}=a$ is not true when a is negative.
(b) The restriction " for any nonnegative real number" is necessary in order to deal with real numbers and it is not true when 'a' is negative.

53. $A=\frac{1}{2}bh,\ b=\sqrt{288},\ h=\sqrt{2}$
$A=\frac{1}{2}\sqrt{288}\cdot\sqrt{2}=\frac{1}{2}\left(12\sqrt{2}\right)\left(\sqrt{2}\right)=12$
The area is 12 square feet

55. $A=\sqrt{12}\sqrt{27}$
$=\sqrt{324}$
$=18$ square feet

Cumulative Review

57. $36x^2-49y^2=(6x)^2-(7y)^2$
$=(6x+7y)(6x-7y)$

59. $m=ak$
$m=\left(\frac{6076}{5280}\right)k$
$m=\left(\frac{1519}{1320}\right)k$
Destroyer is going
$m=\left(\frac{1519}{1320}\right)(35)\approx 40.3$ mph

How Am I Doing? Sections 9.1-9.4

1. $\sqrt{49}=7$

2. $\sqrt{\frac{25}{64}}=\frac{5}{8}$

3. $\sqrt{81}=9$

4. $-\sqrt{100}=-10$

5. $\sqrt{5}=2.236$

6. $-\sqrt{121}=-11$

7. $\sqrt{0.04} = 0.2$

8. $\sqrt{0.36} = 0.6$

9. $\sqrt{x^4} = x^2$

10. $\sqrt{25x^4y^6} = 5x^2y^3$

11. $\sqrt{50} = \sqrt{25 \cdot 2} = 5\sqrt{2}$

12. $\sqrt{x^5} = \sqrt{x^4 x} = x^2\sqrt{x}$

13. $\sqrt{36x^3} = \sqrt{36x^2 x} = 6x\sqrt{x}$

14. $\sqrt{8a^4b^3} = \sqrt{4 \cdot 2a^4b^2b}$
$= 2a^2b\sqrt{2b}$

15. $\sqrt{98} + \sqrt{128} = \sqrt{49 \cdot 2} + \sqrt{64 \cdot 2}$
$= 7\sqrt{2} + 8\sqrt{2}$
$= 15\sqrt{2}$

16. $3\sqrt{2} - \sqrt{8} + \sqrt{18} = 3\sqrt{2} - \sqrt{4 \cdot 2} + \sqrt{9 \cdot 2}$
$= 3\sqrt{2} - 2\sqrt{2} + 3\sqrt{2}$
$= 4\sqrt{2}$

17. $2\sqrt{28} + 3\sqrt{63} - \sqrt{49}$
$= 2\sqrt{4 \cdot 7} + 3\sqrt{9 \cdot 7} - \sqrt{49}$
$= 4\sqrt{7} + 9\sqrt{7} - 7$
$= 13\sqrt{7} - 7$

18. $3\sqrt{8} + \sqrt{12} + \sqrt{50} - 4\sqrt{85}$
$= 3\sqrt{4 \cdot 2} + \sqrt{4 \cdot 3} + \sqrt{25 \cdot 2} - 4\sqrt{25 \cdot 3}$
$= 6\sqrt{2} + 2\sqrt{3} + 5\sqrt{2} - 20\sqrt{3}$

19. $5\sqrt{2x} - 7\sqrt{8x} = 5\sqrt{2x} - 7\sqrt{4 \cdot 2x}$
$= 5\sqrt{2x} - 14\sqrt{2x}$
$= -9\sqrt{2x}$

20. $-4x\sqrt{75} - 2x\sqrt{27}$
$= -4x\sqrt{25 \cdot 3} - 2x\sqrt{9 \cdot 3}$
$= -20x\sqrt{3} - 6x\sqrt{3}$
$= -26x\sqrt{3}$

21. $\left(3\sqrt{5}\right)\left(2\sqrt{10}\right) = 6\sqrt{50}$
$= 6\sqrt{25 \cdot 2}$
$= 30\sqrt{2}$

22. $\left(5x\sqrt{7}\right)\left(3x\sqrt{14}\right) = 15x^2\sqrt{98}$
$= 15x^2\sqrt{49 \cdot 2}$
$= 105x^2\sqrt{2}$

23. $\sqrt{2}\left(\sqrt{6} + 3\sqrt{2}\right) = \sqrt{12} + 3\sqrt{4}$
$= \sqrt{4 \cdot 3} + 3\sqrt{4}$
$= 2\sqrt{3} + 6$

24. $\left(\sqrt{7}+4\right)^2 = \sqrt{49}+2\sqrt{7}(4)+16$
$= 7+8\sqrt{7}+16$
$= 23+8\sqrt{7}$

25. $\left(\sqrt{11}-\sqrt{10}\right)\left(\sqrt{11}+\sqrt{10}\right)$
$= \sqrt{121}-\sqrt{100}$
$= 11-10$
$= 1$

26. $\left(2\sqrt{3}-\sqrt{5}\right)^2$
$= 4\sqrt{9}-2\left(2\sqrt{3}\right)\left(\sqrt{5}\right)+\sqrt{25}$
$= 12-4\sqrt{15}+5$
$= 17-4\sqrt{15}$

27. $\left(\sqrt{x}+3\right)\left(\sqrt{x}-4\right)$
$= \sqrt{x^2}-4\sqrt{x}+3\sqrt{x}-12$
$= x-\sqrt{x}-12$

28. $\left(3\sqrt{7}+4\sqrt{3}\right)\left(5\sqrt{7}-2\sqrt{3}\right)$
$= 15\sqrt{49}-6\sqrt{21}+20\sqrt{21}-8\sqrt{9}$
$= 105+14\sqrt{21}-24$

9.5 Exercises

1. $\dfrac{\sqrt{12}}{\sqrt{3}} = \sqrt{\dfrac{12}{3}} = \sqrt{4} = 2$

3. $\dfrac{\sqrt{7}}{\sqrt{63}} = \sqrt{\dfrac{7}{63}} = \sqrt{\dfrac{1}{9}} = \dfrac{1}{3}$

5. $\dfrac{\sqrt{98}}{\sqrt{2}} = \sqrt{\dfrac{98}{2}} = \sqrt{49} = 7$

7. $\dfrac{\sqrt{6}}{\sqrt{x^4}} = \dfrac{\sqrt{6}}{x^2}$

9. $\dfrac{\sqrt{18}}{\sqrt{a^4}} = \dfrac{\sqrt{9\cdot 2}}{\sqrt{\left(a^2\right)^2}} = \dfrac{3\sqrt{2}}{a^2}$

11. $\dfrac{3}{\sqrt{7}} = \dfrac{3}{\sqrt{7}}\cdot\dfrac{\sqrt{7}}{\sqrt{7}} = \dfrac{3\sqrt{7}}{\sqrt{49}} = \dfrac{3\sqrt{7}}{7}$

13. $\dfrac{5}{\sqrt{13}} = \dfrac{5}{\sqrt{13}}\cdot\dfrac{\sqrt{13}}{\sqrt{13}} = \dfrac{5\sqrt{13}}{13}$

15. $\dfrac{x\sqrt{x}}{\sqrt{2}} = \dfrac{x\sqrt{x}}{\sqrt{2}}\cdot\dfrac{\sqrt{2}}{\sqrt{2}}$
$= \dfrac{x\sqrt{2x}}{\sqrt{4}}$
$= \dfrac{x\sqrt{2x}}{x}$

17. $\dfrac{\sqrt{8}}{\sqrt{x}} = \dfrac{\sqrt{4\cdot 2}}{\sqrt{x}}\cdot\dfrac{\sqrt{x}}{\sqrt{x}} = \dfrac{2\sqrt{2x}}{\sqrt{x^2}} = \dfrac{2\sqrt{2x}}{x}$

19. $\dfrac{3}{\sqrt{12}} = \dfrac{3}{\sqrt{4\cdot 3}} = \dfrac{3}{2\sqrt{3}} = \dfrac{3}{2\sqrt{3}}\cdot\dfrac{\sqrt{3}}{\sqrt{3}}$
$= \dfrac{3\sqrt{3}}{2\sqrt{9}} = \dfrac{\sqrt{3}}{2}$

21. $\frac{6}{\sqrt{a}} = \frac{6}{\sqrt{a}} = \frac{\sqrt{a}}{\sqrt{a}} = \frac{6\sqrt{a}}{\sqrt{a^2}} = \frac{6\sqrt{a}}{a}$

23. $\frac{x}{\sqrt{2x^5}} = \frac{x}{\sqrt{2xx^4}} = \frac{x}{x^2\sqrt{2x}}$
$= \frac{1}{x\sqrt{2x}} \cdot \frac{\sqrt{2x}}{\sqrt{2x}}$
$= \frac{\sqrt{2x}}{x\sqrt{4x^2}} = \frac{\sqrt{2x}}{2x^2}$

25. $\frac{\sqrt{18}}{\sqrt{2x^3}} = \sqrt{\frac{18}{2x^3}} = \sqrt{\frac{9}{x^3}} = \frac{3}{x\sqrt{x}} \cdot \frac{\sqrt{x}}{\sqrt{x}}$
$= \frac{3\sqrt{x}}{x^2}$

27. $\sqrt{\frac{3}{5}} = \frac{\sqrt{3}}{\sqrt{5}} \cdot \frac{\sqrt{5}}{\sqrt{5}} = \frac{\sqrt{15}}{\sqrt{25}} = \frac{\sqrt{5}}{5}$

29. $\sqrt{\frac{13}{14}} = \frac{\sqrt{13}}{\sqrt{14}} = \frac{\sqrt{13}}{\sqrt{14}} \cdot \frac{\sqrt{14}}{\sqrt{14}} = \frac{\sqrt{182}}{\sqrt{196}} = \frac{\sqrt{182}}{14}$

31. $\frac{9}{\sqrt{32x}} = \frac{9}{\sqrt{16 \cdot 2x}} = \frac{9}{4\sqrt{2x}} \cdot \frac{\sqrt{2x}}{\sqrt{2x}}$
$= \frac{9\sqrt{2x}}{4\sqrt{4x^2}}$
$= \frac{9\sqrt{2x}}{8x}$

33. $\frac{4}{\sqrt{3}-1} = \frac{4}{\sqrt{3}-1} \cdot \frac{\sqrt{3}+1}{\sqrt{3}+1}$
$= \frac{4\sqrt{3}+4}{\sqrt{9}-\sqrt{1}} = \frac{4\sqrt{3}+4}{3-1}$
$= \frac{4\left(\sqrt{3}+1\right)}{2} = 2\sqrt{3}+2$

35. $\frac{6}{\sqrt{5}+\sqrt{2}} = \frac{6}{\sqrt{5}+\sqrt{2}} \cdot \frac{\sqrt{5}-\sqrt{2}}{\sqrt{5}-\sqrt{2}}$
$= \frac{6\sqrt{5}-6\sqrt{2}}{\sqrt{25}-\sqrt{4}}$
$= \frac{6\sqrt{5}-6\sqrt{2}}{3}$
$= 2\sqrt{5}-2\sqrt{2}$

37. $\frac{\sqrt{6}}{\sqrt{6}-\sqrt{3}} = \frac{\sqrt{6}}{\sqrt{6}-\sqrt{3}} \cdot \frac{\sqrt{6}+\sqrt{3}}{\sqrt{6}+\sqrt{3}}$
$= \frac{\sqrt{36}+\sqrt{18}}{\left(\sqrt{6}\right)^2-\left(\sqrt{3}\right)^2}$
$= \frac{6+\sqrt{9 \cdot 2}}{6-3}$
$= \frac{6+3\sqrt{2}}{3}$
$= 2+\sqrt{2}$

39. $\frac{\sqrt{7}}{\sqrt{8}+\sqrt{7}} = \frac{\sqrt{7}}{\sqrt{8}+\sqrt{7}} \cdot \frac{\sqrt{8}-\sqrt{7}}{\sqrt{8}-\sqrt{7}}$
$= \frac{\sqrt{56}-\sqrt{49}}{\left(\sqrt{8}\right)^2-\left(\sqrt{7}\right)^2}$
$= \frac{\sqrt{4 \cdot 14}-7}{8-7}$
$= 2\sqrt{14}-7$

41. $$\frac{3x}{2\sqrt{2}-\sqrt{5}} = \frac{3x}{2\sqrt{2}-\sqrt{5}} \cdot \frac{2\sqrt{2}+\sqrt{5}}{2\sqrt{2}+\sqrt{5}}$$
$$= \frac{6x\sqrt{2}+3x\sqrt{5}}{\left(2\sqrt{2}\right)^2-\left(\sqrt{5}\right)^2}$$
$$= \frac{3x\left(2\sqrt{2}+\sqrt{5}\right)}{8-5}$$
$$= x\left(2\sqrt{2}+\sqrt{5}\right)$$

43. $$\frac{\sqrt{x}}{\sqrt{6}+\sqrt{2}} = \frac{\sqrt{x}}{\sqrt{6}+\sqrt{2}} \cdot \frac{\sqrt{6}-\sqrt{2}}{\sqrt{6}-\sqrt{2}}$$
$$= \frac{\sqrt{6x}-\sqrt{2x}}{\sqrt{36}-\sqrt{4}}$$
$$= \frac{\sqrt{6x}-\sqrt{2x}}{4}$$

45. $$\frac{\sqrt{5}-\sqrt{2}}{\sqrt{5}+\sqrt{2}} = \frac{\sqrt{5}-\sqrt{2}}{\sqrt{5}+\sqrt{2}} \cdot \frac{\sqrt{5}-\sqrt{2}}{\sqrt{5}-\sqrt{2}}$$
$$= \frac{\sqrt{25}+2\left(-\sqrt{10}\right)+\sqrt{4}}{\left(\sqrt{5}\right)^2-\left(\sqrt{2}\right)^2}$$
$$= \frac{5-2\sqrt{10}+2}{5-2}$$
$$= \frac{7-2\sqrt{10}}{3}$$

47. $$\frac{4\sqrt{7}+3}{\sqrt{5}-\sqrt{2}} = \frac{4\sqrt{7}+3}{\sqrt{5}-\sqrt{2}} \cdot \frac{\sqrt{5}+\sqrt{2}}{\sqrt{5}+\sqrt{2}}$$
$$= \frac{4\sqrt{35}+4\sqrt{14}+3\sqrt{5}+3\sqrt{2}}{\sqrt{25}-\sqrt{4}}$$
$$= \frac{4\sqrt{35}+4\sqrt{14}+3\sqrt{5}+3\sqrt{2}}{3}$$

49. $$\frac{4\sqrt{3}+2}{\sqrt{8}-\sqrt{6}} = \frac{4\sqrt{3}+2}{\sqrt{8}-\sqrt{6}} \cdot \frac{\sqrt{8}+\sqrt{6}}{\sqrt{8}+\sqrt{6}}$$
$$= \frac{4\sqrt{24}+4\sqrt{18}+2\sqrt{8}+2\sqrt{6}}{\left(\sqrt{8}\right)^2-\left(\sqrt{6}\right)^2}$$
$$= \frac{8\sqrt{6}+12\sqrt{2}+4\sqrt{2}+2\sqrt{6}}{8-6}$$
$$= \frac{10\sqrt{6}+16\sqrt{2}}{2} = 5\sqrt{6}+8\sqrt{2}$$

51. $$\frac{x-25}{\sqrt{x}+5} = \frac{x-25}{\sqrt{x}+5} \cdot \frac{\sqrt{x}-5}{\sqrt{x}-5}$$
$$= \frac{(x-25)\left(\sqrt{x}-5\right)}{x-25}$$
$$= \sqrt{x}-5$$

53. (a) $\left(x+\sqrt{2}\right)\left(x-\sqrt{2}\right) = (x)^2-\left(\sqrt{2}\right)^2$
$= x^2-2$

(b) $x^2-2 = (x)^2-\left(\sqrt{2}\right)^2$
$= \left(x+\sqrt{2}\right)\left(x-\sqrt{2}\right)$

(c) $x^2-12 = (x)^2-\left(2\sqrt{3}\right)^2$
$= \left(x+2\sqrt{3}\right)\left(x-2\sqrt{3}\right)$

55. $s = \sqrt{\frac{3V}{h}}$

a. $s = \frac{\sqrt{3V}}{\sqrt{h}} \cdot \frac{\sqrt{h}}{\sqrt{h}}$

$= \frac{\sqrt{3Vh}}{h}$

b. $V = 24,\ h = 8$

$s = \sqrt{\frac{3(24)}{8}} = 3$ feet

57. $x =$ width, $l = \sqrt{5} + 2$

$A = lw$

$3 = \left(\sqrt{5} + 2\right)x$

$x = \frac{3}{\sqrt{5}+2} = \frac{3}{\sqrt{5}+2} \cdot \frac{\sqrt{5}-2}{\sqrt{5}-2}$

$= \frac{3\sqrt{5}-6}{\sqrt{25}-4} = \frac{3\sqrt{5}-6}{5-4}$

$= \left(3\sqrt{5} - 6\right)$ meters

59. $l = \sqrt{5} + 2 = 2.236 + 2 = 4.236$ meters

$w = 3\sqrt{5} - 6 = 3(2.236) - 6 = 0.708$ meters

Cumulative Review

61. $3(2x-6) = 5x - 7(x-2) + 27$

$6x - 18 = 5x - 7x + 14 + 27$

$6x - 18 = -2x + 41$

$8x - 18 = 41$

$8x = 59$

$x = \frac{59}{8} \approx 7.38$

63. Let $x =$ mg per cup in instant coffee

$x + 0.7x = 120$

$1.7x = 120$

$x \approx 70.6$

Approximately 71 mg per cup in the instant coffee.

9.6 Exercises

1. $c^2 = 5^2 + 12^2$

$c^2 = 25 + 144$

$= 169$

$c = \sqrt{169} = 13$

3. $9^2 = 5^2 + b^2$

$81 = 25 + b^2$

$56 = b^2$

$\sqrt{56} = b$

$\sqrt{4 \cdot 14} = b$

$2\sqrt{14} = b$

5. $a = 7,\ b = 7$

$7^2 + 7^2 = c^2$

$49 + 49 = c^2$

$98 = c^2$

$c = \sqrt{98}$

$c = \sqrt{49 \cdot 2}$

$c = 7\sqrt{2}$

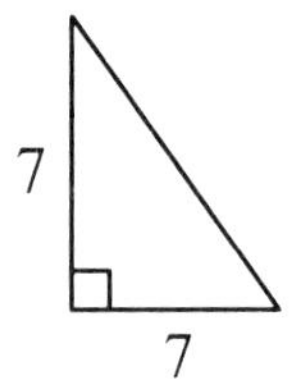

7. $a = \sqrt{14},\ b = 7$

$c^2 = \left(\sqrt{14}\right)^2 + 7^2$

$= 14 + 49$

$= 63$

$c = \sqrt{63} = \sqrt{9 \cdot 7} = 3\sqrt{7}$

9. $c = 20,\ \ b = 18$

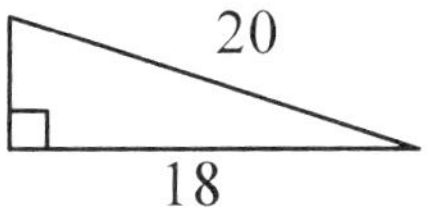

$a^2 + 18^2 = 20^2$

$a^2 + 324 = 400$

$a^2 = 76$

$a = \sqrt{76} = 2\sqrt{19}$

11. $c = \sqrt{82},\ a = 5$

$5^2 + b^2 = \left(\sqrt{82}\right)^2$

$25 + b^2 = 82$

$b^2 = 57$

$b = \sqrt{57}$

13. $c = 12.96,\ \ b = 8.35$

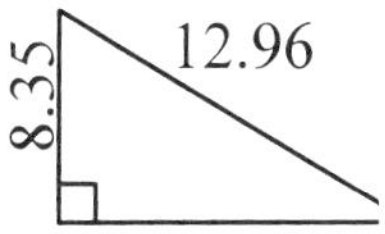

$a^2 + (8.35)^2 = (12.96)^2$

$a^2 + 69.7225 = 167.9616$

$a^2 = 98.2391$

$a \approx 9.91$

15. $c^2 = 8^2 + 5^2$

$c^2 = 64 + 25$

$= 89$

$c = \sqrt{89}$

$c \approx 9.4$ feet

17.

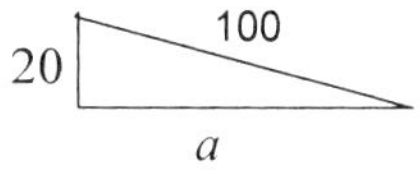

$a^2 + b^2 = c^2$

$a^2 + 20^2 = 100^2$

$a^2 + 400 = 10,000$

$a^2 = 9600$

$a = 98.0$

98 feet

19. Bottom cable

$c^2 = (54)^2 + (130)^2$

$= 2916 + 16,900$

$= 19,816$

$c = \sqrt{19,816} \approx 140.77$ feet

Top cable

$c^2 = (54)^2 + (135)^2$

$= 2916 + 18,225$

$= 21,141$

$c = \sqrt{21,141} \approx 145.40$ feet

21. $\sqrt{x+2} = 3$

$x + 2 = 9$

$x = 7$

Check:

$\sqrt{7+2} \stackrel{?}{=} 3$

$\sqrt{9} \stackrel{?}{=} 3$

$3 = 3$

23. $\sqrt{2x+7}=5$
$$2x+7=5^2$$
$$2x+7=25$$
$$2x=18$$
$$x=9$$

25. $\sqrt{2x+2}=\sqrt{3x-5}$
$$2x+2=3x-5$$
$$2=x-5$$
$$7=x$$

Check: $\sqrt{2(7)+2}\stackrel{?}{=}\sqrt{3(7)-5}$
$$\sqrt{16}\stackrel{?}{=}\sqrt{16}$$
$$4=4$$

27. $\sqrt{2x}-5=4$
$$\sqrt{2x}=9$$
$$2x=9^2$$
$$2x=81$$
$$x=\frac{81}{2}$$

Check: $\sqrt{2\left(\frac{81}{2}\right)}-5\stackrel{?}{=}4$
$$\sqrt{81}-5\stackrel{?}{=}4$$
$$9-5\stackrel{?}{=}4$$
$$4=4$$

29. $\sqrt{2x+15}=x$
$$2x+15=x^2$$
$$x^2-2x-15=0$$
$$(x-5)(x+3)=0$$

Check:
$$\sqrt{2(5)+15}\stackrel{?}{=}5 \qquad \sqrt{2(-3)+15}\stackrel{?}{=}-3$$
$$\sqrt{25}\stackrel{?}{=}5 \qquad \sqrt{9}\stackrel{?}{=}-3$$
$$5=5 \qquad 3\neq-3$$

$x-5$ only

31. $\sqrt{5y+1}=y+1$
$$5y+1=(y+1)^2$$
$$5y+1=y^2+2y+1$$
$$y^2-3y=0$$
$$y(y-3)=0$$

$$y=0,\quad y-3=0$$
$$y=3$$

Check:
$$\sqrt{5(0)+1}\stackrel{?}{=}0+1 \qquad \sqrt{5(3)+1}\stackrel{?}{=}3+1$$
$$\sqrt{1}=1 \qquad \sqrt{16}\stackrel{?}{=}4$$
$$1=1 \qquad 4=4$$

$y=0, 3$

33. $\sqrt{x+3}=3x-1$
$$x+3=(3x-1)^2$$
$$x+3=9x^2-6x+1$$
$$9x^2-7x-2=0$$
$$(9x+2)(x-1)=0$$
$$9x+2=0 \qquad x-1=0$$
$$9x=-2 \qquad x=1$$
$$x=-\frac{2}{9}$$

Check:

$$\sqrt{-\frac{2}{9}+3} \stackrel{?}{=} 3\left(-\frac{2}{9}\right)-1$$

$$\sqrt{\frac{25}{9}} \stackrel{?}{=} -\frac{5}{3}$$

$$\frac{5}{3} \neq -\frac{5}{3} \text{ Doens't check}$$

$$\sqrt{1+3} \stackrel{?}{=} 3(1)-1$$

$$\sqrt{4} \stackrel{?}{=} 2$$

$$2 = 2$$

The only solution is $x = 1$.

35. $\sqrt{3y+1} - y = 1$

$$\sqrt{3y+1} = y+1$$

$$3y+1 = (y+1)^2$$

$$3y+1 = y^2+2y+1$$

$$y^2 - y = 0$$

$$y(y-1) = 0$$

$$y = 0 \qquad y-1 = 0$$

$$y = 1$$

Check:

$$\sqrt{3(0)+1} - 0 \stackrel{?}{=} 1 \qquad \sqrt{3(1)+1} - 1 \stackrel{?}{=} 1$$

$$\sqrt{1} - 0 \stackrel{?}{=} 1 \qquad \sqrt{4} - 1 \stackrel{?}{=} 1$$

$$1 = 1 \qquad 2 - 1 \stackrel{?}{=} 1$$

$$1 = 1$$

$y = 0, 1$

37. $\sqrt{6y+1} - 3y = y$

$$\sqrt{6y+1} = 4y$$

$$6y+1 = 16y^2$$

$$0 = 16y^2 - 6y - 1$$

$$0 = (8y+1)(2y-1)$$

$$8y+1 = 0 \qquad 2y-1 = 0$$

$$y = -\frac{1}{8} \qquad y = \frac{1}{2}$$

Check:

$$\sqrt{6\left(-\frac{1}{8}\right)+1} - 3\left(-\frac{1}{8}\right) \stackrel{?}{=} -\frac{1}{8}$$

$$\sqrt{\frac{1}{4}} + \frac{3}{8} \stackrel{?}{=} -\frac{1}{8}$$

$$\frac{1}{2} + \frac{3}{8} \stackrel{?}{=} -\frac{1}{8}$$

$$\frac{7}{8} \neq -\frac{1}{8}$$

Check:

$$\sqrt{6\left(\frac{1}{2}\right)+1} - 3\left(\frac{1}{2}\right) \stackrel{?}{=} \frac{1}{2}$$

$$\sqrt{4} - \frac{3}{2} \stackrel{?}{=} \frac{1}{2}$$

$$2 - \frac{3}{2} \stackrel{?}{=} \frac{1}{2}$$

$$\frac{1}{2} = \frac{1}{2}$$

$$y = \frac{1}{2} \text{ only}$$

39. $\sqrt{12x+1} = 2\sqrt{6x} - 1$

$$12x+1 = \left(2\sqrt{6x}-1\right)^2$$
$$12x+1 = 4\sqrt{36x^2} - 2\left(2\sqrt{6x}\right) + 1$$
$$12x+1 = 24x - 4\sqrt{6x} + 1$$
$$-12x+1 = -4\sqrt{6x} + 1$$
$$-12x = -4\sqrt{6x}$$
$$3x = \sqrt{6x}$$
$$(3x)^2 = 6x$$
$$9x^2 - 6x = 0$$
$$3x(3x-2) = 0$$

$3x = 0$ $\quad$ $3x - 2 = 0$

$x = 0$ $\quad$ $3x = 2$

$x = \frac{2}{3}$

Check:

$$\sqrt{12(0)+1} \stackrel{?}{=} 2\sqrt{6(0)} - 1$$
$$\sqrt{1} \stackrel{?}{=} -1$$
$$1 \neq -1$$
$$x = \frac{2}{3} \text{ only}$$
$$\sqrt{12\left(\frac{2}{3}\right)+1} \stackrel{?}{=} 2\sqrt{6\left(\frac{2}{3}\right)} - 1$$
$$\sqrt{9} \stackrel{?}{=} 2(2) - 1$$
$$3 = 3$$

Cumulative Review

41. $\frac{5x}{x-4} = 5 + \frac{4x}{x-4}$

$$(x-4)\left(\frac{5x}{x-4}\right) = 5(x-4) + (x-4)\left(\frac{4x}{x-4}\right)$$
$$5x = 5x - 20 + 4x$$
$$5x = 9x - 20$$
$$-4x = -20$$
$$x = 5$$

43. $f(x) = 2x^2 - 3x + 6$

$$f(-2) = 2(-2)^2 - 3(-2) + 6$$
$$= 2(4) + 6 + 6 = 20$$

45. x = number of birds at the start

$$x + 50(25) = 1700$$
$$x + 1250 = 1700$$
$x = 450$ birds at the start

9.7 Exercises

1. $y = kx, \quad y = 9, \quad x = 2$

$$9 = 2k \Rightarrow k = \frac{9}{2}$$
$$y = \frac{9}{2}x, \quad x = 16$$
$$y = \frac{9}{2}(16) = 72$$

3. $y = kx^3, \quad y = 12, \quad x = 2$

$$12 = k(2)^3 \Rightarrow k = \frac{12}{8} = \frac{3}{2}$$

$$y = \frac{3}{2}x^3, \quad x = 7$$

$$y = \frac{3}{2}(7)^3 = \frac{3}{2}(343) = \frac{1029}{2}$$

5. $y = kx^2, \quad y = 900, \quad x = 25$

$$900 = k(25)^2 \Rightarrow k = \frac{900}{625} = \frac{36}{25}$$

$$y = \frac{36}{25}x^2, \quad x = 30$$

$$y = \frac{36}{25}(30)^2 = 1296$$

7. $c = kw, \ c = 200, \ w = 50$

$$200 = k(50)$$

$$4 = k$$

$$c = 4w, \ w = 70$$

$$c = 4(70)$$

$$c = 280 \text{ calories}$$

9. $T = \text{time}, \quad s = \text{length of side}$

$$T = ks^3, \quad T = 7, \quad s = 2.0$$

$$7 = k(2.0)^3 \Rightarrow k = \frac{7}{8}$$

$$T = \frac{7}{8}s^3, \quad s = 4.0$$

$$T = \frac{7}{8}(4.0)^3 = 56 \text{ min.}$$

11. $y = \frac{k}{x}, \quad y = 18, \quad x = 5$

$$18 = \frac{k}{5} \Rightarrow k = 90$$

$$y = \frac{90}{x}, \quad x = 8$$

$$y = \frac{90}{8} = \frac{45}{4}$$

13. $y = \frac{k}{x}, \quad y = \frac{1}{4}, \quad x = 8$

$$\frac{1}{4} = \frac{k}{8} \Rightarrow k = 2$$

$$y = \frac{2}{x}, \quad x = 1$$

$$y = \frac{2}{1} = 2$$

15. $y = \frac{k}{x^2}, \quad y = 30, \quad x = 2$

$$30 = \frac{k}{(2)^2} \Rightarrow k = 120$$

$$y = \frac{120}{x^2}, \quad x = 9$$

$$y = \frac{120}{(9)^2} = \frac{120}{81} = \frac{40}{27}$$

17. $T = \text{time}, F = \text{temperature}$

$$T = \frac{k}{F}$$

$$2.3 = \frac{k}{60}$$

$$138 = k$$

$$T = \frac{138}{F}$$

$$T = \frac{138}{40}$$
$$T = 3.45$$
3.45 minutes

19. W = weight, D = distance from center

$$W = \frac{k}{D^2},\quad W = 1000,\quad D = 4000$$
$$1000 = \frac{k}{(4000)^2} \Rightarrow k = 16,000,000,000$$
$$W = \frac{16,000,000,000}{D^2},\quad D = 6000$$
$$W = \frac{16,000,000,000}{(6000)^2} = \frac{4000}{9}$$
$$= 444\frac{4}{9} \text{ pounds}$$

21. $d = \frac{15}{11}\sqrt{h}$

(a) $h = 121;\ d = \frac{15}{11}\sqrt{121} = 15$ km

$h = 36;\ d = \frac{15}{11}\sqrt{36} \approx 8.2$ km

$h = 4;\ d = \frac{15}{11}\sqrt{4} \approx 2.7$ km

(b) $h = 81;\ d = \frac{15}{11}\sqrt{81} \approx 12.3$ km

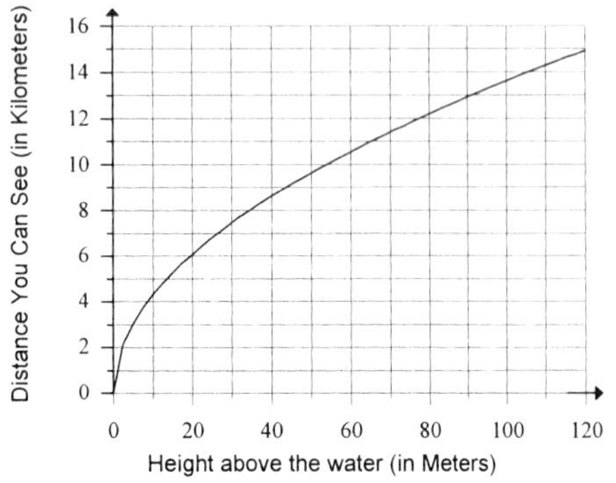

23. $t^2 = kd^3$

a. $1^1 = k(1)^3$
$$1 = k$$

b. $t^2 = 1(1.5)^3$
$$t^2 = 3.375$$
$$t = \sqrt{3.375}$$
$$\approx 1.837 \text{ earth years}$$

Cumulative Review

25. $$\frac{80,000}{320} = \frac{120,000}{320+a}$$
$$\frac{2}{320} = \frac{3}{320+a}$$
$$2(320+a) = 3(320)$$
$$640 + 2a = 960$$
$$2a = 320$$
$$a = 160$$

27. $12x^2 - 20x - 48,$
$$= 4(3x^2 - 5x - 12)$$
Grouping number $= 36$
$$= 4(3x^2 - 9x + 4x - 12)$$
$$= 4[3x(x-3) + 4(x-3)]$$
$$= 4(x-3)(3x+4)$$

29. $x =$ second side

$x + 5 =$ first side

$2x =$ third side

$$x + x + 5 + 2x = 69$$
$$4x + 5 = 69$$
$$4x = 64$$
$$x = 16$$

First side: $x + 5 = 16 + 5 = 21$ feet

Second side: $x = 16$ feet

Third side: $2x = 2(16) = 32$ feet

Putting Your Skills to Work

1. $A = 112(39) = 4368$ square feet

2. $\dfrac{3750}{4} = 937.5$ watts

3. $\dfrac{937.5 \text{ watts}}{4368 \text{ sq. ft}} = 0.21$ watts/sq. ft.

4. $A = 150(39) = 5850$ sq. ft.

$4(5850)(0.21) = 4914$ watts

Increase $= 4914 - 3750 = 1164$ watts

5. $p = k\sqrt{l}$, $l = 130$, $p = 400$

$$400 = k\sqrt{130}$$
$$\frac{400}{\sqrt{130}} = k$$
$$p = \frac{400}{\sqrt{130}}\sqrt{l},\ l = 190$$
$$p = \frac{400}{\sqrt{130}} \cdot \sqrt{190}$$
$$= 400\sqrt{\frac{190}{130}}$$
$$= 484 \text{ watts}$$

6. $p = k\sqrt{w}$, $w = 30$, $p = 4400$

$$4400 = k\sqrt{30}$$
$$\frac{4400}{\sqrt{30}} = k$$
$$p = \frac{4400}{\sqrt{30}}\sqrt{w},\ w = 50$$
$$p = 4400 \cdot \sqrt{\frac{50}{30}}$$
$$= 5680 \text{ watts}$$

Chapter 9 Review Problems

1. $\sqrt{36} = 6$

2. $\sqrt{25} = 5$

3. $\sqrt{169} = 13$

4. $\sqrt{196} = 14$

5. $-\sqrt{81} = -9$

6. $\sqrt{64} = 8$

7. $\sqrt{49} = 7$

8. $-\sqrt{121} = -11$

9. $-\sqrt{144} = -12$

10. $\sqrt{289} = 17$

11. $\sqrt{0.04} = 0.2$

12. $\sqrt{0.49} = 0.7$

13. $\sqrt{\frac{1}{25}} = \frac{1}{5}$

14. $\sqrt{\frac{36}{49}} = \frac{6}{7}$

15. $\sqrt{105} \approx 10.247$

16. $\sqrt{198} \approx 14.071$

17. $\sqrt{77} = 8.775$

18. $\sqrt{88} \approx 9.381$

19. $\sqrt{27} = \sqrt{9 \cdot 3} = 3\sqrt{3}$

20. $\sqrt{52} = \sqrt{4 \cdot 13} = 2\sqrt{13}$

21. $\sqrt{28} = \sqrt{4 \cdot 7} = 2\sqrt{7}$

22. $\sqrt{125} = \sqrt{25 \cdot 5} = 5\sqrt{5}$

23. $\sqrt{40} = \sqrt{4 \cdot 10} = 2\sqrt{10}$

24. $\sqrt{80} = \sqrt{16 \cdot 5} = 4\sqrt{5}$

25. $\sqrt{x^8} = \sqrt{\left(x^4\right)^2} = x^4$

26. $\sqrt{y^{10}} = \sqrt{\left(y^5\right)^2} = y^5$

27. $\sqrt{x^5 y^6} = \sqrt{x^4 x y^6} = x^2 y^3 \sqrt{x}$

28. $\sqrt{a^3 b^4} = \sqrt{a^2 \cdot a \cdot \left(b^2\right)^2} = ab^2\sqrt{a}$

29. $\sqrt{16x^3 y^5} = \sqrt{16x^2 x y^4 y} = 4xy^2\sqrt{xy}$

30. $\sqrt{98x^4 y^6} = \sqrt{49 \cdot 2 \cdot \left(x^2\right)^2 \left(y^3\right)^2}$
$= 7x^2 y^3 \sqrt{2}$

31. $\sqrt{12x^5} = \sqrt{4 \cdot 3x^4 x} = 2x^2\sqrt{3x}$

32. $\sqrt{27x^7} = \sqrt{9 \cdot 3(x^3)^2 x} = 3x^3\sqrt{3x}$

33. $\sqrt{75x^{10}} = \sqrt{25 \cdot 3x^{10}} = 5x^5\sqrt{3}$

34. $\sqrt{125x^{12}} = \sqrt{25 \cdot 5(x^6)^2} = 5x^6\sqrt{5}$

35. $\sqrt{120a^3b^4c^5} = \sqrt{4 \cdot 30a^2ab^4c^4c}$
$= 2ab^2c^2\sqrt{30ac}$

36. $\sqrt{121a^6b^4c} = \sqrt{11^2(a^3)^2(b^2)^2 c}$
$= 11a^3b^2\sqrt{c}$

37. $\sqrt{56x^7y^9} = \sqrt{4 \cdot 14x^6xy^8y}$
$= 2x^3y^4\sqrt{14xy}$

38. $\sqrt{99x^{13}y^7} = \sqrt{9 \cdot 11(x^6)^2 x(y^3)^2 y}$
$= 3x^6y^3\sqrt{11xy}$

39. $\sqrt{63} - \sqrt{7} + \sqrt{28} = \sqrt{9 \cdot 7} - \sqrt{7} + \sqrt{4 \cdot 7}$
$= 3\sqrt{7} - \sqrt{7} + 2\sqrt{7}$
$= 4\sqrt{7}$

40. $5\sqrt{6} - \sqrt{24} + 2\sqrt{54}$
$= 5\sqrt{6} - \sqrt{4 \cdot 6} + 2\sqrt{9 \cdot 6}$
$= 5\sqrt{6} - 2\sqrt{6} + 6\sqrt{6}$
$= 9\sqrt{6}$

41. $x\sqrt{3} + 3x\sqrt{3} + \sqrt{27x^2} = 4x\sqrt{3} + \sqrt{9 \cdot 3x^2}$
$= 4x\sqrt{3} + 3x\sqrt{3}$
$= 7x\sqrt{3}$

42. $a\sqrt{2} + \sqrt{12a^2} + a\sqrt{98}$
$= a\sqrt{2} + 2a\sqrt{3} + 7a\sqrt{2}$
$= 8a\sqrt{2} + 2a\sqrt{3}$

43. $5\sqrt{5} - 6\sqrt{20} + 2\sqrt{10}$
$= 5\sqrt{5} - 6\sqrt{4 \cdot 5} + 2\sqrt{10}$
$= 5\sqrt{5} - 12\sqrt{5} + 2\sqrt{10}$
$= -7\sqrt{5} + 2\sqrt{10}$

44. $3\sqrt{6} - 5\sqrt{18} + 3\sqrt{24}$
$= 3\sqrt{6} - 15\sqrt{2} + 6\sqrt{6}$
$= 9\sqrt{6} - 15\sqrt{2}$

45. $(2\sqrt{x})(3\sqrt{x^3}) = 6\sqrt{x^4} = 6x^2$

46. $(-5\sqrt{a})(2\sqrt{ab}) = -10\sqrt{a^2b} = -10a\sqrt{b}$

47. $(\sqrt{2a^3})(\sqrt{8b^2}) = \sqrt{16a^2ab^2}$
$= 4ab\sqrt{a}$

48. $(5x\sqrt{x})(-3x^2\sqrt{x}) = -15x^3(x) = -15x^4$

49. $\sqrt{7}\left(\sqrt{28}-3\sqrt{7}\right)=\sqrt{196}-3\sqrt{49}$
$=14-21$
$=-7$

50. $\sqrt{8}\left(5\sqrt{2}-3\sqrt{8}\right)=5\sqrt{16}-3\sqrt{64}$
$=20-24$
$=-4$

51. $\sqrt{2}\left(\sqrt{5}-\sqrt{3}-2\sqrt{2}\right)=\sqrt{10}-\sqrt{6}-2\sqrt{4}$
$=\sqrt{10}-\sqrt{6}-4$

52. $\sqrt{5}\left(\sqrt{6}-2\sqrt{5}+\sqrt{10}\right)$
$=\sqrt{30}-2(5)+\sqrt{50}$
$=\sqrt{30}-10+5\sqrt{2}$

53. $\left(\sqrt{11}+2\right)\left(2\sqrt{11}-1\right)$
$=2\sqrt{121}-\sqrt{11}+4\sqrt{11}-2$
$=22+3\sqrt{11}-2$
$=20+3\sqrt{11}$

54. $\left(\sqrt{10}+3\right)\left(3\sqrt{10}-1\right)$
$=3(10)-\sqrt{10}+9\sqrt{10}-3$
$=30+8\sqrt{10}-3=27+8\sqrt{10}$

55. $\left(2+3\sqrt{6}\right)\left(4-2\sqrt{3}\right)$
$=8-4\sqrt{3}+12\sqrt{6}-6\sqrt{18}$
$=8-4\sqrt{3}+12\sqrt{6}-18\sqrt{2}$

56. $\left(5-\sqrt{2}\right)\left(3-\sqrt{12}\right)$
$=\left(5-\sqrt{2}\right)\left(3-2\sqrt{3}\right)$
$=15-10\sqrt{3}-3\sqrt{2}+2\sqrt{6}$

57. $\left(2\sqrt{3}+3\sqrt{6}\right)^2=4\sqrt{9}+2(6)\sqrt{18}+9\sqrt{36}$
$=12+36\sqrt{2}+54$
$=66+36\sqrt{2}$

58. $\left(5\sqrt{2}-2\sqrt{6}\right)^2$
$=\left(5\sqrt{2}\right)^2-2\left(5\sqrt{2}\right)\left(2\sqrt{6}\right)+\left(2\sqrt{6}\right)^2$
$=25(2)-20\sqrt{12}+4(6)$
$=50-40\sqrt{3}+24$
$=74-40\sqrt{3}$

59. $\dfrac{1}{\sqrt{3x}}=\dfrac{1}{\sqrt{3x}}\cdot\dfrac{\sqrt{3x}}{\sqrt{3x}}=\dfrac{\sqrt{3x}}{3x}$

60. $\dfrac{2y}{\sqrt{5}}=\dfrac{2y}{\sqrt{5}}\cdot\dfrac{\sqrt{5}}{\sqrt{5}}=\dfrac{2y\sqrt{5}}{5}$

61. $\dfrac{x^2y}{\sqrt{8}}=\dfrac{x^2y}{2\sqrt{2}}=\dfrac{x^2y}{2\sqrt{2}}\cdot\dfrac{\sqrt{2}}{\sqrt{2}}=\dfrac{x^2y\sqrt{2}}{4}$

62. $\dfrac{3ab}{\sqrt{2b}}=\dfrac{3ab}{\sqrt{2b}}\cdot\dfrac{\sqrt{2b}}{\sqrt{2b}}=\dfrac{3ab\sqrt{2b}}{2b}=\dfrac{3a\sqrt{2b}}{2}$

63. $\sqrt{\dfrac{5}{6}}=\dfrac{\sqrt{5}}{\sqrt{6}}\cdot\dfrac{\sqrt{6}}{\sqrt{6}}=\dfrac{\sqrt{30}}{\sqrt{36}}=\dfrac{\sqrt{30}}{6}$

64. $\sqrt{\frac{6}{11}} = \frac{\sqrt{6}}{\sqrt{11}} \cdot \frac{\sqrt{11}}{\sqrt{11}} = \frac{\sqrt{66}}{\sqrt{121}} = \frac{\sqrt{66}}{11}$

65. $\frac{\sqrt{a^5}}{\sqrt{2a}} = \frac{\sqrt{a^5}}{\sqrt{2a}} \cdot \frac{\sqrt{2a}}{\sqrt{2a}} = \frac{\sqrt{2a^6}}{2a} = \frac{a^3\sqrt{2}}{2a}$
$= \frac{a^2\sqrt{2}}{2}$

66. $\frac{\sqrt{x^3}}{\sqrt{3x}} = \sqrt{\frac{x^3}{3x}} = \sqrt{\frac{x^2}{3}} = \frac{\sqrt{x^2}}{\sqrt{3}} = \frac{x}{\sqrt{3}}$
$= \frac{x}{\sqrt{3}} \cdot \frac{\sqrt{3}}{\sqrt{3}} = \frac{x\sqrt{3}}{3}$

67. $\frac{3}{\sqrt{5}+\sqrt{2}} = \frac{3}{\sqrt{5}+\sqrt{2}} \cdot \frac{\sqrt{5}-\sqrt{2}}{\sqrt{5}-\sqrt{2}}$
$= \frac{3\sqrt{5}-3\sqrt{2}}{\left(\sqrt{5}\right)^2-\left(\sqrt{2}\right)^2}$
$= \frac{3\sqrt{5}-3\sqrt{2}}{3}$
$= \sqrt{5}-\sqrt{2}$

68. $\frac{2}{\sqrt{6}-\sqrt{3}} = \frac{2}{\sqrt{6}-\sqrt{3}} \cdot \frac{\sqrt{6}+\sqrt{3}}{\sqrt{6}+\sqrt{3}}$
$= \frac{2\left(\sqrt{6}+\sqrt{3}\right)}{6-3} = \frac{2\left(\sqrt{6}+\sqrt{3}\right)}{3}$

69. $\frac{1-\sqrt{5}}{2+\sqrt{5}} = \frac{1-\sqrt{5}}{2+\sqrt{5}} \cdot \frac{2-\sqrt{5}}{2-\sqrt{5}}$
$= \frac{2-3\sqrt{5}+5}{2^2-\left(\sqrt{5}\right)^2}$
$= \frac{7-3\sqrt{5}}{-1}$
$= -7+3\sqrt{5}$

70. $\frac{1-\sqrt{3}}{3+\sqrt{3}} = \frac{1-\sqrt{3}}{3+\sqrt{3}} \cdot \frac{3-\sqrt{3}}{3-\sqrt{3}} = \frac{3-4\sqrt{3}+3}{9-3}$
$= \frac{6-4\sqrt{3}}{6} = \frac{2\left(3-2\sqrt{3}\right)}{6} = \frac{3-2\sqrt{3}}{3}$

71. $c^2 = a^2 + b^2, \quad a = 5, \quad b = 8$
$c^2 = 5^2 + 8^2$
$c = \sqrt{25+64}$
$c = \sqrt{89}$

72. $a^2 + b^2 = c^2$
$a^2 + 3^2 = \left(\sqrt{11}\right)^2$
$a^2 + 9 = 11$
$a^2 = 2$
$a = \sqrt{2}$

73. $c^2 = a^2 + b^2, \quad c = 5, \quad a = 3.5$

$$5^2 = (3.5)^2 + b^2$$
$$25 = 12.25 + b^2$$
$$12.75 = b^2$$
$$\sqrt{12\frac{3}{4}} = b$$
$$\sqrt{\frac{51}{4}} = b$$
$$\frac{\sqrt{51}}{2} = b$$

74. $a^2 + b^2 = c^2$

$$(2.400)^2 + (2.000)^2 = c^2$$
$$5.760 + 4.000 = c^2$$
$$9.760 = c^2$$
$$3.124 = c$$

75. $x^2 = 18^2 + 24^2$

$$x^2 = 324 + 576$$
$$x^2 = 900$$
$$x = 30$$

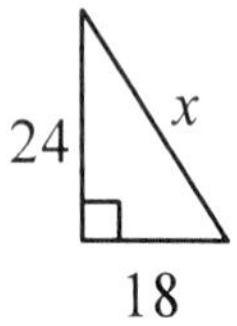

It is 30 meters from his feet to the top of the pole.

76. $a^2 + b^2 = c^2$

$$(20.0)^2 + b^2 = (50.0)^2$$
$$400.0 + b^2 = 2500.0$$
$$b^2 = 2100.0$$
$$b \approx 45.8$$

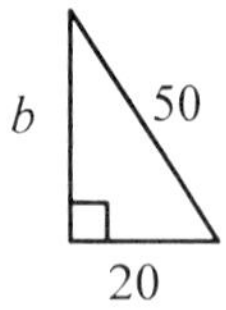

It is 45.8 miles

77. x = length of a side

$$x^2 + x^2 = \left(\sqrt{455}\right)^2$$
$$2x^2 = 455$$
$$x^2 = 227.5$$
$$x - 15.1 \text{ feet}$$

78. $\sqrt{x-4} = 5$

$$x - 4 = 25$$
$$x = 29$$

Check:

$$\sqrt{29-4} \stackrel{?}{=} 5$$
$$\sqrt{25} = 5$$
$$5 = 5$$

79. $\sqrt{3x+1} = 8$

$$3x + 1 = 64$$
$$3x = 63$$
$$x = 21$$

Check:

$$\sqrt{3(21)+1} \stackrel{?}{=} 8$$
$$\sqrt{64} \stackrel{?}{=} 8$$
$$8 = 8$$

80. $\sqrt{1-3x} = \sqrt{5+x}$

$\left(\sqrt{1-3x}\right)^2 = \left(\sqrt{5+x}\right)^2$

$1-3x = 5+x$

$1 = 5+4x$

$-4 = 4x$

$-1 = x$

Check:

$\sqrt{1-3(-1)} \stackrel{?}{=} \sqrt{5+(-1)}$

$\sqrt{1+3} = \sqrt{4}$

$x = -1$

81. $\sqrt{-5+2x} = \sqrt{1+x}$

$-5+2x = 1+x$

$x = 6$

Check:

$\sqrt{-5+2(6)} \stackrel{?}{=} \sqrt{1+6}$

$\sqrt{-5+12} \stackrel{?}{=} \sqrt{7}$

$\sqrt{7} = \sqrt{7}$

$x = 6$

82. $\sqrt{10x+9} = -1+2x$

$\left(\sqrt{10x+9}\right)^2 = (-1+2x)^2$

$10x+9 = 1-4x+4x^2$

$0 = 4x^2-14x-8$

$0 = 2\left(2x^2-7x-4\right)$

$0 = 2(2x+1)(x-4)$

$2x+1=0 \qquad x-4=0$

$x = -\frac{1}{2} \qquad x = 4$

Check: $x = -\frac{1}{2}$

$\sqrt{10\left(-\frac{1}{2}\right)+9} \stackrel{?}{=} -1+2\left(-\frac{1}{2}\right)$

$\sqrt{4} \stackrel{?}{=} -2$

$2 \neq -2$

Check: $x = 4$

$\sqrt{10(4)+9} \stackrel{?}{=} -1+2(4)$

$\sqrt{49} \stackrel{?}{=} 7$

$7 = 7$

$x = 4$

83. $\sqrt{2x-5} = 10-x$

$2x-5 = (10-x)^2$

$2x-5 = 100-20x+x^2$

$0 = x^2-22x+105$

$0 = (x-7)(x-15)$

$x-7=0 \qquad x-15=0$

$x=7 \qquad x=15$

Check: $x = 7$

$\sqrt{2(7)-5} \stackrel{?}{=} 10-7$

$\sqrt{9} \stackrel{?}{=} 3$

$3 = 3$

Check: $x = 15$

$\sqrt{2(15)-5} \stackrel{?}{=} 10-5$

$\sqrt{25} \stackrel{?}{=} -5$

$5 \neq -5$

$x = 7$ only

84. $6-\sqrt{5x-1}=x+1$

$-\sqrt{5x-1}=x-5$

$\left(-\sqrt{5x-1}\right)^2=(x-5)^2$

$5x-1=x^2-10x+25$

$0=x^2-15x+26$

$0=(x-13)(x-2)$

$x-13=0 \qquad x-2=0$

$x=13 \qquad x=2$

Check: $x=13$

$6-\sqrt{5(13)-1}\stackrel{?}{=}13+1$

$6-\sqrt{64}\stackrel{?}{=}14$

$6-8\neq 14$

Check: $x=2$

$6-\sqrt{5(2)-1}\stackrel{?}{=}2+1$

$6-\sqrt{9}\stackrel{?}{=}3$

$6-3=3$

$x=2$

85. $4x+\sqrt{x+2}=5x-4$

$\sqrt{x+2}=x-4$

$x+2=(x-4)^2$

$x+2=x^2-8x+16$

$0=x^2-9x+14$

$0=(x-2)(x-7)$

$x-2=0 \qquad x-7=0$

$x=2 \qquad x=7$

Check: $x=7$

$4(7)+\sqrt{7+2}\stackrel{?}{=}5(7)-4$

$28+\sqrt{9}\stackrel{?}{=}31$

$31=31$

$x=7$ only

86. $y=k\sqrt{x}$

$35=k\sqrt{25}$

$35=5k$

$7=k$

$y=7\sqrt{x}=7\sqrt{121}=7(11)=77$

87. $y=\frac{k}{x^2},\quad y=\frac{6}{5},\quad x=5$

$\frac{6}{5}=\frac{k}{5^2}\Rightarrow k=30$

$y=\frac{30}{x^2},\quad x=15$

$y=\frac{30}{(15)^2}=\frac{2}{15}$

88. $y=\frac{k}{x^3}$

$4=\frac{k}{2^3}$

$4=\frac{k}{8}$

$32=k$

$y=\frac{32}{x^3}=\frac{32}{4^3}=\frac{32}{64}=\frac{1}{2}$

89. $I = \frac{k}{d^2}, \quad I = I_1, \quad d = 1$

$I_1 = \frac{k}{1} \Rightarrow k = I_1$

$I = \frac{I_1}{d^2}, \quad d = 4$

$I_4 = \frac{I_1}{4^2} = \frac{I_1}{16}$

$\frac{I_4}{I_1} = \frac{\frac{I_1}{16}}{I_1} = \frac{1}{16}$

The intensity is $\frac{1}{16}$ as great.

90. l = length of skid mark, s = speed of car

$l = ks^2$

$40 = k(30)^2$

$40 = 900k$

$\frac{2}{45} = k$

$l = \frac{2}{45}s^2 = \frac{2}{45}(55)^2 = 134$

134 feet

91. H = horsepower, s = speed

$H = ks^3, \quad s = s_m, \quad H = H_m$

$H_1 = k(s_m)^3 \Rightarrow k = \frac{H_1}{(s_m)^3}$

$H_1 = \frac{H_m}{s_m^3}s^3, \quad s = 2s_m$

$H = \frac{H_m}{s_m^3}(25_m)^3 = 8H_m$

The horsepower is 8 times as much.

92. $\sqrt{\frac{36}{121}} = \frac{6}{11}$

93. $\sqrt{0.0004} = 0.02$

94. $\sqrt{98x^6y^3} = \sqrt{49 \cdot 2x^6y^2y}$

$= 7x^3y\sqrt{2y}$

95. $3\sqrt{27} - 2\sqrt{75} + \sqrt{48}$

$= 3\sqrt{9 \cdot 3} - 2\sqrt{25 \cdot 3} + \sqrt{16 \cdot 3}$

$= 9\sqrt{3} - 10\sqrt{3} + 4\sqrt{3}$

$= 3\sqrt{3}$

96. $(2\sqrt{3} + \sqrt{5})(3\sqrt{3} - \sqrt{10})$

$= 6\sqrt{9} - 2\sqrt{30} + 3\sqrt{15} - \sqrt{50}$

$= 18 - 2\sqrt{30} + 3\sqrt{15} - \sqrt{25 \cdot 2}$

$= 18 - 2\sqrt{30} + 3\sqrt{15} - 5\sqrt{2}$

97. $(3\sqrt{5} + 2)^2 = 9\sqrt{25} + 2(3\sqrt{5})(2) + 4$

$= 45 + 12\sqrt{5} + 4$

$= 49 + 12\sqrt{5}$

98. $\frac{5}{\sqrt{12}} = \frac{5}{\sqrt{4 \cdot 3}} = \frac{5}{2\sqrt{3}} = \frac{5}{2\sqrt{3}} \cdot \frac{\sqrt{3}}{\sqrt{3}} = \frac{5\sqrt{3}}{6}$

99. $\frac{\sqrt{3}+\sqrt{6}}{2\sqrt{3}+\sqrt{2}}=\frac{\sqrt{3}+\sqrt{6}}{2\sqrt{3}+\sqrt{2}}\cdot\frac{2\sqrt{3}-\sqrt{2}}{2\sqrt{3}-\sqrt{2}}$

$$=\frac{2\sqrt{9}-\sqrt{6}+2\sqrt{18}-\sqrt{12}}{4\sqrt{9}-\sqrt{4}}$$
$$=\frac{6-\sqrt{6}+2\sqrt{9\cdot 2}-\sqrt{4\cdot 3}}{12-2}$$
$$=\frac{6-\sqrt{6}+6\sqrt{2}-2\sqrt{3}}{10}$$

100. $\sqrt{2x-3}=9$

$$2x-3=81$$
$$2x=84$$
$$x=42$$

101. $\sqrt{10x+5}=2x+1$

$$10x+5=4x^2+4x+1$$
$$0=4x^2-6x-4$$
$$0=2(2x^2-3x-2)$$
$$0=2(2x+1)(x-2)$$

$2x+1=0$ $\quad x-2=0$

$2x=-1$ $\quad x=2$

$x=-\frac{1}{2}$

How Am I Doing? Chapter 9 Test

1. $\sqrt{121}=11$

2. $\sqrt{\frac{9}{100}}=\frac{3}{10}$

3. $\sqrt{48x^2y^7}=\sqrt{3\cdot 16x^2y^6y}$

$$=4xy^3\sqrt{3y}$$

4. $\sqrt{100x^3yz^4}=\sqrt{100\cdot x^2\cdot xy(z^2)^2}$

$$=10xz^2\sqrt{xy}$$

5. $8\sqrt{3}+5\sqrt{27}-5\sqrt{48}=8\sqrt{3}+5\sqrt{9\cdot 3}-5\sqrt{16\cdot 3}$

$$=8\sqrt{3}+15\sqrt{3}-20\sqrt{3}$$
$$=3\sqrt{3}$$

6. $\sqrt{4a}+\sqrt{8a}+\sqrt{36a}+\sqrt{18a}$

$$=2\sqrt{a}+2\sqrt{2a}+6\sqrt{a}+3\sqrt{2a}$$
$$=8\sqrt{a}+5\sqrt{2a}$$

7. $(2\sqrt{a})(3\sqrt{b})(2\sqrt{ab})=12\sqrt{a^2b^2}$

$$=12ab$$

8. $\sqrt{5}(\sqrt{10}+2\sqrt{3}-3\sqrt{5})$

$$=\sqrt{50}+2\sqrt{15}-3\sqrt{25}$$
$$=\sqrt{25\cdot 2}+2\sqrt{15}-15$$
$$=5\sqrt{2}+2\sqrt{15}-15$$

9. $(2\sqrt{3})^2=4\sqrt{9}+2(2\sqrt{3})(5)+25$

$$=12+20\sqrt{3}+25$$
$$=37+20\sqrt{3}$$

10. $\left(4\sqrt{2}-\sqrt{5}\right)\left(3\sqrt{2}+\sqrt{5}\right)$
$=12(2)+4\sqrt{10}-3\sqrt{10}-5$
$=24+\sqrt{10}-5=19+\sqrt{10}$

11. $\sqrt{\dfrac{x}{5}}=\dfrac{\sqrt{x}}{\sqrt{5}}=\dfrac{\sqrt{x}}{\sqrt{5}}\cdot\dfrac{\sqrt{5}}{\sqrt{5}}=\dfrac{\sqrt{5x}}{5}$

12. $\dfrac{3}{\sqrt{12}}=\dfrac{3}{2\sqrt{3}}=\dfrac{3}{2\sqrt{3}}\cdot\dfrac{\sqrt{3}}{\sqrt{3}}=\dfrac{3\sqrt{3}}{2(3)}=\dfrac{\sqrt{3}}{2}$

13. $\dfrac{\sqrt{3}+4}{5+\sqrt{3}}=\dfrac{\sqrt{3}+4}{5+\sqrt{3}}\cdot\dfrac{5-\sqrt{3}}{5-\sqrt{3}}$
$=\dfrac{5\sqrt{3}-\sqrt{9}+20-4\sqrt{3}}{25-\sqrt{9}}$
$=\dfrac{\sqrt{3}-3+20}{25-3}$
$=\dfrac{17+\sqrt{3}}{22}$

14. $\dfrac{3a}{\sqrt{5}+\sqrt{2}}=\dfrac{3a}{\sqrt{5}+\sqrt{2}}\cdot\dfrac{\sqrt{5}-\sqrt{2}}{\sqrt{5}-\sqrt{2}}$
$=\dfrac{3a\left(\sqrt{5}-\sqrt{2}\right)}{5-2}=\dfrac{3a\left(\sqrt{5}-\sqrt{2}\right)}{3}$
$=a\left(\sqrt{5}-\sqrt{2}\right)$

15. $\sqrt{156}=12.49$

16. $c^2=a^2+b^2$
$13^2=6^2+x^2$
$169=36+x^2$
$133=x^2$
$\sqrt{133}=x$

17. $x^2=\left(3\sqrt{2}\right)^2+3^2=9\sqrt{4}+9=27$
$x=\sqrt{27}=\sqrt{9\cdot 3}=3\sqrt{3}$

18. $6-\sqrt{2x+1}=0$
$6=\sqrt{2x+1}$
$6^2=\left(\sqrt{2x+1}\right)^2$
$36=2x+1$
$35=2x$
$\dfrac{35}{2}=x$

Check:
$6-\sqrt{2\left(\dfrac{35}{2}\right)+1}\stackrel{?}{=}0$
$6-\sqrt{36}\stackrel{?}{=}0$
$6-6=0$
$x=\dfrac{35}{2}$

19. $x=5+\sqrt{x+7}$
$x-5=\sqrt{x+7}$
$x^2-10x+25=x+7$
$x^2-11x+18=0$
$(x-2)(x-9)=0$

$x - 2 = 0 \qquad x - 9 = 0$

$x = 2 \qquad x = 9$

2 is an extraneous solution. $x = 9$

20. i = illumination, d = distance

$i = \frac{k}{d^2}$

$i = \frac{k}{8^2}$

$64i = k$

$i = \frac{64i}{d^2}$

$\frac{1}{4}i = \frac{64i}{d^2}$

$d^2 = 256$

$d = 16$

16 inches

21. $c = ks$, $c = 23.40$, $s = 780$

$23.40 = k(780)$

$0.03 = k$

$c = 0.03s$, $s = 2859$

$c = 0.03(2859)$

$c = \$85.77$

22. A = area, p = perimeter

$A = kp^2$

$6.93 = k(12)^2$

$6.93 = 144k$

$0.048125 = k$

$A = 0.048125p^2$

$= 0.048125(21)^2 = 21.2 \text{ cm}^2$

Cumulative Test Chapters 0 - 9

1. $\frac{28}{42} = \frac{2 \cdot 2 \cdot 7}{2 \cdot 3 \cdot 7} = \frac{2}{3}$

2. $3\frac{1}{4} + 5\frac{2}{3} = 3\frac{3}{12} + 5\frac{8}{12} = 8\frac{11}{12}$

3. $6\frac{1}{4} \div 6\frac{2}{3} = \frac{25}{4} \div \frac{20}{3}$

$= \frac{25}{4} \cdot \frac{3}{20}$

$= \frac{5(5)}{4} \cdot \frac{3}{4(5)}$

$= \frac{15}{16}$

4. $0.07\% = 0.0007$

5. $-11 - 16 + 8 + 4 - 13 + 31$

$= -27 + 8 + 4 - 13 + 31$

$= -19 + 4 - 13 + 31$

$= -15 - 13 + 31$

$= -28 + 31$

$= 3$

6. $(-3)^2 \cdot 4 - 8 \div 2 - (3 - 2)^3$

$= 9 \cdot 4 - 8 \div 2 - 1^3$

$= 9 \cdot 4 - 8 \div 2 - 1$

$= 36 - 4 - 1 = 31$

7. $4a^2bc^3 - 2abc^2 - 7a^2bc^3 = -3a^2bc^3 - 2abc^2$

8. $(x+3)(x-2)(x+2) = (x+3)(x^2-4)$
$= x^3 - 4x + 3x^2 - 12$
$= x^3 + 3x^2 - 4x - 12$

9. $(4x-5)^2 = 16x^2 - 2(4x)(5) + 25$
$= 16x^2 - 40x + 25$

10. $(3x-11)(3x+11) = (3x)^2 - 11^2$
$= 9x^2 - 121$

11. $\frac{4x}{x+2} + \frac{4}{x+2} = \frac{4x+4}{x+2}$

12. $\frac{x^2-5x+6}{x^2-5x-6} \div \frac{x^2-4}{x^2+2x+1}$
$= \frac{x^2-5x+6}{x^2-5x-6} \cdot \frac{x^2+2x+1}{x^2-4}$
$= \frac{(x-2)(x-3)}{(x-6)(x+1)} \cdot \frac{(x+1)^2}{(x-2)(x+2)}$
$= \frac{(x-3)(x+1)}{(x-6)(x+2)}$

13. $\sqrt{98x^5y^6} = \sqrt{49 \cdot 2x^4xy^6}$
$= 7x^2y^3\sqrt{2x}$

14. $\sqrt{50} - \sqrt{98} + \sqrt{162}$
$= \sqrt{25 \cdot 2} - \sqrt{49 \cdot 2} + \sqrt{81 \cdot 2}$
$= 5\sqrt{2} - 7\sqrt{2} + 9\sqrt{2}$
$= 7\sqrt{2}$

15. $(\sqrt{6} - \sqrt{3})^2 = \sqrt{36} + 2(-\sqrt{3})(\sqrt{6}) + \sqrt{9}$
$= 6 - 2\sqrt{18} + 3$
$= 9 - 2\sqrt{9 \cdot 2}$
$= 9 - 6\sqrt{2}$

16. $\frac{\sqrt{3}+\sqrt{2}}{\sqrt{3}-\sqrt{2}} = \frac{\sqrt{3}+\sqrt{2}}{\sqrt{3}-\sqrt{2}} \cdot \frac{\sqrt{3}+\sqrt{2}}{\sqrt{3}+\sqrt{2}}$
$= \frac{3+2\sqrt{6}}{3-2} = 5 + 2\sqrt{6}$

17. $(3\sqrt{6} - \sqrt{2})(\sqrt{6} + 4\sqrt{2})$
$= 3\sqrt{36} + 12\sqrt{12} - \sqrt{12} - 4\sqrt{4}$
$= 18 + 11\sqrt{12} - 8$
$= 10 + 11\sqrt{4 \cdot 3}$
$= 10 + 22\sqrt{3}$

18. $\frac{1}{3-\sqrt{2}} = \frac{1}{3-\sqrt{2}} \cdot \frac{3+\sqrt{2}}{3+\sqrt{2}}$
$= \frac{3+\sqrt{2}}{9-2} = \frac{3+\sqrt{2}}{7}$

19. $-\sqrt{16} = -4$

20. Not a real number

21. $c^2 = a^2 + b^2,\ a = 4, b = \sqrt{7}$
$c^2 = 4^2 + (\sqrt{7})^2 = 16 + 7 = 23$
$c = \sqrt{23}$

22. $a^2 + b^2 = c^2$

$19^2 + b^2 = 21^2$

$361 + b^2 = 441$

$b^2 = 80$

$b = 4\sqrt{5}$

23. $\sqrt{4x+5} = x$

$4x + 5 = x^2$

$0 = x^2 - 4x - 5$

$0 = (x-5)(x+1)$

$x - 5 = 0 \qquad x + 1 = 0$

$x = 5 \qquad x = -1$

-1 is an extraneous solution. $x = 5$

24. $\sqrt{3y-2} + 2 = y$

$\sqrt{3y-2} = y - 2$

$\left(\sqrt{3y-2}\right)^2 = (y-2)^2$

$3y - 2 = y^2 - 4y + 4$

$0 = y^2 - 7y + 6$

$0 = (y-6)(y-1)$

$y - 6 = 0 \qquad y - 1 = 0$

$y = 6 \qquad y = 1$

Check: $y = 6$

$\sqrt{3(6)-2} + 2 \stackrel{?}{=} 6$

$\sqrt{16} + 2 \stackrel{?}{=} 6$

$4 + 2 = 6$

Check: $y = 1$

$\sqrt{3(1)-2} + 2 \stackrel{?}{=} 1$

$\sqrt{1} + 2 \stackrel{?}{=} 1$

$1 + 2 \neq 1$

$y = 6$

25. $y = \dfrac{k}{x}, \quad y = 4, \quad x = 5$

$4 = \dfrac{k}{5} \Rightarrow k = 20$

$y = \dfrac{20}{x}, \quad x = 2$

$y = \dfrac{20}{2} = 10$

26. A = surface area, r = radius

$A = kr^2$

$1256 = k(2)^2$

$1256 = 4k$

$314 = k$

$A = 314r^2 = 314(0.5)^2$

$= 314(0.25) = 78.5$

78.5 m^2

Chapter 10

10.1 Exercises

1. $x^2 + 8x + 7 = 0$

$a = 1, \quad b = 8, \quad c = 7$

3. $8x^2 - 11x = 0$

$a = 8, \quad b = -11, \quad c = 0$

5. $x^2 + 15x - 7 = 12x + 8$

$x^2 + 3x - 15 = 0$

$a = 1, \quad b = 3, \quad c = -15$

7. $27x^2 - 9x = 0$

$9x(3x - 1) = 0$

$9x = 0 \qquad 3x - 1 = 0$

$x = 0 \qquad x = \frac{1}{3}$

9. $5x^2 - 7x = 3x$

$5x^2 - 10x = 0$

$5x(x - 2) = 0$

$5x = 0 \qquad x - 2 = 0$

$x = 0 \qquad x = 2$

11. $11x^2 - 13x = 8x - 3x^2$

$14x^2 - 21x = 0$

$7x(2x - 3) = 0$

$7x = 0 \qquad 2x - 3 = 0$

$x = 0 \qquad x = \frac{3}{2}$

13. $x^2 - 3x - 28 = 0$

$(x + 4)(x - 7) = 0$

$x + 4 = 0 \qquad x - 7 = 0$

$x = -4 \qquad x = 7$

15. $2x^2 + 11x - 6 = 0$

$(2x - 1)(x + 6) = 0$

$2x - 1 = 0 \qquad x + 6 = 0$

$2x = 1 \qquad x = -6$

$x = \frac{1}{2}$

17. $3x^2 - 16x + 5 = 0$

$(3x - 1)(x - 5) = 0$

$3x - 1 = 0 \qquad x - 5 = 0$

$3x = 1 \qquad x = 5$

$x = \frac{1}{3}$

19. $x^2 = 3x + 18$

$x^2 - 3x - 18 = 0$

$(x + 3)(x - 6) = 0$

$x + 3 = 0 \qquad x - 6 = 0$

$x = -3 \qquad x = 6$

21. $12x^2+17x+6=0$

$(4x+3)(3x+2)=0$

$4x+3=0 \qquad 3x+2=0$

$4x=-3 \qquad 3x=-2$

$x=-\frac{3}{4} \qquad x=-\frac{2}{3}$

23. $m^2+13m+24=2m-6$

$m^2+11m+30=0$

$(m+6)(m+5)=0$

$m+6=0 \qquad m+5=0$

$m=-6 \qquad m=-5$

25. $8y^2=14y-3$

$8y^2-14y+3=0$

$(4y-1)(2y-3)=0$

$4y-1=0 \qquad 2y-3=0$

$4y=1 \qquad 2y=3$

$y=\frac{1}{4} \qquad y=\frac{3}{2}$

27. $25x^2-60x+36=0$

$(5x-6)(5x-6)=0$

$5x-6=0$

$5x=6$

$x=\frac{6}{5}$

29. $(x-5)(x+2)=-10$

$x^2-3x-10=-10$

$x^2-3x=0$

$x(x-3)=0$

$x=0 \qquad x-3=0$

$x=3$

31. $3x^2+8x-10=-6x-10$

$3x^2+14x=0$

$x(3x+14)=0$

$x=0 \qquad 3x+14=0$

$3x=-14$

$x=-\frac{14}{3}$

33. $y(y-7)=2(5-2y)$

$y^2-7y-=10-4y$

$y^2-3y=10$

$y^2-3y-10=0$

$(y+2)(y-5)=0$

$y+2=0 \qquad y-5=0$

$y=-2 \qquad y=5$

35. $(x-6)(x-4)=3$

$x^2-10x+24=3$

$x^2-10x+21=0$

$(x-3)(x-7)=0$

$x-3=0 \qquad x-7=0$

$x=3 \qquad x=7$

37. $2x(x+3) = (3x+1)(x+1)$

$$2x^2 + 6x = 3x^2 + 4x + 1$$
$$0 = x^2 - 2x + 1$$
$$0 = (x-1)(x-1)$$
$$0 = x - 1$$
$$1 = x$$

39. $x + \frac{8}{x} = 6$

$$x^2 + 8 = 6x$$
$$x^2 - 6x + 8 = 0$$
$$(x-2)(x-4) = 0$$
$$x - 2 = 0 \qquad x - 4 = 0$$
$$x = 2 \qquad x = 4$$

Check: $2 + \frac{8}{2} \stackrel{?}{=} 6 \qquad 4 + \frac{8}{4} \stackrel{?}{=} 6$

$$6 = 6 \qquad 6 = 6$$

41. $\frac{x}{4} - \frac{7}{x} = -\frac{3}{4}$

$$x^2 - 28 = -3x$$
$$x^2 + 3x - 28 = 0$$
$$(x+7)(x-4) = 0$$
$$x + 7 = 0 \qquad x - 4 = 0$$
$$x = -7 \qquad x = 4$$

Check: $\frac{-7}{4} - \frac{7}{-7} \stackrel{?}{=} -\frac{3}{4} \qquad \frac{4}{4} - \frac{7}{4} \stackrel{?}{=} -\frac{3}{4}$

$$-\frac{3}{4} = -\frac{3}{4} \qquad -\frac{3}{4} = -\frac{3}{4}$$

43. $\frac{3y-8}{2} = \frac{2}{y+1}$

$$(y+1)(3y-8) = 4$$
$$3y^2 - 5y - 8 = 4$$
$$3y^2 - 5y - 12 = 0$$
$$(3y+4)(y-3) = 0$$
$$3y + 4 = 0 \qquad y - 3 = 0$$
$$3y = -4 \qquad y = 3$$
$$y = -\frac{4}{3}$$

Check: $\frac{3\left(-\frac{4}{3}\right) - 8}{2} \stackrel{?}{=} \frac{2}{-\frac{4}{3} + 1}$

$$-6 = -6$$

$$\frac{3(3) - 8}{2} \stackrel{?}{=} \frac{2}{3+1}$$
$$\frac{1}{2} = \frac{1}{2}$$

45. $\frac{4}{x} + \frac{3}{x+5} = 2$

$$(x+5)(4) + 3x = 2x(x+5)$$
$$4x + 20 + 3x = 2x^2 + 10x$$
$$2x^2 + 3x - 20 = 0$$
$$(2x-5)(x+4) = 0$$
$$2x - 5 = 0 \qquad x + 4 = 0$$
$$x = \frac{5}{2} \qquad x = -4$$

47. $\frac{24}{x^2-4}=1+\frac{2x-6}{x-2}$

$$24=(x^2-4)(1)+(x+2)(2x-6)$$
$$24=x^2-4+2x^2-2x-12$$
$$0=3x^2-2x-40$$
$$0=(3x+10)(x-4)$$

$3x+10=0 \qquad x-4=0$

$x=-\frac{10}{3} \qquad x=4$

Check:

$$\frac{24}{(-10/3)^2-4}\stackrel{?}{=}1+\frac{2(-10/3)-6}{-10/3-2}$$
$$\frac{27}{8}=\frac{27}{8}$$
$$\frac{24}{4^2-4}\stackrel{?}{=}1+\frac{2(4)-6}{4-2}$$
$$2=2$$

49. You can always factor out x.

51. $ax^2-7x+c=0$

$x=-\frac{3}{2}\Rightarrow\frac{9}{4}a+\frac{21}{2}+c=0 \qquad (1)$

$x=6\Rightarrow 36a-42+c=0 \qquad (2)$

Multiply (1) by -1 and add it to (2)

$$\frac{135}{4}a-\frac{105}{2}=0$$
$$a=\frac{14}{9}$$

Substitute $\frac{14}{9}$ for a in (2)

$$36\left(\frac{14}{9}\right)-42+c=0$$
$$c=-14$$
$$a=\frac{14}{9},\quad c=-14$$

53. $t=\frac{n^2-n}{2},\quad n=20$

$$t=\frac{(20)^2-20}{2}=190$$

190 truck routes.

55. $t=\frac{n^2-n}{2},\quad t=15$

$$15=\frac{n^2-n}{2}$$
$$0=n^2-n-30$$
$$0=(n-6)(n+5)$$

$n-6=0 \qquad n+5=0$

$n=6 \qquad n=-5$, has no meaning

6 cities can be serviced.

57. $t=\frac{n^2-n}{2},\quad t=45$

$$45=\frac{n^2-n}{2}$$
$$0=n^2-n-90$$
$$0=(n-10)(n+9)$$

$n-10=0 \qquad n+9=0$

$n=10 \qquad n=-9$

has no meaning

10 cities can be serviced.

For Exercises 59-63 use

$R=-2.5n^2+80n-280$

59. $320=-2.5n^2+80n-280$

$0=-2.5n^2+80n-600$

$0=-2.5(n^2-32n+240)$

$0=-2.5(n-12)(n-20)$

$n-12=0 \qquad n-20=0$

$n=12 \qquad n=20$

They may produce 1200 or2000 labels.

61. $\frac{12+20}{2}=16$, average $=1600$ labels

$R=-2.5(16)^2+80(16)-280=360$

The revenue for producing 1600 labels is \$360, which is more than producing either 1200 or 2000 labels.

63. $R=-2.5(15)^2+80(15)-280=\$357.5$

$R=-2.5(17)^2+80(17)-280=\$357.5$

It appears that the maximum revenue is at 1600 labels.

Cumulative Review

65. $$\frac{\frac{1}{x}+\frac{3}{x-2}}{\frac{4}{x^2-4}}=\frac{\frac{1}{x}+\frac{3}{x-2}}{\frac{4}{(x+2)(x-2)}}\cdot\frac{x(x+2)(x-2)}{x(x+2)(x-2)}$$

$$=\frac{(x+2)(x-2)+3x(x+2)}{4x}$$

$$=\frac{x^2-4+3x^2+6x}{4x}$$

$$=\frac{4x^2+6x-4}{4x}$$

$$=\frac{2x^2-3x-2}{2x}$$

$$=\frac{(2x-1)(x+2)}{2x}$$

67. Dust $=0.12(36)=4.32$ mg

10.2 Exercises

1. $x^2=64$

$x=\pm\sqrt{64}$

$x=\pm 8$

3. $x^2=98$

$x=\pm\sqrt{98}$

$x=\pm 7\sqrt{2}$

5. $x^2 - 28 = 0$
$x^2 = 28$
$x = \pm\sqrt{28}$
$x = \pm 2\sqrt{7}$

7. $5x^2 = 45$
$x^2 = 9$
$x = \pm\sqrt{9}$
$x = \pm 3$

9. $6x^2 = 120$
$x^2 = 20$
$x = \pm\sqrt{20}$
$x = \pm 2\sqrt{5}$

11. $3x^2 - 375 = 0$
$3x^2 = 375$
$x^2 = 125$
$x = \pm\sqrt{125}$
$x = \pm 5\sqrt{5}$

13. $5x^2 + 13 = 73$
$5x^2 = 60$
$x^2 = 12$
$x = \pm\sqrt{12}$
$x = \pm 2\sqrt{3}$

15. $13x^2 + 17 = 82$
$13x^2 = 65$
$x^2 = 5$
$x = \pm\sqrt{5}$

17. $(x-7)^2 = 16$
$x - 7 = \pm 4$
$x = 7 \pm 4$
$x = 7 + 4 = 11$
$x = 7 - 4 = 3$

19. $(x+4)^2 = 6$
$x + 4 = \pm\sqrt{6}$
$x = -4 \pm \sqrt{6}$

21. $(2x+5)^2 = 2$
$2x + 5 = \pm\sqrt{2}$
$2x = -5 \pm \sqrt{2}$
$x = \dfrac{-5 \pm \sqrt{2}}{2}$

23. $(3x-1)^2 = 7$
$3x - 1 = \pm\sqrt{7}$
$3x = 1 \pm \sqrt{7}$
$x = \dfrac{1 \pm \sqrt{7}}{3}$

25. $(7x+2)^2 = 12$
$7x + 2 = \pm\sqrt{12}$
$7x = -2 \pm 2\sqrt{3}$
$x = \dfrac{-2 \pm 2\sqrt{3}}{7}$

27. $(4x-5)^2 = 54$

$$4x-5 = \pm\sqrt{54}$$
$$4x = 5 \pm 3\sqrt{6}$$
$$x = \frac{5 \pm 3\sqrt{6}}{4}$$

29. $x^2 - 6x = 11, \left[\frac{1}{2}(6)\right]^2 = 3^2 = 9$

$$x^2 - 6x + 9 = 11 + 9$$
$$(x-3)^2 = 20$$
$$x - 3 = \pm\sqrt{20}$$
$$x = 3 \pm 2\sqrt{5}$$

31. $x^2 + 6x - 7 = 0$

$$x^2 + 6x = 7$$
$$x^2 + 6x + 9 = 7 + 9$$
$$(x+3)^2 = 16$$
$$x + 3 = \pm 4$$
$$x = -3 \pm 4$$
$$x = -3 + 4 = 1$$
$$x = -3 - 4 = -7$$

33. $x^2 - 12x - 5 = 0$

$$x^2 - 12x = 5, \left[\frac{1}{2}(-12)\right]^2 = 6^2 = 36$$
$$x^2 - 12x + 36 = 5 + 36$$
$$(x-6)^2 = 41$$
$$x - 6 = \pm\sqrt{41}$$
$$x = 6 \pm \sqrt{41}$$

35. $x^2 - 7x = 0$

$$x^2 - 7x + \frac{49}{4} = \frac{49}{4}$$
$$\left(x - \frac{7}{2}\right)^2 = \frac{49}{4}$$
$$x - \frac{7}{2} = \pm\sqrt{\frac{49}{4}}$$
$$x - \frac{7}{2} = \pm\frac{7}{2}$$
$$x = \frac{7}{2} \pm \frac{7}{2}$$
$$x = \frac{7}{2} + \frac{7}{2} = 7$$
$$x = \frac{7}{2} - \frac{7}{2} = 0$$

37. $5x^2 - 25x = 0$

$$x^2 - 5x = 0$$
$$x^2 - 5x + \frac{25}{4} = \frac{25}{4}$$
$$\left(x - \frac{5}{2}\right)^2 = \frac{25}{4}$$
$$x - \frac{5}{2} = \pm\sqrt{\frac{25}{4}}$$
$$x - \frac{5}{2} = \pm\frac{5}{2}$$
$$x = \frac{5}{2} \pm \frac{5}{2}$$
$$x = \frac{5}{2} + \frac{5}{2} = 5$$
$$x = \frac{5}{2} - \frac{5}{2} = 0$$

39. $2x^2 - 7x = 9$

$x^2 - \frac{7}{2}x = \frac{9}{2}, \quad \left[\frac{1}{2}\left(-\frac{7}{2}\right)\right]^2 = \frac{49}{16}$

$x^2 - \frac{7}{2}x + \frac{49}{16} = \frac{9}{2} + \frac{49}{16}$

$\left(x - \frac{7}{4}\right)^2 = \frac{121}{16}$

$x - \frac{7}{4} = \pm\sqrt{\frac{121}{16}}$

$x = \frac{7}{4} \pm \frac{11}{4}$

$x = \frac{7}{4} - \frac{11}{4} \qquad x = \frac{7}{4} + \frac{11}{4}$

$x = -\frac{4}{4} = -1 \qquad x = \frac{18}{4} = \frac{9}{2}$

41. $x^2 + bx - 7 = 0$

$x^2 + bx = 7, \quad \left(\frac{b}{2}\right)^2 = \frac{b^2}{4}$

$x^2 + bx + \frac{b^2}{4} = 7 + \frac{b^2}{4}$

$\left(x + \frac{b}{2}\right)^2 = \frac{28 + b^2}{4}$

$x + \frac{b}{2} = \pm\sqrt{\frac{28 + b^2}{4}}$

$x = -\frac{b}{2} \pm \frac{\sqrt{28 + b^2}}{2}$

$x = \frac{-b \pm \sqrt{28 + b^2}}{2}$

Cumulative Review

43. $3a - 5b = 8 \quad (1)$

$5a - 7b = 8 \quad (2)$

Multiply (1) by 5 and (2) by -3

$15a - 25b = 40$

$-15a + 21b = -24$

$-4b = 16$

$b = -4$

Substitute -4 for b in (1)

$3a - 5(-4) = 8$

$3a = -12$

$a = -4$

$a = -4, \ b = -4$

45. $P = \frac{F}{A}, \quad A = LW$

$F = 16, \ L = 8, \ W = 0.01$

$P = \frac{16}{8(0.01)}$

$= \frac{16}{0.08}$

$= 200$

The pressure is 200 pounds per square inch.

47. $21 - 7 = 14$ single men

$\frac{14}{24} \approx 0.583 = 58.3\%$ were single

10.3 Exercises

* Use $x = \dfrac{-b \pm \sqrt{b^2 - 4ac}}{2a}$ in Exercises 1 - 31.

1. $3x^2 + 4x - 7 = 0$

   $a = 3, \quad b = 4, \quad c = -7$

   $\sqrt{b^2 - 4ac} = \sqrt{16 + 84}$

   $= \sqrt{100}$

   $= 10$

   There are two rational roots.

3. $4x^2 = -5x + 6$, no

   $4x^2 + 5x - 6 = 0$

   $a = 4, \quad b = 5, \quad c = -6$

5. $x^2 + 3x - 10 = 0$

   $a = 1, \quad b = 3, \quad c = -10$

   $x = \dfrac{-3 \pm \sqrt{(3)^2 - 4(1)(-10)}}{2(1)}$

   $= \dfrac{-3 \pm \sqrt{49}}{2} = \dfrac{-3 \pm 7}{2}$

   $x = \dfrac{-3 \pm 7}{2}$

   $x = \dfrac{-3 + 7}{2} = 2$

   $x = \dfrac{-3 - 7}{2} = -5$

7. $x^2 - 3x - 8 = 0$

   $a = 1, \quad b = -3, \quad c = -8$

   $x = \dfrac{-(-3) \pm \sqrt{(-3)^2 - 4(1)(-8)}}{2(1)}$

   $= \dfrac{3 \pm \sqrt{41}}{2}$

9. $4x^2 - 5x - 6 = 0$

   $a = 4, \quad b = -5, \quad c = -6$

   $x = \dfrac{-(-5) \pm \sqrt{(-5)^2 - 4(4)(-6)}}{2(4)}$

   $= \dfrac{5 \pm \sqrt{121}}{8} = \dfrac{5 \pm 11}{8}$

11. $2x^2 = 3x + 20$

    $a = 2, \quad b = -3, \quad c = -20$

    $x = \dfrac{-(-3) \pm \sqrt{(-3)^2 - 4(2)(-20)}}{2(2)}$

    $x = \dfrac{3 \pm \sqrt{169}}{4}$

    $x = \dfrac{3 \pm 13}{4}$

    $x = \dfrac{3 \pm 13}{4}$

    $x = \dfrac{3 + 13}{4} = 4$

    $x = \dfrac{3 - 13}{4} = -\dfrac{5}{2}$

13. $6x^2 - 3x = 1$

$6x^2 - 3x - 1 = 0$

$a = 6, \quad b = -3, \quad c = -1$

$$x = \frac{-(-3) \pm \sqrt{(-3)^2 - 4(6)(-1)}}{2(6)}$$

$$= \frac{3 \pm \sqrt{33}}{12}$$

15. $x + \frac{3}{2} = 3x^2$

$3x^2 - x - \frac{3}{2} = 0$

$6x^2 - 2x - 3 = 0$

$a = 6, \quad b = -2, \quad c = -3$

$$x = \frac{-(-2) \pm \sqrt{(-2)^2 - 4(6)(-3)}}{2(6)}$$

$$= \frac{2 \pm \sqrt{76}}{12} = \frac{2 \pm 2\sqrt{19}}{12} = \frac{1 \pm \sqrt{19}}{6}$$

17. $\frac{x}{2} + \frac{5}{x} = \frac{7}{2}$

$x^2 + 10 = 7x$

$x^2 - 7x + 10 = 0$

$a = 1, \quad b = -7, \quad c = 10$

$$x = \frac{-(-7) \pm \sqrt{(-7)^2 - 4(1)(10)}}{2(1)}$$

$$= \frac{7 \pm \sqrt{9}}{2} = \frac{7 \pm 3}{2}$$

$$x = \frac{7+3}{2} = 5$$

$$x = \frac{7-3}{2} = 2$$

$(x-7)^2 = 16$

$x - 7 = \pm 4$

$x = 7 \pm 4$

$x = 7 + 4 = 11$

$x = 7 - 4 = 3$

19. $5x^2 + 6x + 2 = 0$

$a = 5, \quad b = 6, \quad c = 2$

$$x = \frac{-(6) \pm \sqrt{(6)^2 - 4(5)(2)}}{2(5)} = \frac{-6 \pm \sqrt{-4}}{10}$$

No real solution.

21. $3y^2 = 8y - 3$

$3y^2 - 8y + 3 = 0$

$a = 3, \quad b = -8, \quad c = 3$

$$y = \frac{-(-8) \pm \sqrt{(-8)^2 - 4(3)(3)}}{2(3)}$$

$$y = \frac{8 \pm \sqrt{28}}{6}$$

$$y = \frac{8 \pm 2\sqrt{7}}{6}$$

$$y = \frac{4 \pm \sqrt{7}}{3}$$

23. $\frac{d^2}{2} + \frac{5d}{6} - 2 = 0$

$3d^2 + 5d - 12 = 0$

$a = 3, \quad b = 5, \quad c = -12$

$$d = \frac{-(5) \pm \sqrt{(5)^2 - 4(3)(-12)}}{2(3)}$$

$$= \frac{-5 \pm \sqrt{169}}{6} = \frac{-5 \pm 13}{6}$$

$$d = \frac{-5-13}{6} \qquad d = \frac{-5+13}{6}$$

$$d = -3 \qquad d = \frac{4}{3}$$

25. $4x^2 + 8x + 11 = 0$

$a = 4, \quad b = 8, \quad c = 11$

$$x = \frac{-8 \pm \sqrt{(8)^2 - 4(4)(11)}}{2(4)}$$

$$x = \frac{-8 \pm \sqrt{-24}}{8}$$

$\sqrt{-24}$ is not real

There is no real solution.

27. $x^2 - 10x + 25 = 0$

$a = 1, \quad b = -10, \quad c = 25$

$$x = \frac{-(-10) \pm \sqrt{(-10)^2 - 4(1)(25)}}{2(1)}$$

$$x = \frac{10 \pm \sqrt{0}}{2}$$

$x = 5$

29. $x^2 + 5x - 2 = 0$

$a = 1, \quad b = 5, \quad c = -2$

$$x = \frac{-(5) \pm \sqrt{(5)^2 - 4(1)(-2)}}{2(1)} = \frac{-5 \pm \sqrt{33}}{2}$$

$x = 0.372, \quad -5.372$

31. $2x^2 - 7x - 5 = 0$

$a = 2, \quad b = -7, \quad c = -5$

$$x = \frac{-(-7) \pm \sqrt{(-7)^2 - 4(2)(-5)}}{2(2)}$$

$$x = \frac{7 \pm \sqrt{89}}{4}$$

$$x = \frac{7 + \sqrt{89}}{4} \approx 4.108$$

$$x = \frac{7 - \sqrt{89}}{4} \approx -0.608$$

33. $5x^2 + 10x + 1 = 0$

$a = 5, \quad b = 10, \quad c = 1$

$$x = \frac{-(10) \pm \sqrt{(10)^2 - 4(5)(1)}}{2(5)}$$

$$= \frac{-10 \pm \sqrt{80}}{10} = \frac{-10 \pm 4\sqrt{5}}{10} = \frac{-5 \pm 2\sqrt{5}}{5}$$

$x = -0.106, \quad -1.894$

35. $6x^2 - 13x + 6 = 0$

$(3x-2)(2x-3) = 0$

$3x - 2 = 0 \qquad 2x - 3 = 0$

$3x = 2 \qquad 2x = 3$

$x = \frac{2}{3} \qquad x = \frac{3}{2}$

37.
$$3\left(x^2+1\right)=10x$$
$$3x^2+3=10x$$
$$3x^2-10x+3=0$$
$$(3x-1)(x-3)=0$$
$$3x-1=0 \qquad x-3=0$$
$$3x=1 \qquad x=3$$
$$x=\frac{1}{3}$$

39.
$$(t+5)(t-3)=7$$
$$t^2+2t-15=7$$
$$t^2+2t=22$$
$$t^2+2t+1=22+1$$
$$(t+1)^2=23$$
$$t+1=\pm\sqrt{23}$$
$$t=-1\pm\sqrt{23}$$

41.
$$y^2-\frac{2}{5}y=2$$
$$5y^2-2y=10$$
$$5y^2-2y-10=0$$
$$a=5,\quad b=-2,\quad c=-10$$
$$y=\frac{-(-2)\pm\sqrt{(-2)^2-4(5)(-10)}}{2(5)}$$
$$=\frac{2\pm\sqrt{204}}{10}=\frac{2\pm2\sqrt{51}}{10}=\frac{1\pm\sqrt{51}}{5}$$

43.
$$3x^2-13=0$$
$$3x^2=13$$
$$x^2=\frac{13}{3}$$
$$x=\pm\sqrt{\frac{13}{3}}=\pm\frac{\sqrt{13}}{\sqrt{3}}=\pm\frac{\sqrt{39}}{3}$$

45.
$$x(x-2)=7$$
$$x^2-2x=7$$
$$x^2-2x-7=0$$
$$a=1,\quad b=-2,\quad c=-7$$
$$x=\frac{-(-2)\pm\sqrt{(-2)^2-4(1)(-7)}}{2(1)}$$
$$=\frac{2\pm\sqrt{32}}{2}=\frac{2\pm4\sqrt{2}}{2}=1\pm2\sqrt{2}$$

47.
Total area − Pool area = Tile area
$$(2x+30)(2x+20)-20(30)=216$$
$$4x^2+100x+600-600=216$$
$$4x^2+100x-216=0$$
$$x^2+25x-54=0$$
$$(x+27)(x-2)=0$$
$$x+27=0 \qquad x-2=0$$
$$x=-27 \qquad x=2$$

Width can't be negative.

Width is 2 feet.

Cumulative Review

49. $\frac{3x}{x-2}+\frac{4}{x+2}-\frac{x+22}{x^2-4}$

$$=\frac{3x}{x-2}\cdot\frac{x+2}{x+2}+\frac{4}{x+2}\cdot\frac{x-2}{x-2}-\frac{x+22}{x^2-4}$$

$$=\frac{3x^2+6x+4x-8-x-22}{(x-2)(x+2)}$$

$$=\frac{3x^2+9x-30}{(x-2)(x+2)}$$

$$=\frac{3(x+5)(x-2)}{(x-2)(x+2)}$$

$$=\frac{3(x+5)}{x+2}$$

51.
$$\begin{array}{r} x^2+5x+2 \\ x+3\overline{)x^3+8x^2+17x+6} \\ \underline{x^3+3x^2} \\ 5x^2+17x \\ \underline{5x^2+15x} \\ 2x+6 \\ \underline{2x+6} \end{array}$$

$(x^3+8x^2+17x+6)\div(x+3)=x^2+5x+2$

53. $V=\pi r^2h$

$\approx 3.14(2.40)^2(5.35)$

≈ 96.76 cubic feet

How Am I Doing? Sections 10.1-10.3

1. $x^2-13x-48=0$

$(x-16)(x+3)=0$

$x-16=0 \qquad x+3=0$

$x=16 \qquad x=-3$

2. $5x^2+7x=14x$

$5x^2-7x=0$

$x(5x-7)=0$

$5x-7=0 \qquad x=0$

$5x=7$

$x=\frac{7}{5}$

3. $5x^2=22x-8$

$5x^2-22x+8=0$

$(5x-2)(x-4)=0$

$5x-2=0 \qquad x-4=0$

$5x=2 \qquad x=4$

$x=\frac{2}{5}$

4. $-2x+1=8x^2$

$8x^2+2x-1=0$

$(4x-1)(2x+1)=0$

$4x-1=0 \qquad 2x+1=0$

$4x=1 \qquad 2x=-1$

$x=\frac{1}{4} \qquad x=-\frac{1}{2}$

5. $x(x+9)=4(x+6)$
$x^2+9x=4x+24$
$x^2+5x-24=0$
$(x+8)(x-3)=0$
$x+8=0 \qquad x-3=0$
$x=-8 \qquad x=3$

6. $\frac{5}{x+2}=\frac{2x-1}{5}$
$25=(x+2)(2x-1)$
$25=2x^2+3x-2$
$0=2x^2+3x-27$
$0=(2x+9)(x-3)$
$2x+9=0 \qquad x-3=0$
$2x=-9 \qquad x=3$
$x=-\frac{9}{2}$

7. $x^2-18=0$
$x^2=18$
$x=\pm\sqrt{18}$
$x=\pm 3\sqrt{2}$

8. $3x^2+1=0$
$3x^2=75$
$x^2=25$
$x=\pm\sqrt{25}$
$x=\pm 5$

9. $x^2+8x+5=0$
$x^2+8x=-5$
$x^2+8x+16=-5+16$
$(x+4)^2=11$
$x+4=\pm\sqrt{11}$
$x=-4\pm\sqrt{11}$

10. $2x^2+3x-7=0$
$2x^2+3x=7$
$x^2+\frac{3}{2}x=\frac{7}{2}$

$x^2+\frac{3}{2}x+\frac{9}{16}=\frac{7}{2}+\frac{9}{16}$
$\left(x+\frac{3}{4}\right)^2=\frac{65}{16}$
$x+\frac{3}{4}=\pm\sqrt{\frac{65}{16}}$
$x=-\frac{3}{4}\pm\sqrt{\frac{65}{4}}$
$x=\frac{-3\pm\sqrt{65}}{4}$

11. $2x^2+4x-5=0$
$a=2, \quad b=4, \quad c=-5$
$x=\frac{-4\pm\sqrt{(4)^2-4(2)(-5)}}{2(2)}$
$x=\frac{-4\pm\sqrt{56}}{4}$
$x=\frac{-4\pm 2\sqrt{14}}{4}$
$x=\frac{-2\pm\sqrt{14}}{2}$

12. $2x^2 = 7x - 4$

$2x^2 - 7x + 4 = 0$

$a = 2, \quad b = -7, \quad c = 4$

$$x = \frac{-(-7) \pm \sqrt{(-7)^2 - 4(2)(4)}}{2(2)}$$

$$x = \frac{7 \pm \sqrt{17}}{4}$$

13. $3x^2 + 8x + 1 = 0$

$a = 3, \quad b = 8, \quad c = 1$

$$x = \frac{-8 \pm \sqrt{(8)^2 - 4(3)(1)}}{2(3)}$$

$$x = \frac{-8 \pm \sqrt{52}}{6}$$

$$x = \frac{-8 \pm 2\sqrt{13}}{6}$$

$$x = \frac{-4 \pm \sqrt{13}}{3}$$

14. $5x^2 + 3 = 4x$

$5x^2 - 4x + 3 = 0$

$a = 5, \quad b = -4, \quad c = 3$

$$x = \frac{-(-4) \pm \sqrt{(-4)^2 - 4(5)(3)}}{2(5)}$$

$$x = \frac{4 \pm \sqrt{-44}}{10}$$

$\sqrt{-44}$ is not a real number

There is no real solution.

10.4 Exercises

1. It may have 2, 1, or none.

3. If $b = 0$, then the parabola always has a vertex at $(0, c)$ that is on the y-axis.

5. $y = x^2 + 2$

x	y
−2	6
−1	3
0	2
1	3
2	6

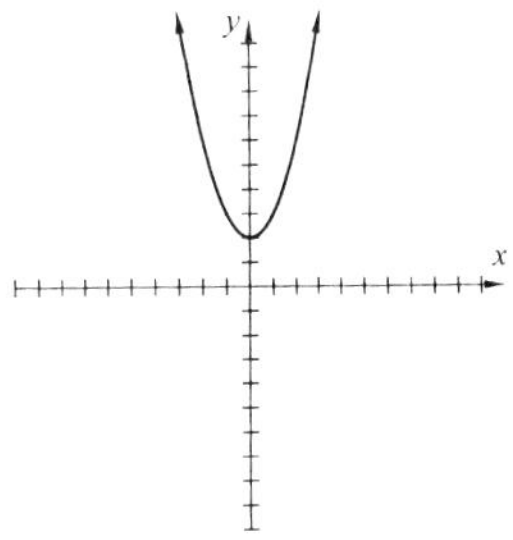

7. $y = -\frac{1}{3}x^2$

x	y
−6	−12
−3	−3
0	0
3	−3
6	−12

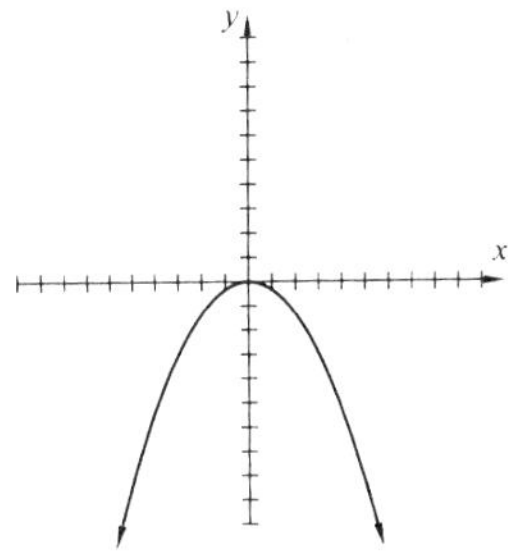

9. $y = 2x^2 - 3$

x	y
−2	5
−1	−1
0	−3
1	−1
2	5

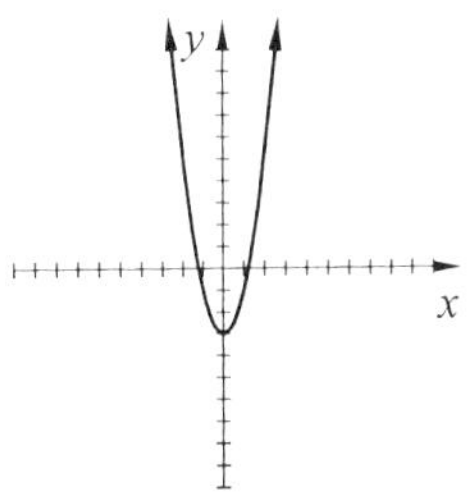

11. $y = (x-2)^2$

x	y
0	4
1	1
2	0
3	1
4	4

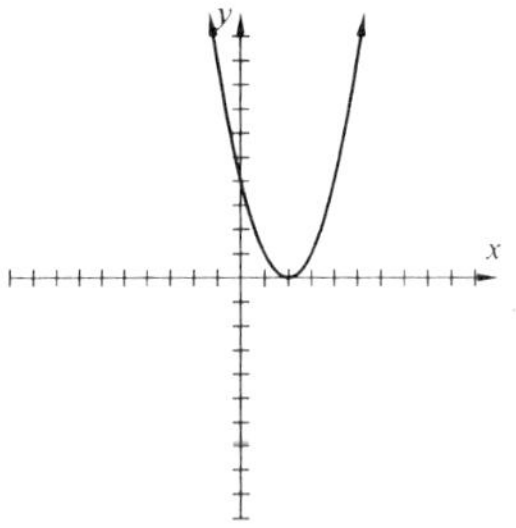

13. $y = -\frac{1}{2}x^2 + 4$

x	y
−4	−4
−2	2
0	4
2	2
4	−4

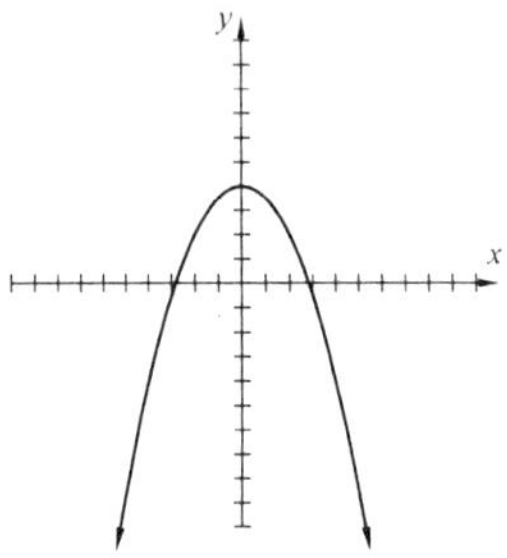

15. $y = \frac{1}{2}(x-3)^2$

x	y
−1	8
1	2
3	0
5	2
7	8

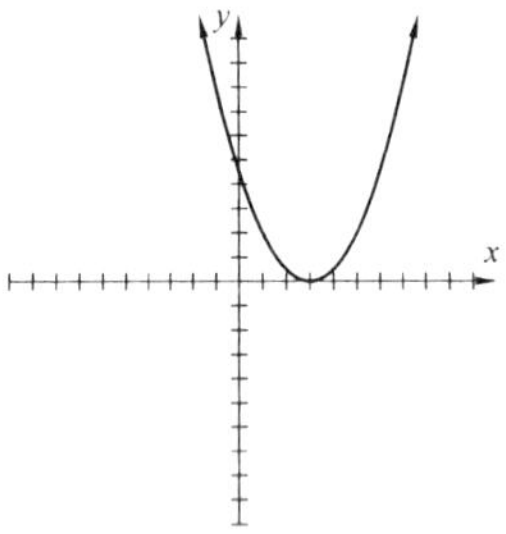

17. $y = x^2 + 4x, \quad a = 1, \quad b = 4$

$a > 0$: opens up

Vertex: $x = -\frac{b}{2a} = -\frac{4}{2(1)} = -2$

$$y = (-2)^2 + 4(-2) = -4$$

$V(-2, -4)$

x	y
−4	0
−3	−3
−2	−4
−1	−3
0	0

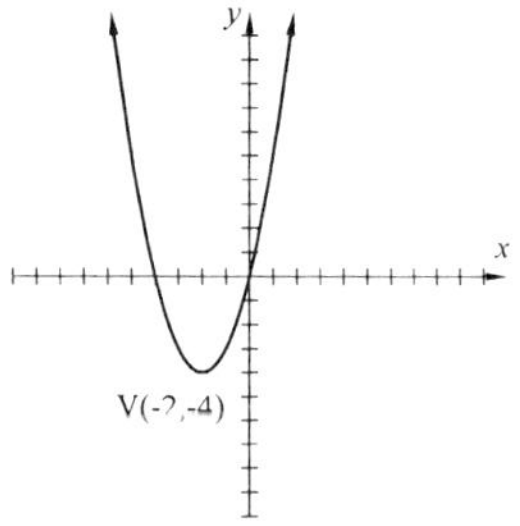

19. $y = -x^2 - 4x, \quad a = -1, \quad b = -4$

$a < 0$: opens down

Vertex: $x = -\frac{(-4)}{2(-1)} = -2$

$$y = -(-2)^2 - 4(-2) = 4$$

$V(-2, 4)$

x	y
−4	0
−3	3
−2	4
−1	3
0	0

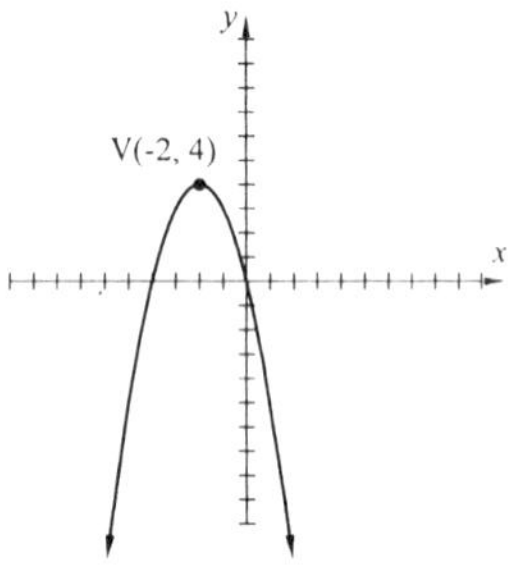

21. $y = -2x^2 + 8x, \quad a = -2, \quad b = 8$

$a < 0$: opens down

Vertex: $x = -\frac{8}{2(-2)} = 2$

$$y = -2(2)^2 + 8(2) = 8$$

$V(2, 8)$

x	y
−1	−10
0	0
1	8
4	0
5	−10

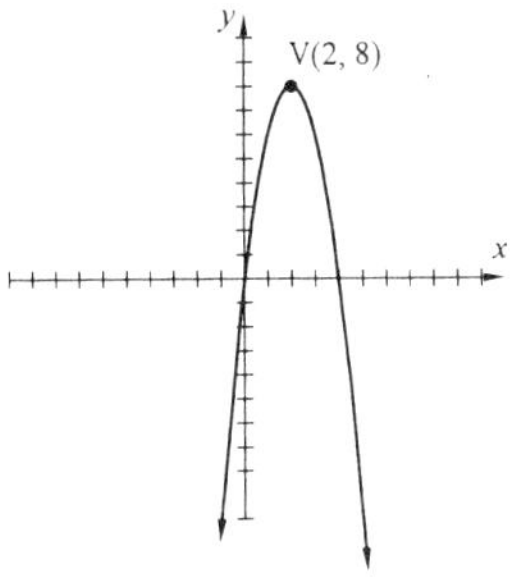

23. $y = x^2 + 2x - 3,\quad a = 1,\quad b = 2$

Vertex: $x = -\dfrac{2}{2(1)} = -1$

$y = (-1)^2 + 2(-1) - 3 = -4$

$V(-1, -4)$

y – intercept: $y = 0^2 + 2(0) - 3 = -3$

x – intercepts: $0 = x^2 + 2x - 3$

$0 = (x+3)(x-1)$

$x = 1, -3$

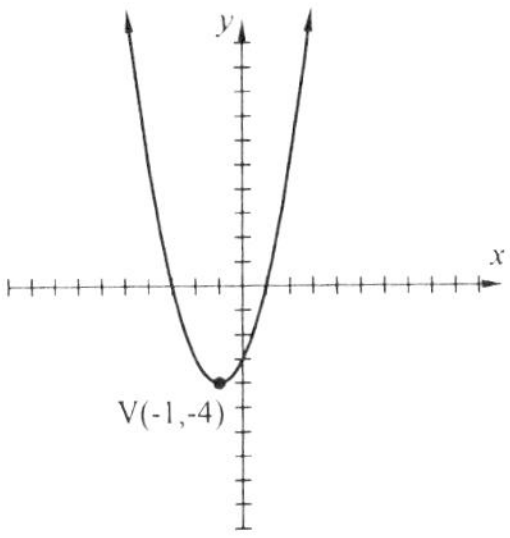

25. $y = -x^2 - 6x - 5,\quad a = -1,\quad b = -6$

Vertex: $x = -\dfrac{(-6)}{2(-1)} = -3$

$y = -(-3)^2 - 6(-3) - 5 = 4$

$V(-3, 4)$

y – intercept: $y = -0^2 - 6(0) - 5 = -5$

x – intercepts: $0 = -x^2 - 6x - 5$

$0 = -(x+5)(x+1)$

$x = -5, -1$

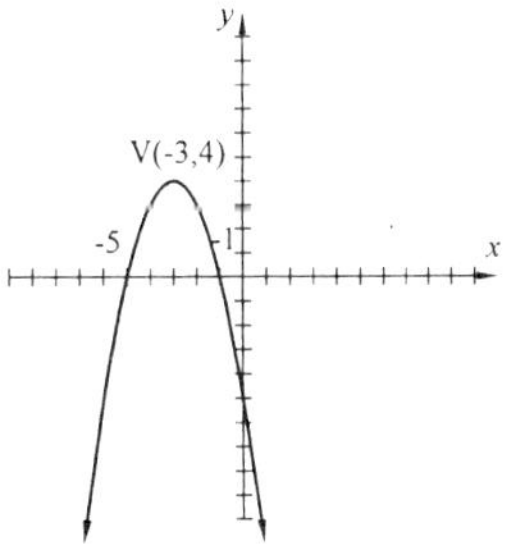

27. $y = x^2 + 6x + 9,\quad a = 1,\quad b = 6$

Vertex: $x = -\dfrac{6}{2(1)} = -3$

$y = (-3)^2 + 6(-3) + 9 = 0$

$V(-3, 0)$

y – intercept: $y = 0^2 + 6(0) + 9 = 9$

x – intercepts: $0 = x^2 + 6x + 9$

$0 = (x+3)^2$

$x = -3$

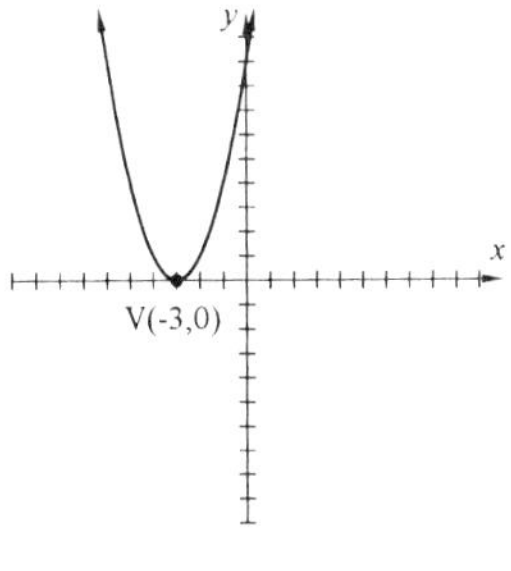

$h = -4.9t^2 + 39.2t + 4,$

$a = -4.9, \ b = 39.2$

Vertex: $t = -\dfrac{39.2}{2(-4.9)} = 4$

29. $h = -4.9(4)^2 + 39.2(4) + 4 = 82.4$

$V(4, 82.4)$

h – intercept: $h = 4$

t – intercept: $0 = -4.9t^2 + 39.2t + 10$

$t \approx -0.1, 8.1$

(a)

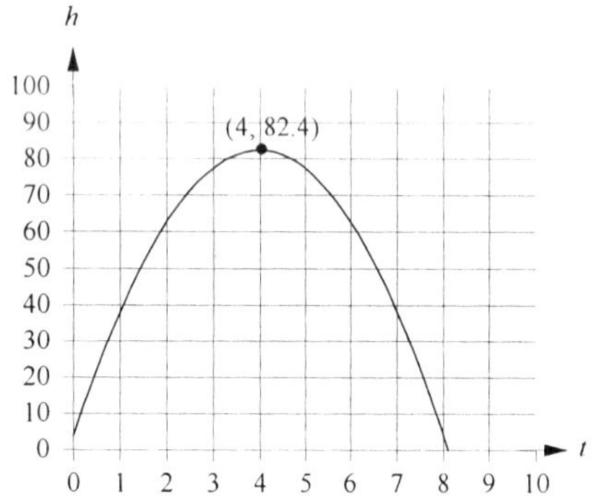

(b) $t = 2$

$h = -4.9(2)^2 + 39.2(2) + 4 = 62.8$

It is 62.8 m high.

(c) $V(4, 82.4)$

The maximum height is 82.4 m.

(d) The x – intercept is 8.1

After $8.1s$ it will strike the earth.

31. $N = 9x - x^2, \ a = -1, \ b = 9$

(a) Vertex: $x = -\dfrac{9}{2(-1)} = 4.5$

$N = 9(4.5) - (4.5)^2 = 20.25$

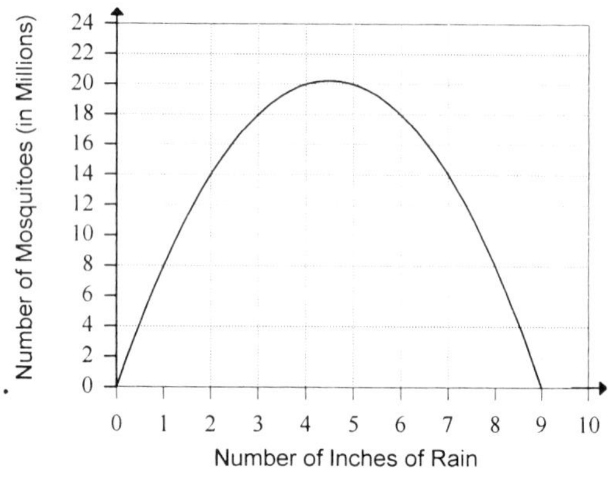

(a) Maximum is at the vertex: 4.5 inches

(b) Maximum is at the vertex: 20.25 million

33. $y = x^2 + 10.6x - 212.16$

Vertex: $x = -\dfrac{10.6}{2(1)} = -5.3$

$y = (-5.3)^2 + 10.6(-5.3) - 212.16$

$= -240.25$

$V(-5.3, -240.25)$

y-intercept: $(0, -212.16)$

x-intercepts:

$$x = \frac{-10.6 \pm \sqrt{(10.6)^2 - 4(1)(-212.16)}}{2(1)}$$

$= -20.8, \ 10.2$

$(-20.8, 0)$ and $(10.2, 0)$

35. $y = -301x^2 - 167x + 1528$

Vertex: $x = -\dfrac{-167}{2(-301)} \approx -0.277$

$y = -301(-0.277)^2 - 167(-0.277) + 1528$

≈ 1551.164

$V(-0.277, 1551.164)$

y-intercept: $(0, 1528)$

x-intercepts:

$$x = \frac{167 \pm \sqrt{(-167)^2 - 4(-301)(1528)}}{2(-301)}$$

$= -2.548,\ 1.993$

$(-2.548, 0)$ and $(1.993, 0)$

Cumulative Review

37. $y = kx^2,\quad y = 12,\quad x = 2$

$12 = k(2)^2 \Rightarrow k = 3$

$y = 3x^2,\quad x = 5$

$y = 3(5)^2 = 75$

39. $a = \dfrac{V_2 - V_1}{T_2 - T_1} = \dfrac{360 - 355}{49 - 40} = \dfrac{5}{9}$

$V = aT + b$

$355 = \dfrac{5}{9}(40) + b$

$3195 = 200 + 9b$

$2995 = 9b$

$V = \dfrac{5}{9}T + \dfrac{2995}{9}$

10.5 Exercises

1. $A = \dfrac{1}{2}ba,\ x = \text{altitude},\ x - 5 = \text{base}$

$88 = \dfrac{1}{2}(x-5)(x)$

$176 = x^2 - 5x$

$0 = x^2 - 5x - 176$

$0 = (x-16)(x+11)$

$x - 16 = 0 \qquad x + 11 = 0$

$x = 16 \qquad x = -11$

Length can't be negative.

$x - 5 = 16 - 5 = 11$

The altitude is 16 centimeters.

The base is 11 centimeters

3. x = length of the sides of the old garden

New garden area $= (x+7)(x+11)$

$x^2 + 18x + 77 = 396$

$x^2 + 18x - 319 = 0$

$(x-11)(x+29) = 0$

$x - 11 = 0 \qquad x + 29 = 0$

$x = 11, \qquad x = -29,$

Length can't be negative.

The old garden was 11 feet by 11 feet.

5. s = original number of people

c = cost per person

$$s \cdot c = 420 \qquad (1)$$

$$(s+7)(c-5) = 420 \qquad (2)$$

Substitute $\frac{420}{5}$ for c in (2)

$$(s+7)\left(\frac{420}{s}-5\right) = 420$$

$$(s+7)(420-5s) = 420s$$

$$(s+7)(84-s) = 84s$$

$$s^2+7s-588 = 0$$

$$(s+28)(s-21) = 0$$

$s = -28$, $s = 21$

Number of people can't be negative.

Original number of people was 21.

7. s = number of members

c = cost

$$s \cdot c = 400 \qquad (1)$$

$$(s+20)(c-1) = 400 \qquad (2)$$

Substitute $\frac{400}{s}$ for c in (2)

$$(s+20)\left(\frac{400}{s}-1\right) = 400$$

$$(s+20)(400-s) = 400s$$

$$s^2-380s-8000 = -400s$$

$$s^2+20s-8000 = 0$$

$$(s+100)(s-80) = 0$$

$s+100 = 0 \qquad s-80 = 0$

$s = -100 \qquad s = 80$

Number of members can't be negative.

$s+20 = 100$

80 were expected, 100 attended.

9. x = width, y = length

Length: $2x+y = 100 \qquad (1)$

Area: $xy = 1250 \qquad (2)$

Substitute $\frac{1250}{y}$ for x in (1)

$$2\left(\frac{1250}{y}\right)+y = 100$$

$$2500+y^2 = 100y$$

$$y^2-100y+2500 = 0$$

$$(y-50)(y-50) = 0$$

$y-50 = 0$

$y = 50$

$$x = \frac{1250}{y} = \frac{1250}{50} = 25$$

The length is 50 feet, width is 25 feet.

11. x = actual speed

$x+200$ = proposed speed

Actual test: time $= \frac{2400}{x}$

Proposed test: time $= \frac{2400}{x+200}$

$$\frac{2400}{x+200} = \frac{2400}{x}-1$$

$$2400x = 2400(x+200)-x(x+200)$$

$$2400x = 2400x+480{,}000-x^2-200x$$

$$0 = -x^2-200x+480{,}000$$

$$0 = -(x+800)(x-600)$$

$x = -800 \qquad x = 600$

Speed can't be negative.

The actual speed was 600 mph.

Use $y = -2.5x^2 + 22.5x + 50$ in Exercises 13 – 17.

13. $x = 1991 - 1990 = 1$

$$y = -2.5(1)^2 + 22.5(1) + 50 = 70$$

15. $x = 1992 - 1990 = 2$

$$y = -2.5(2)^2 + 22.5(2) + 50 = 85$$

$$x = 1993 - 1990 = 3$$

$$y = -2.5(3)^2 + 22.5(3) + 50 = 95$$

$95 - 85 = 10$ more centers

17. $-2.5x^2 + 22.5x + 50 = 100$

$$2.5x^2 - 22.5x + 50 = 0$$

$$x^2 - 9x + 20 = 0$$

$$(x-5)(x-4) = 0$$

$x - 5 = 0 \qquad x - 4 = 0$

$x = 5 \qquad x = 4$

1995 and 1994

19. $y = -0.018x^2 + 1.336x,\ y = 0$

$$0 = -0.018x^2 + 1.336x$$

$$0 = x(-0.018x + 1.336)$$

$$x = 0$$

$$-0.018x + 1.336 = 0$$

$$-0.018x = -1.336$$

$$x \approx 74.2$$

21. $\dfrac{n(n+1)}{2} = s, \quad s = 91$

$$\frac{n(n+1)}{2} = 91$$

$$n^2 + n = 182$$

$$n^2 + n - 182 = 0$$

$$(n+14)(x-13) = 0$$

$n = -14 \qquad n = 13$

Counting numbers can't be negative.

$n = 13$

Cumulative Review

23. $3x + 11 \ge 9x - 4$

$$-6x \ge -15$$

$$x \le \frac{5}{2}$$

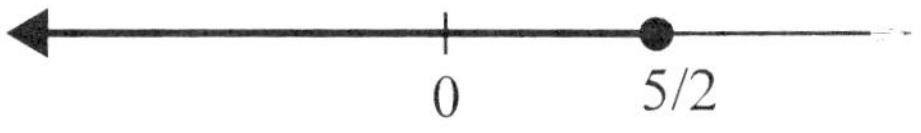

25. x = length of section 1 and of section 2.

l = length of section 3

Volume of 3 = $\left(\dfrac{2}{3}\right)$ volume of 1

$$2(2)l = \frac{2}{3}(2)(2)x$$

$$l = \frac{2}{3}x$$

Total length $= x + x + \dfrac{2}{3}x = 3$

$$\frac{8}{3}x = 3$$

$$x = \frac{9}{8}$$

$$l=\frac{2}{3}\left(\frac{9}{8}\right)=\frac{3}{4}$$

Section 3 is 0.75 feet long

Putting Your Skills to Work

In Exercises 1-5 use $h=16t^2+80t+4$.

1.

t	0	1	2	3	4	5
h	4	68	100	100	68	4

* 2.

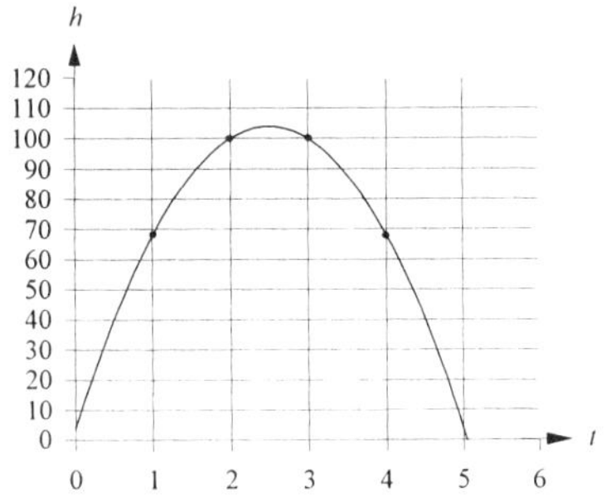

3. $t=2$

$h=-16(2)^2+80(2)+4$

$h=100$ feet

4. Maximum height occurs at 2.5 seconds

$h=-16(2.5)^2+80(2.5)+4$

$h=104$ feet

5. $h=0$

$0=-16t^2+80t+4$

$0=-4t^2+20t+1$

$a=-4,\ b=20,\ c=1$

$$t=\frac{-20\pm\sqrt{(20)^2-4(-4)(1)}}{2(-4)}$$

$$t=\frac{-20\pm\sqrt{416}}{-8}$$

$$t=\frac{-20+\sqrt{416}}{-8}\approx-0.0495$$

Time cannot be negative.

$$t=\frac{-20-\sqrt{416}}{-8}\approx 5.05$$

It lands after about 5.05 seconds.

6. $h=-16t^2+90t+5$

$a=-16,\ b=90,\ c=5$

$$t_{\text{max}}=-\frac{b}{2a}=\frac{-90}{2(-16)}=2.8125 \text{ sec}$$

$$h_{\text{max}}=-16(2.8125)^2+90(2.8125)+5$$

$=131.6$ feet

Chapter 10 Review Problems

1. $6x^2 = 5x - 8$

$6x^2 - 5x + 8 = 0$

$a = 6,\ b = -5,\ c = 8$

2. $9x^2 + 3x = -5x^2 + 16$

$14x^2 + 3x - 16 = 0$

$a = 14,\ b = 3,\ c = -16$

3. $(x-4)(2x-1) = x^2 + 7$

$2x^2 - 9x + 4 = x^2 + 7$

$x^2 - 9x - 3 = 0$

$a = 1,\ b = -9,\ c = -3$

4. $2x(5x-2) = x(3-x)$

$10x^2 - 4x = 3x - x^2$

$11x^2 - 7x = 0$

$a = 11,\ b = -7,\ c = 0$

5. $\frac{3}{x^2} - \frac{6}{x} - 2 = 0$

$3 - 6x - 2x^2 = 0$

$-2x^2 - 6x + 3 = 0$

$a = -2,\ b = -6,\ c = 3$

6. $\frac{x}{x+2} - \frac{3}{x-2} = 5$

$x(x-2) - 3(x+2) = 5(x+2)(x-2)$

$x^2 - 2x - 3x - 6 = 5x^2 - 20$

$-4x^2 - 5x + 14 = 0$

$4x^2 + 5x - 14 = 0$

$a = 4,\ b = 5,\ c = -14$

7. $x^2 + 26x + 25 = 0$

$(x+25)(x+1) = 0$

$x + 25 = 0 \qquad x + 1 = 0$

$x = -25 \qquad x = -1$

8. $x^2 + 16x + 64 = 0$

$(x+8)(x+8) = 0$

$x + 8 = 0$

$x = -8$

9. $x^2 + 6x - 55 = 0$

$(x+11)(x-5) = 0$

$x + 11 = 0 \qquad x - 5 = 0$

$x = -11 \qquad x = 5$

10. $4x^2 - 16x + 15 = 0$

$(2x-3)(2x-5) = 0$

$2x - 3 = 0 \qquad 2x - 5 = 0$

$2x = 3 \qquad 2x = 5$

$x = \frac{3}{2} \qquad x = \frac{5}{2}$

11. $9x^2 - 24x + 16 = 0$

$(3x-4)(3x-4) = 0$

$3x - 4 = 0$

$3x = 4$

$x = \frac{4}{3}$

12. $15x^2 = 26x - 8$

$15x^2 - 26x + 8 = 0$

$(5x-2)(3x-4) = 0$

$5x - 2 = 0 \qquad 3x - 4 = 0$

$5x = 2 \qquad 3x = 4$

$x = \frac{2}{5} \qquad x = \frac{4}{3}$

13. $10x^2 = 19x - 7$

$10x^2 - 19x + 7 = 0$

$(2x-1)(5x-7) = 0$

$5x - 2 = 0 \qquad 3x - 4 = 0$

$5x = 2 \qquad 3x = 4$

$x = \frac{2}{5} \qquad x = \frac{4}{3}$

14. $3x^2 = -14x - 11$

$3x^2 + 14x + 11 = 0$

$(3x+11)(x+1) = 0$

$3x + 11 = 0 \qquad x + 1 = 0$

$3x = -11$

$x = -\frac{11}{3} \qquad x = -1$

15. $x^2 + \frac{1}{6}x - 2 = 0$

$6x^2 + x - 12 = 0$

$(2x+3)(3x-4) = 0$

$2x + 3 = 0 \qquad 3x - 4 = 0$

$2x = -3 \qquad 3x = 4$

$x = -\frac{3}{2} \qquad x = \frac{4}{3}$

16. $\frac{1}{2}x^2 = \frac{3}{4}x - \frac{1}{4}$

$2x^2 = 3x - 1$

$2x^2 - 3x + 1 = 0$

$(2x-1)(x-1) = 0$

$2x - 1 = 0 \qquad x - 1 = 0$

$2x = 1 \qquad x = 1$

$x = \frac{1}{2}$

17. $x^2 + \frac{2}{15}x = \frac{1}{15}$

$15x^2 + 2x = 1$

$15x^2 + 2x - 1 = 0$

$(5x-1)(3x+1) = 0$

$5x - 1 = 0 \qquad 3x + 1 = 0$

$5x = 1 \qquad 3x = -1$

$x = \frac{1}{5} \qquad x = -\frac{1}{3}$

18. $1 + \frac{13}{12x} - \frac{1}{3x^2} = 0$

$12x^2 + 13x - 4 = 0$

$(3x+4)(4x-1) = 0$

$3x + 4 = 0 \qquad 4x - 1 = 0$

$3x = -4 \qquad 4x = 1$

$x = -\frac{4}{3} \qquad x = \frac{1}{4}$

19. $x^2 + 8x - 2 = 3x - 2$
$$x^2 + 5x = 0$$
$$x(x+5) = 0$$
$$x + 5 = 0 \qquad x = 0$$
$$x = -5$$

20. $x^2 - 6x + 5 = 2x + 5$
$$x^2 - 8x = 0$$
$$x(x-8) = 0$$
$$x - 8 = 0 \qquad x = 0$$
$$x = 8$$

21. $$1 + \frac{2}{3x+4} = \frac{3}{3x+2}$$
$$(3x+4)(3x+2) + 2(3x+2) = 3(3x+4)$$
$$9x^2 + 18x + 8 + 6x + 4 = 9x + 12$$
$$9x^2 + 15x = 0$$
$$3x(3x+5) = 0$$
$$3x = 0 \qquad 3x + 5 = 0$$
$$x = 0 \qquad 3x = -5$$
$$x = -\frac{5}{3}$$

22. $$2 - \frac{5}{x+1} = \frac{3}{x-1}$$
$$2(x+1)(x-1) - 5(x-1) = 3(x+1)$$
$$2x^2 - 2 - 5x + 5 = 3x + 3$$
$$2x^2 - 8x = 0$$
$$2x(x-4) = 0$$
$$2x = 0 \qquad x - 4 = 0$$
$$x = 0 \qquad x = 4$$

23. $$5 + \frac{24}{2-x} = \frac{24}{2+x}$$
$$5(2-x)(2+x) + 24(2+x) = 24(2-x)$$
$$20 - 5x^2 + 48 + 24x = 48 - 24x$$
$$-5x^2 + 48x + 20 = 0$$
$$5x^2 - 48x - 20 = 0$$
$$(5x+2)(x-10) = 0$$
$$5x + 2 = 0 \qquad x - 10 = 0$$
$$5x = -2 \qquad x = 10$$
$$x = -\frac{2}{5}$$

24. $$\frac{4}{9} = \frac{5x}{3} - x^2$$
$$4 = 15x - 9x^2$$
$$9x^2 - 15x + 4 = 0$$
$$(3x-1)(3x-4) = 0$$
$$3x - 1 = 0 \qquad 3x - 4 = 0$$
$$3x = 1 \qquad 3x = 4$$
$$x = \frac{1}{3} \qquad x = \frac{4}{3}$$

25. $x^2 - 8 = 41$
$$x^2 = 49$$
$$x = \pm\sqrt{49}$$
$$x = \pm 7$$

26. $x^2 + 11 = 92$
$$x^2 = 81$$
$$x = \pm\sqrt{81}$$
$$x = \pm 9$$

27. $x^2 - 5 = 17$
$$x^2 = 22$$
$$x = \pm\sqrt{22}$$

28. $x^2 + 11 = 50$
$$x^2 = 39$$
$$x = \pm\sqrt{39}$$

29. $2x^2 - 1 = 15$
$$2x^2 = 16$$
$$x^2 = 8$$
$$x = \pm\sqrt{8}$$
$$x = \pm 2\sqrt{2}$$

30. $3x^2 + 4 = 154$
$$3x^2 = 150$$
$$x^2 = 50$$
$$x = \pm\sqrt{50}$$
$$x = \pm 5\sqrt{2}$$

31. $3x^2 + 6 = 60$
$$3x^2 = 54$$
$$x^2 = 18$$
$$x = \pm\sqrt{18}$$
$$x = \pm 3\sqrt{2}$$

32. $2x^2 - 5 = 43$
$$2x^2 = 48$$
$$x^2 = 24$$
$$x = \pm\sqrt{24}$$
$$x = \pm 2\sqrt{6}$$

33. $(x-4)^2 = 7$
$$x - 4 = \pm\sqrt{7}$$
$$x = 4 \pm \sqrt{7}$$

34. $(x-2)^2 = 3$
$$x - 2 = \pm\sqrt{3}$$
$$x = 2 \pm \sqrt{3}$$

35. $(4x+3)^2 = 24$
$$4x + 3 = \pm\sqrt{24}$$
$$4x = -3 \pm 2\sqrt{6}$$
$$x = \frac{-3 \pm 2\sqrt{6}}{4}$$

36. $(2x+7)^2 = 45$
$$2x + 7 = \pm\sqrt{45}$$
$$2x = -7 \pm 3\sqrt{5}$$
$$x = \frac{-7 \pm 3\sqrt{5}}{2}$$

37. $x^2 + 10x - 11 = 0$
$$x^2 + 10x = 11, \quad \left[\frac{1}{2}(10)\right]^2 = 5^2 = 25$$
$$x^2 + 10x + 25 = 11 + 25$$
$$(x+5)^2 = 36$$
$$x + 5 = \pm\sqrt{36}$$
$$x = -5 \pm 6$$
$$x = -5 - 6 \qquad x = -5 + 6$$
$$x = -11 \qquad x = 1$$

38. $x^2+12x+11=0$

$$x^2+12x=-11,\quad \left[\frac{1}{2}(12)\right]^2=6^2=36$$
$$x^2+12x+36=-11+36$$
$$(x+6)^2=25$$
$$x+6=\pm\sqrt{25}$$
$$x=-6\pm 5$$
$$x=-6-5 \qquad x=-6+5$$
$$x=-11 \qquad x=-1$$

39. $2x^2-8x-90=0$

$$2x^2-8x=90$$
$$x^2-4x=45,\quad \left[\frac{1}{2}(-4)\right]^2=(-2)^2=4$$
$$x^2-4x+4=45+4$$
$$(x-2)^2=49$$
$$x-2=\pm\sqrt{49}$$
$$x=2\pm 7$$
$$x=2+7 \qquad x=2-7$$
$$x=9 \qquad x=-5$$

40. $-5x^2+30x-35=0$

$$x^2-6x+7=0$$
$$x^2-6x=-7$$
$$x^2-6x+9=-7+9$$
$$(x-3)^2=2$$
$$x-3=\pm\sqrt{2}$$
$$x=3\pm\sqrt{2}$$

41. $3x^2+6x-6=0$

$$3x^2+6x=6$$
$$x^2+2x=2,\quad \left[\frac{1}{2}(2)\right]^2=1^2=1$$
$$x^2+2x+1=2+1$$
$$(x+1)^2=3$$
$$x+1=\pm\sqrt{3}$$
$$x=-1\pm\sqrt{3}$$

42. $2x^2+10x-3=0$

$$2x^2+10x=3$$
$$x^2+5x=\frac{3}{2},\quad \left[\frac{1}{2}(5)\right]^2=\left(\frac{5}{2}\right)^2=\frac{25}{4}$$
$$x^2+5x+\frac{25}{4}=\frac{3}{2}+\frac{25}{4}$$
$$\left(x+\frac{5}{2}\right)^2=\frac{31}{4}$$
$$x+\frac{5}{2}=\pm\sqrt{\frac{31}{4}}$$
$$x=-\frac{5}{2}\pm\frac{\sqrt{31}}{2}$$
$$x=\frac{-5\pm\sqrt{31}}{2}$$

Use $x=\dfrac{-b\pm\sqrt{b^2-4ac}}{2a}$ in Exercises 43-67.

43. $x^2+4x-6=0$

$a=1,\ b=4,\ c=-6$

$$x=\frac{-4\pm\sqrt{(4)^2-4(1)(-6)}}{2(1)}$$

$$= \frac{-4 \pm \sqrt{40}}{2}$$

$$= \frac{-4 \pm 2\sqrt{10}}{2}$$

$$x = -2 \pm \sqrt{10}$$

44. $x^2 + 4x - 8 = 0$

$a = 1, \quad b = 4, \quad c = -8$

$$x = \frac{-4 \pm \sqrt{(4)^2 - 4(1)(-8)}}{2(1)}$$

$$= \frac{-4 \pm \sqrt{48}}{2}$$

$$= \frac{-4 \pm 4\sqrt{3}}{2}$$

$$x = -2 \pm 2\sqrt{3}$$

45. $2x^2 - 7x + 4 = 0$

$a = 2, \quad b = -7, \quad c = 4$

$$x = \frac{-(-7) \pm \sqrt{(-7)^2 - 4(2)(4)}}{2(2)}$$

$$x = \frac{7 \pm \sqrt{17}}{4}$$

46. $2x^2 + 5x - 6 = 0$

$a = 2, \quad b = 5, \quad c = -6$

$$x = \frac{-(5) \pm \sqrt{(5)^2 - 4(2)(-6)}}{2(2)}$$

$$x = \frac{-5 \pm \sqrt{73}}{4}$$

47. $3x^2 - 8x - 4 = 0$

$a = 3, \quad b = -8, \quad c = -4$

$$x = \frac{-(-8) \pm \sqrt{(-8)^2 - 4(3)(-4)}}{2(3)}$$

$$x = \frac{8 \pm \sqrt{112}}{6}$$

$$x = \frac{8 \pm 4\sqrt{7}}{6}$$

$$x = \frac{4 \pm 2\sqrt{7}}{3}$$

48. $4x^2 - 2x - 11 = 0$

$a = 4, \quad b = -2, \quad c = -11$

$$x = \frac{-(-2) \pm \sqrt{(-2)^2 - 4(4)(-11)}}{2(4)}$$

$$x = \frac{2 \pm \sqrt{180}}{8}$$

$$x = \frac{2 \pm 6\sqrt{5}}{8}$$

$$x = \frac{1 \pm 3\sqrt{5}}{4}$$

49. $3x^2 - 5x = 4$

$3x^2 - 5x - 4 = 0$

$a = 3, \quad b = -5, \quad c = -4$

$$x = \frac{-(-5) \pm \sqrt{25 - 4(3)(-4)}}{2(3)}$$

$$= \frac{5 \pm \sqrt{73}}{6}$$

50. $4x^2 + 3x = 2$

$4x^2 + 3x - 2 = 0$

$a = 4, \ b = 3, \ c = -2$

$$x = \frac{-(3) \pm \sqrt{(3)^2 - 4(4)(-2)}}{2(4)}$$

$$x = \frac{-3 \pm \sqrt{41}}{8}$$

51. $2x^2 - 9x + 10 = 0$

$(2x - 5)(x - 2) = 0$

$2x - 5 = 0 \qquad x - 2 = 0$

$x = \frac{5}{2} \qquad x = 2$

52. $4x^2 - 4x - 3 = 0$

$(2x + 1)(2x - 3) = 0$

$2x + 1 = 0 \qquad 2x - 3 = 0$

$x = \frac{1}{2} \qquad x = \frac{3}{2}$

53. $25x^2 + 10x + 1 = 0$

$(5x + 1)^2 = 0$

$5x + 1 = 0$

$x = -\frac{1}{5}$

54. $2x^2 - 11x + 12 = 0$

$(x - 4)(2x - 3) = 0$

$x - 4 = 0 \qquad 2x - 3 = 0$

$x = 4 \qquad x = \frac{3}{2}$

55. $3x^2 - 6x + 2 = 0$

$a = 3, \ b = -6, \ c = 2$

$$x = \frac{-(-6) \pm \sqrt{(-6)^2 - 4(3)(2)}}{2(3)}$$

$$= \frac{6 \pm \sqrt{12}}{6}$$

$$= \frac{6 \pm 2\sqrt{3}}{6}$$

$$= \frac{3 \pm \sqrt{3}}{3}$$

56. $5x^2 - 7x = 8$

$5x^2 - 7x - 8 = 0$

$a = 5, \ b = -7, \ c = -8$

$$x = \frac{-(-7) \pm \sqrt{(-7)^2 - 4(5)(-8)}}{2(5)}$$

$$= \frac{7 \pm \sqrt{209}}{10}$$

57. $4x^2 + 4x = x^2 + 5$

$3x^2 + 4x - 5 = 0$

$a = 3, \ b = 4, \ c = -5$

$$x = \frac{-4 \pm \sqrt{4^2 - 4(3)(-5)}}{2(3)}$$

$$= \frac{-4 \pm \sqrt{76}}{6}$$

$$= \frac{-4 \pm 2\sqrt{19}}{6}$$

$$= \frac{-2 \pm \sqrt{19}}{3}$$

58. $5x^2 + 7x + 1 = 0$

$a = 5,\ b = 7,\ c = 1$

$$x = \frac{-(7) \pm \sqrt{(7)^2 - 4(5)(1)}}{2(5)}$$

$$= \frac{-7 \pm \sqrt{29}}{10}$$

59. $x^2 = 9x + 3$

$x^2 - 9x - 3 = 0$

$a = 1,\ b = -9,\ c = -3$

$$x = \frac{-(-9) \pm \sqrt{(-9)^2 - 4(1)(3)}}{2(1)}$$

$$x = \frac{9 \pm \sqrt{93}}{2}$$

60. $3x^2 = 6 - 7x$

$3x^2 + 7x - 6 = 0$

$(x + 3)(3x - 2) = 0$

$x + 3 = 0 \qquad 3x - 2 = 0$

$x = -3 \qquad x = \frac{2}{3}$

61. $-27x^2 + 3x = -12x^2 - 12$

$-15x^2 + 3x + 12 = 0$

$-3(5x^2 - x - 4) = 0$

$5x^2 - x - 4 = 0$

$(5x + 4)(x - 1) = 0$

$5x + 4 = 0 \qquad x - 1 = 0$

$5x = -4 \qquad x = 1$

$x = -\frac{4}{5}$

62. $2x^2 - 1 = 35$

$2x^2 = 36$

$x^2 = 18$

$x = \pm\sqrt{18}$

$= \pm 3\sqrt{2}$

63. $\frac{(y-2)^2}{20} + 3 + y = 0$

$(y-2)^2 + 60 + 20y = 0$

$y^2 - 4y + 4 + 60 + 20y = 0$

$y^2 + 16y + 64 = 0$

$(y+8)^2 = 0$

$y + 8 = 0$

$y = -8$

64. $\frac{(y+2)^2}{5} + 2y = -9$

$(y+2)^2 + 10y = -45$

$y^2 + 4y + 4 + 10y + 45 = 0$

$y^2 + 14y + 49 = 0$

$(y+7)^2 = 0$

$y + 7 = 0$

$y = -7$

65. $3x^2 + 1 = 6 - 8x$

$3x^2 + 8x - 5 = 0$

$$x = \frac{-8 \pm \sqrt{8^2 - 4(3)(-5)}}{2(3)}$$

$$= \frac{-8 \pm \sqrt{124}}{6}$$

$= \frac{-8 \pm 2\sqrt{31}}{6}$

$= \frac{-4 \pm \sqrt{31}}{3}$

66. $2x^2 + 10x = 2x - 7$

$2x^2 + 8x + 7 = 0$

$x = \frac{-8 \pm \sqrt{8^2 - 4(2)(7)}}{2(2)}$

$= \frac{-8 \pm \sqrt{8}}{4}$

$= \frac{-8 \pm 2\sqrt{2}}{4}$

$= \frac{-4 \pm \sqrt{2}}{2}$

67. $2y - 10 = 10y(y - 2)$

$2y - 10 = 10y^2 - 20y$

$0 = 10y^2 - 22y + 10$

$0 = 5y^2 - 11y + 5$

$y = \frac{-(-11) \pm \sqrt{(-11)^2 - 4(5)(5)}}{2(5)}$

$= \frac{11 \pm \sqrt{21}}{10}$

68. $\frac{3y - 2}{4} = \frac{y^2 - 2}{y}$

$y(3y - 2) = 4(y^2 - 2)$

$3y^2 - 2y = 4y^2 - 8$

$0 = y^2 + 2y - 8$

$0 = (y + 4)(y - 2)$

$y + 4 = 0 \qquad y - 2 = 0$

$y = -4 \qquad y = 2$

69. $\frac{y^2 + 5}{2y} = \frac{2y - 1}{3}$

$3(y^2 + 5) = 2y(2y - 1)$

$3y^2 + 15 = 4y^2 - 2y$

$0 = y^2 - 2y - 15$

$0 = (y - 5)(y + 3)$

$y - 5 = 0 \qquad y + 3 = 0$

$y = 5 \qquad y = -3$

70. $\frac{5x^2}{2} = x - \frac{7x^2}{2}$

$5x^2 = 2x - 7x^2$

$12x^2 - 2x = 0$

$2x(6x - 1) = 0$

$2x = 0 \qquad 6x - 1 = 0$

$x = 0 \qquad x = \frac{1}{6}$

71. $y = 2x^2$

x	y
−2	8
−1	2
0	0
1	2
2	8

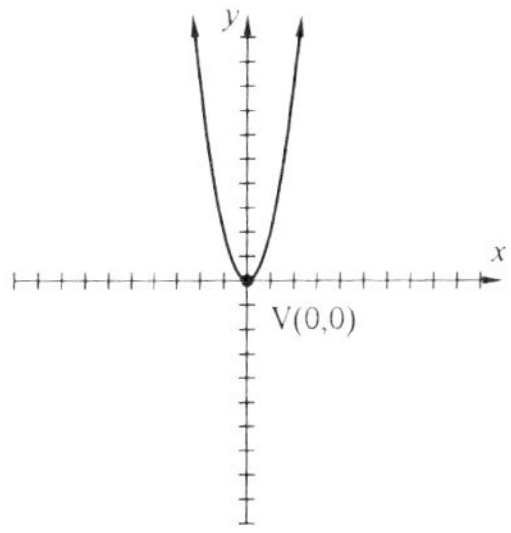

72. $y = x^2 + 4$

x	y
-2	8
-1	5
0	4
1	5
2	8

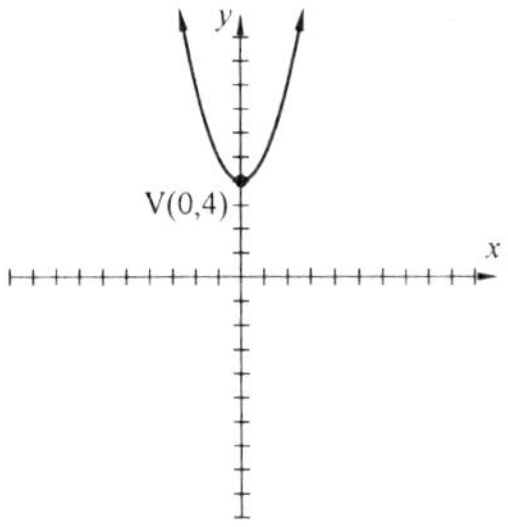

73. $y = x^2 - 3$

x	y
-2	1
-1	-2
0	-3
1	-2
2	1

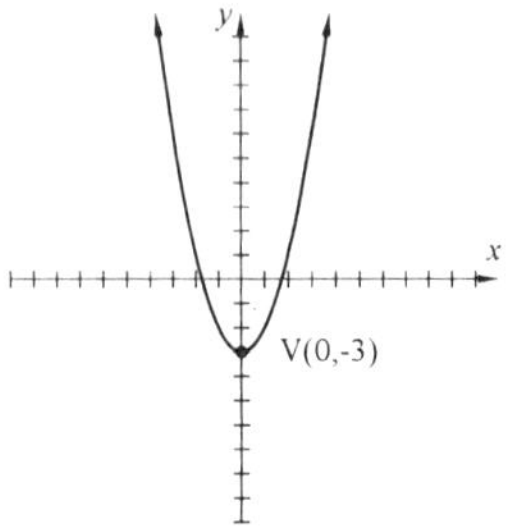

74. $y = -\frac{1}{2}x^2$

x	y
-2	-2
-1	$-\frac{1}{2}$
0	0
1	$-\frac{1}{2}$
2	-2

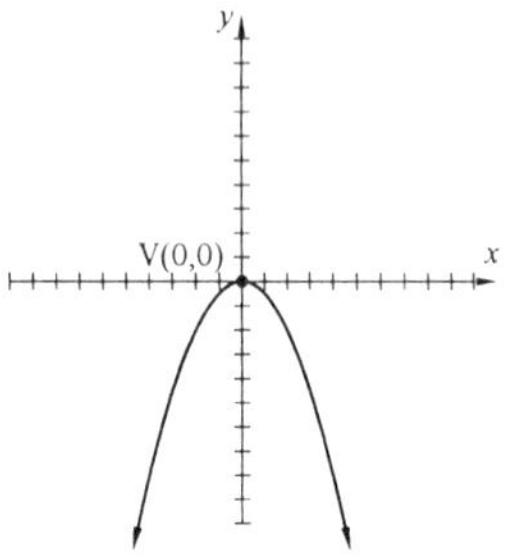

75. $y = x^2 - 3x - 4$

Vertex: $x = \frac{-(-3)}{2(1)} = \frac{3}{2}$

$$y = \left(\frac{3}{2}\right)^2 - 3\left(\frac{3}{2}\right) - 4 = -\frac{25}{4}$$

$V\left(\frac{3}{2}, -\frac{25}{4}\right)$

x	y
-1	0
0	-4
1.5	-6.25
3	-4
4	0

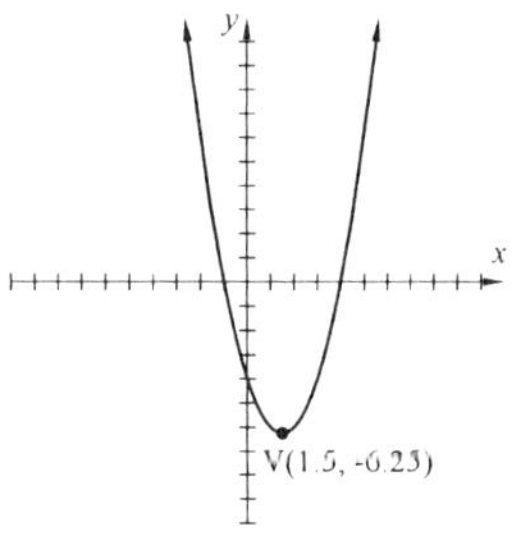

76. $y = \frac{1}{2}x^2 - 2$

x	y
-2	0
-1	-1.5
0	-2
1	-1.5
2	0

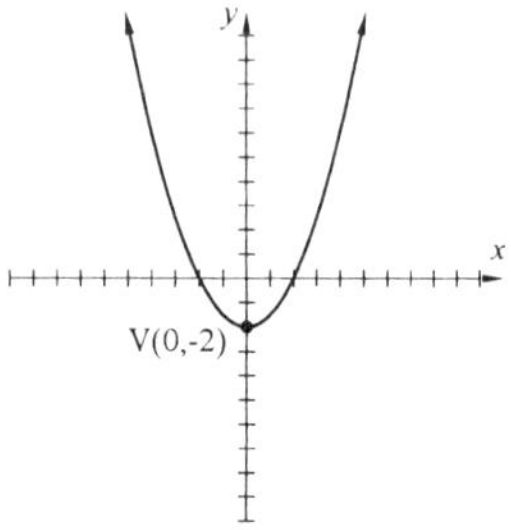

77. $y = -2x^2 + 12x - 17$

Vertex: $x = \frac{-12}{2(-2)} = 3$

$$y = -2(3)^2 + 12(3) - 17 = 1$$

$V(3, 1)$

x	y
1	-7
2	-1
3	1
4	-1
5	-7

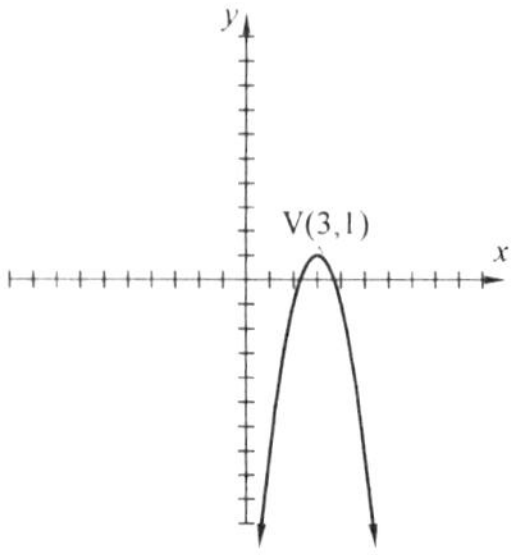

78. $y = -3x^2 - 2x + 4$

Vertex: $x = -\frac{-2}{2(-3)} = -\frac{1}{3}$

$y = -3\left(-\frac{1}{3}\right)^2 - 2\left(-\frac{1}{3}\right) + 4 = \frac{13}{3}$

$V\left(-\frac{1}{3}, \frac{13}{3}\right)$

x	y
−2	−4
−1	3
0	4
1	−1

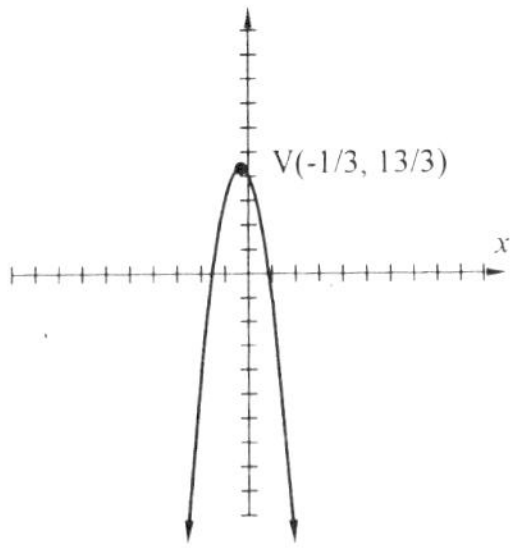

79. L = length

w = width

length: $L + 2w = 11$ (1)

area: $Lw = 15$ (2)

Substitute $\frac{15}{w}$ for L in (1)

$\frac{15}{w} + 2w = 11$

$15 + 2w^2 = 11w$

$2w^2 - 11w + 15 = 0$

$(2w - 5)(w - 3) = 0$

$2w - 5 = 0 \qquad w - 3 = 0$

$w = \frac{5}{2} \qquad w = 3$

$L = \frac{15}{w} = 6 \qquad L = \frac{15}{w} = 5$

The dimensions are 5 feet by 3 feet or 6 feet by 2.5 feet.

80. x = first leg

$x - 4$ = second leg

$x^2 + (x - 4)^2 = 20^2$

$x^2 + x^2 - 8x + 16 = 400$

$2x^2 - 8x - 384 = 0$

$x^2 - 4x - 192 = 0$

$(x - 16)(x + 12) = 0$

$x - 16 = 0 \qquad x + 12 = 0$

$x = 16 \qquad x = -12$

Negative length isn't allowed

$x - 4 = 16 - 4 = 12$

The legs are 16 cm and 12 cm

81. x = width

$x + 6$ = length

$x^2 + (x + 6)^2 = 30^2$

$x^2 + x^2 + 12x + 36 = 900$

$2x^2 + 12x - 864 = 0$

$2(x^2 + 6x - 432) = 0$

$x^2 + 6x - 432 = 0$

$(x + 24)(x - 18) = 0$

$x + 24 = 0 \qquad x - 18 = 0$

$x = -24 \qquad x = 18$

Negative length isn't allowed

width $= x = 18$ inches

length $= x + 6 = 18 + 6 = 24$ inches

82. $h = -225t^2 + 291,600,\ h = 0$

$0 = -225t^2 + 291,600$

$225t^2 = 291,600$

$t^2 = 1296$

$t = 36$ seconds

83. x = members last year.

$x - 4$ = members this year.

$$\frac{720}{x} + 6 = \frac{720}{x-4}$$

$(x-4)720 + 6x(x-4) = 720x$

$720x - 2880 + 6x^2 = 24x = 720x$

$6x^2 - 24x - 2880 = 0$

$6(x^2 - 4x - 480) = 0$

$x^2 - 4x - 480 = 0$

$(x-24)(x+20) = 0$

$x - 24 = 0 \qquad x + 20 = 0$

$x = 24 \qquad x = -20$

Negative number of members isn't allowed.

24 members last year.

84. r = speed on the trip to there

$r + 15$ = speed on the trip back

$d = r \cdot t$

$\frac{d}{r} = t$

$$3.5 = \frac{90}{r} + \frac{90}{r+15}$$

$3.5r(r+15) = 90(r+15) + 90r$

$3.5r^2 + 52.5r = 90r + 1350 + 90r$

$3.5r^2 - 127.5r - 1350 = 0$

$7r^2 - 255r - 2700 = 0$

$(7r+60)(r-45) = 0$

$7r + 60 = 0 \qquad r - 45 = 0$

$r = -\frac{60}{7} \qquad r = 45$

Speed must be positive

$r + 15 = 45 + 15 = 60$

Speed going was 45 mph.

Speed returning was 60 mph.

Use $y = -0.05x^2 + 2.5x + 60$ in Exer. 85-88.

85. $x = 2030 - 1980 = 50$

$y = -0.05(50)^2 + 2.5(50) + 60$

$= 60$ pizza stores

86. $x = 2040 - 1980 = 60$

$y = -0.05(60)^2 + 2.5(60) + 60 = 30$

Difference $= 30 - 60 = -30$

30 fewer pizza stores

87. $-0.05x^2 + 2.5x + 60 = 80$

$-0.05x^2 + 2.5x - 20 = 0$

$x^2 - 50x + 400 = 0$

$(x-10)(x-40) = 0$

$x - 10 = 0 \qquad x - 40 = 0$

$x = 10 \qquad x = 40$

Years 1990 and 2020.

88. $-0.05x^2 + 2.5x + 60 = 90$

$$-0.05x^2 + 2.5x - 30 = 0$$
$$x^2 - 50x + 600 = 0$$
$$(x-20)(x-30) = 0$$
$$x - 20 = 0 \qquad x - 30 = 0$$
$$x = 20 \qquad x = 30$$

Years 2000 and 2010.

How Am I Doing? Chapter 10 Test

1. $5x^2 + 7x = 4$

$$5x^2 + 7x - 4 = 0$$
$$a = 5,\ b = 7,\ c = -4$$
$$x = \frac{-7 \pm \sqrt{7^2 - 4(5)(-4)}}{2(5)}$$
$$= \frac{-7 \pm \sqrt{129}}{10}$$

2. $3x^2 + 13x = 10$

$$3x^2 + 13x - 10 = 0$$
$$(x+5)(3x-2) = 0$$
$$x + 5 = 0 \qquad 3x - 2 = 0$$
$$x = -5 \qquad x = \frac{2}{3}$$

3. $2x^2 = 2x - 5$

$$2x^2 - 2x + 5 = 0$$
$$a = 2,\ b = -2,\ c = 5$$
$$x = \frac{-(-2) \pm \sqrt{(-2)^2 - 4(2)(5)}}{2(2)}$$
$$= \frac{2 \pm \sqrt{-36}}{4}$$

There is no real solution.

4. $4x^2 - 19x = 4x - 15$

$$4x^2 - 23x + 15 = 0$$
$$(4x + -3)(x-5) = 0$$
$$4x - 3 = 0 \qquad x - 5 = 0$$
$$4x = 3 \qquad x = 5$$
$$x = \frac{3}{4}$$

5. $12x^2 + 11x = 5$

$$12x^2 + 11x - 5 = 0$$
$$(4x+5)(3x-1) = 0$$
$$4x + 5 = 0 \qquad 3x - 1 = 0$$
$$4x = -5 \qquad 3x = 1$$
$$x = -\frac{5}{4} \qquad x = \frac{1}{3}$$

6. $$18x^2 + 32 = 48x$$
$$18x^2 - 48x + 32 = 0$$
$$9x^2 - 24x + 16 = 0$$
$$(3x - 4)^2 = 0$$
$$3x - 4 = 0$$
$$3x = 4$$
$$x = \frac{4}{3}$$

7. $2x^2 - 11x + 3 = 5x + 3$
$$2x^2 - 16x = 0$$
$$2x(x - 8) = 0$$
$$2x = 0 \qquad x - 8 = 0$$
$$x = 0 \qquad x = 8$$

8. $5x^2 + 7 = 52$
$$5x^2 = 45$$
$$x^2 = 9$$
$$x = \pm\sqrt{9}$$
$$x = \pm 3$$

9. $$2x(x - 6) = 6 - x$$
$$2x^2 - 12x = 6 - x$$
$$2x^2 - 11x - 6 = 0$$
$$(2x + 1)(x - 6) = 0$$
$$2x + 1 = 0 \qquad x - 6 = 0$$
$$x = -\frac{1}{2} \qquad x = 6$$

10. $$x^2 - x = \frac{3}{4}$$
$$4x^2 - 4x = 3$$
$$4x^2 - 4x - 3 = 0$$
$$(2x + 1)(2x - 3) = 0$$
$$2x + 1 = 0 \qquad 2x - 3 = 0$$
$$x = -\frac{1}{2} \qquad x = -\frac{3}{2}$$

11. $y = 3x^2 - 6x$

Vertex: $x = -\frac{-6}{2(3)} = 1$
$$y = 3(1)^2 - 6(1) = -3$$
V(1, − 3)

x	y
−1	9
0	0
1	−3
2	0
3	9

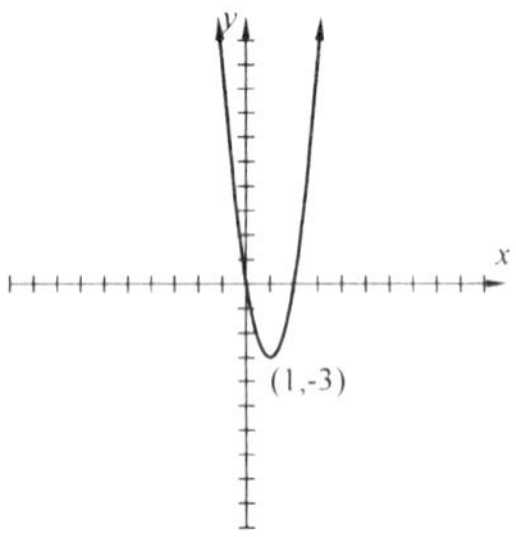

12. $y = -x^2 + 8x - 12$

Vertex: $x = -\frac{8}{2(-1)} = 4$
$$y = -(4)^2 + 8(4) - 12 = 4$$
V(4, 4)

x	y
2	0
3	3
4	4
5	3
6	0

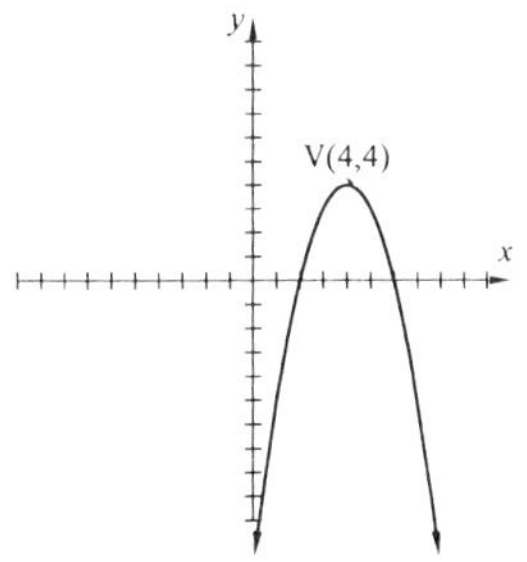

13. x = length of one leg
$x+3$ = length of other leg

$$x^2+(x+3)^2=(15)^2$$
$$x^2+x^2+6x+9=225$$
$$2x^2+6x-216=0$$
$$x^2+3x-108=0$$
$$(x+12)(x-9)=0$$
$$x+12=0 \qquad x-9=0$$
$$x=-12 \qquad x=9$$

Length can't be negative, so
$x=9$ and $x+3=12$.
The lengths are 9 m and 12 m.

14. $S=-5t^2+vt+h$
$$S=-5t^2+33t+14$$
$$0=-5t^2+33t+14$$
$$0=5t^2-33t-14$$
$$0=(5t+2)(t-7)$$
$$5t-2=0 \qquad t-7=0$$
$$t=-\frac{2}{5} \qquad t=7$$

Time must be positive.
It will strike the ground after 7 sec.

Cumulative Test for Chapters 0 - 10

1. $3x\{2y-3[x+2(x+2y)]\}$
$$=3x\{2y-3[x+2x+4y]\}$$
$$=3x[2y-3(3x+4y)]$$
$$=3x[2y-9x-12y]$$
$$=3x(-9x-10y)$$
$$=-27x^2-30y$$

2. For $x=-1$, $y=2$
$$\frac{3x-2xy}{5x}=\frac{3(-1)-2(-1)(2)}{5(-1)}$$
$$=\frac{-3+4}{-5}$$
$$=-\frac{1}{5}$$

3. $\frac{1}{2}(x-2)=\frac{1}{3}(x+10)-2x$
$$3(x-2)=2(x+10)-6(2x)$$
$$3x-6=2x+20-12x$$
$$13x=26$$
$$x=2$$

4. $16x^4-1=(4x^2+1)(4x^2-1)$
$$=(4x^2+1)(2x+1)(2x-1)$$

5. $(x-3)(2x+5)(x+2)$
$$=(x-3)(2x^2+9x+10)$$
$$=2x^3+9x^2+10x-6x^2-27x-30$$
$$=2x^3+3x^2-17x-30$$

6. $$\frac{3x}{x^2-4} = \frac{2}{x+2} + \frac{4}{2-x}$$
$$\frac{3x}{(x+2)(x-2)} = \frac{2}{x+2} - \frac{4}{x-2}$$
$$3x = 2(x-2) - 4(x+2)$$
$$3x = 2x - 4 - 4x - 8$$
$$3x = -2x - 12$$
$$5x = -12$$
$$x = -\frac{12}{5}$$

7. $y = -\frac{3}{4}x + 2$

slope $= -\frac{3}{4}$

y − intercept = 2

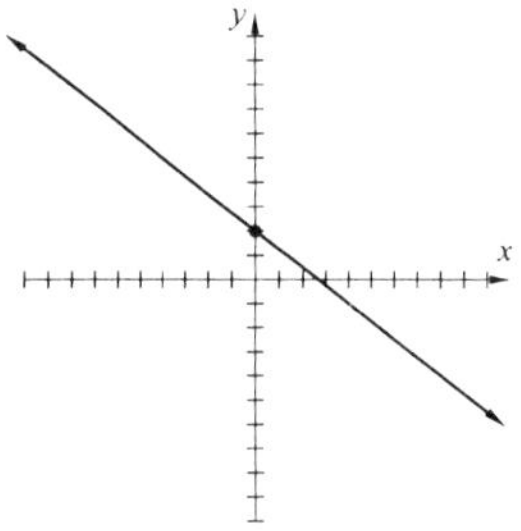

8. $m = \frac{4}{3}$, $(-6, -2)$
$$y = mx + b$$
$$-2 = \frac{4}{3}(-6) + b$$
$$-2 = -8 + b$$
$$6 = b$$
$$y = \frac{4}{3}x + 6$$

9. $3x + 2y = 5$ (1)

$7x + y = 19$ (2)

Multiply (2) by -2
$$3x + 2y = 5$$
$$\underline{-14x - 2y = -38}$$
$$-11x = -33$$
$$x = 3$$

Substitute 3 for x in (1)
$$3(3) + 2y = 5$$
$$2y = -4$$
$$y = -2$$
$$(3, -2)$$

10. $(-3x^3y^2)^4 = (-3)^4(x^3)^4(y^2)^4 = 81x^{12}y^8$

11. $\sqrt{18x^5y^6z^3} = \sqrt{9 \cdot 2x^4xy^6z^2z}$
$$= 3x^2y^3z\sqrt{2xz}$$

12. $(\sqrt{2} + \sqrt{3})(2\sqrt{2} - 4\sqrt{3})$
$$= 2(2) - 4\sqrt{6} + 2\sqrt{6} - 4(3)$$
$$= 4 - 2\sqrt{6} - 12$$
$$= -8 - 2\sqrt{6}$$

13. $\frac{\sqrt{5}-3}{\sqrt{5}+2} = \frac{\sqrt{5}-3}{\sqrt{5}+2} \cdot \frac{\sqrt{5}-2}{\sqrt{5}-2}$
$$= \frac{\sqrt{25} - 2\sqrt{5} - 3\sqrt{5} + 6}{(\sqrt{5})^2 - 2^2}$$
$$= \frac{11 - 5\sqrt{5}}{5 - 4}$$
$$= 11 - 5\sqrt{5}$$

14. $3(t+b)=2a-5t$

$$3t+3b=2a-5t$$
$$8t+3b=2a$$
$$8t=2a-3b$$
$$t=\frac{2a-3b}{8}$$

15. $H=25b^2-6$

$$H+6=25b^2$$
$$\pm\sqrt{H+6}=\sqrt{5^2b^2}$$
$$\pm\sqrt{H+6}=5b$$
$$\frac{\pm\sqrt{H+6}}{5}=b$$

16. $2x^2+3x=35$

$$2x^2+3x-35=0$$
$$(x+5)(2x-7)=0$$
$$x+5=0 \qquad 2x-7=0$$
$$x=-5 \qquad x=\frac{7}{2}$$

17. $(2x+1)^2=20$

$$2x+1=\pm\sqrt{20}$$
$$2x=-1\pm2\sqrt{5}$$
$$x=\frac{-1\pm2\sqrt{5}}{2}$$

18. $\frac{60}{R}=\frac{60}{R+8}+10$

$$60(R+8)=60R+10R(R+8)$$
$$60R+480=60R+10R^2+80R$$
$$0=10R^2+80R-480$$
$$0=R^2+8R-48$$
$$0=(R+12)(R-4)$$
$$R+12=0 \qquad R-4=0$$
$$R=-12 \qquad R=4$$

19. $3x^2+11x+2=0$

$$a=3,\ b=11,\ c=2$$
$$x=\frac{-11\pm\sqrt{11^2-4(3)(2)}}{2(3)}$$
$$=\frac{-11\pm\sqrt{97}}{6}$$

20. $6x^2+11x-10=0$

$$(3x-2)(2x+5)=0$$
$$3x-2=0 \qquad 2x+5=0$$
$$3x=2 \qquad 2x=-5$$
$$x=\frac{2}{3} \qquad x=-\frac{5}{2}$$

21. $x^2=72$

$$x=\pm\sqrt{72}$$
$$x=\pm\sqrt{36\cdot2}$$
$$x=\pm6\sqrt{2}$$

22. $3x^2 + 4 = 79$

$$3x^2 = 75$$
$$x^2 = 25$$
$$x = \pm\sqrt{25}$$
$$x = \pm 5$$

23. $y = x^2 + 6x + 10$

Vertex: $x = \dfrac{-6}{2(1)} = -3$

$$y = (-3)^2 + 6(-3) + 10 = 1$$

$V(-3, 1)$

x	y
-5	5
-4	2
-3	1
-2	2
-1	5

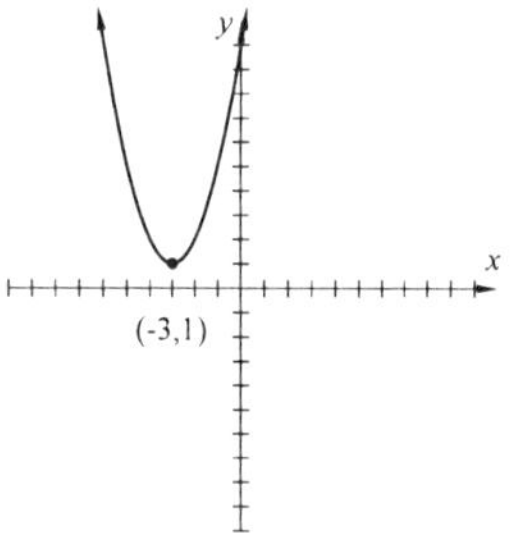

24. x = width, $3x - 2$ = length

$$x(3x - 2) = 96$$
$$3x^2 - 2x = 96$$
$$3x^2 - 2x - 96 = 0$$
$$(3x + 16)(x - 6) = 0$$
$$3x + 16 = 0 \qquad x - 6 = 0$$
$$x = -\frac{16}{3} \qquad x = 6$$

Length must be positive

$3x - 2 = 3(6) - 2 = 16$

The width is 6 m and the length is 16 m.

Practice Final Examination

1. $-2x + 3y\{7 - 2[x - (4x + y)]\}$

$$= -2x + 3y[7 - 2(x - 4x - y)]$$
$$= -2x + 3y[7 - 2(-3x - y)]$$
$$= -2x + 3y(7 + 6x + 2y)$$
$$= -2x + 21y + 18xy + 6y^2$$

2. $2x^2 - 3xy - 4y,\ x = -2,\ y = 3$

$$2(-2)^2 - 3(-2)(3) - 4(3)$$
$$= 8 + 18 - 12 = 14$$

3. $(-3x^2y)(-6x^3y^4) = 18x^5y^5$

4. $5x^2y - 6xy + 8xy - 3x^2y - 10xy$

$$= 2x^2y - 8xy$$

5. $\frac{1}{2}(x + 4) - \frac{2}{3}(x - 7) = 4x$

$$3(x + 4) - 4(x - 7) = 6(4x)$$
$$3x + 12 - 4x + 28 = 24x$$
$$40 = 25x$$
$$\frac{8}{5} = x$$

6. $p = 2a + 2b$

$$p - 2a = 2b$$
$$\frac{p - 2a}{2} = b$$

7. $5x+3-(4x-2)\le 6x-8$

$5x+3-4x+2\le 6x-8$

$-5x\le -13$

$x\ge \frac{13}{5}=2.6$

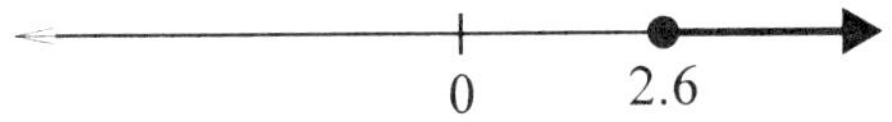

8. $(2x+y)(x^2-3xy+2y^2)$

$=2x^3-6x^2y+4xy^2+x^2y-3xy^2+2y^3$

$=2x^3-5x^2y+xy^2+2y^3$

9. $4x^2-18x-10=2(2x^2-9x-5)$

$=2(2x+1)(x-5)$

10. $3x^3-9x^2-30x=3x(x^2-3x-10)$

$=3x(x-5)(x+2)$

11. $\frac{2}{x-3}-\frac{3}{x^2-x-6}+\frac{4}{x+2}$

$=\frac{2}{x-3}\cdot\frac{x+2}{x+2}-\frac{3}{(x-3)(x+2)}+\frac{4}{x+2}\cdot\frac{x-3}{x-3}$

$=\frac{2x+4-3+4x-12}{(x+2)(x-3)}$

$=\frac{6x-11}{(x+2)(x-3)}$

12. $\frac{\frac{3}{x}+\frac{5}{2x}}{1+\frac{2}{x+2}}=\frac{\frac{3}{x}+\frac{5}{2x}}{1+\frac{2}{x+2}}\cdot\frac{2x(x+2)}{2x(x+2)}$

$=\frac{6(x+2)+5(x+2)}{2x(x+2)+4x}$

$=\frac{11(x+2)}{2x(x+2+2)}$

$=\frac{11(x+2)}{2x(x+4)}$

13. $\frac{2}{x+2}=\frac{4}{x-2}+\frac{3x}{x^2-4}$

$(x-2)(2)=4(x+2)+3x$

$2x-4=4x+8+3x$

$-12=5x$

$-\frac{12}{5}=x$

14. $5x-2y-3=0$

$-2y=-5x+3$

$y=\frac{-5}{-2}x+\frac{3}{-2}$

$y=\frac{5}{2}x-\frac{3}{2}$

$m=\frac{5}{2}$

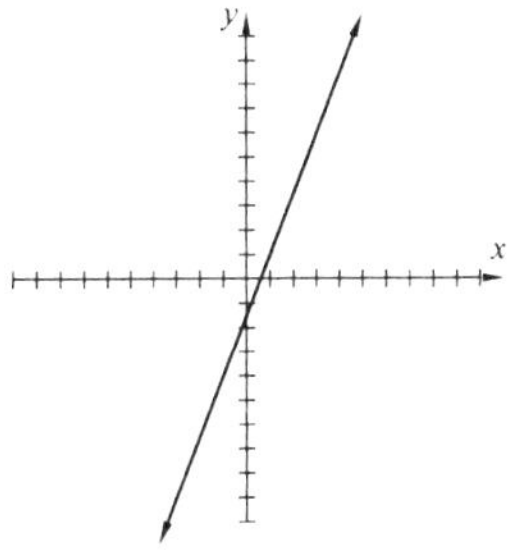

15. $m = -\frac{3}{4},\ (-2, 5)$

$y = mx + b$

$5 = -\frac{3}{4}(-2) + b \Rightarrow b = \frac{7}{2}$

$y = -\frac{3}{4}x + \frac{7}{2}$ or $3x + 4y = 14$

16. $A = \frac{1}{2}\pi r^2 + wl,\ r = 3,\ w = 3,\ l = 8$

$A = \frac{1}{2}(3.13)(3)^2 + (3)(8)$

$= 38.13$ sq in.

17. $2x + 7y = 4$ (1)

$-3x - 5y = 5$ (2)

Multiply (1) by 3 and (2) by 2

$$\begin{aligned} 6x + 21y &= 12 \\ -6x - 10y &= 10 \\ \hline 11y &= 22 \\ y &= 2 \end{aligned}$$

Substitute 2 for y in (1)

$$\begin{aligned} 2x + 7(2) &= 4 \\ 2x + 14 &= 4 \\ 2x &= -10 \\ x &= -5 \end{aligned}$$

$(-5, 2)$

18. $a - \frac{3}{4}b = \frac{1}{4}$ (1)

$\frac{3}{2}a + \frac{1}{2}b = -\frac{9}{2}$ (2)

Multiply (1) by 4 and (2) by 6

$$\begin{aligned} 4a - 3b &= 1 \\ 9a + 3b &= -27 \\ \hline 13a \quad &= -26 \\ a &= -2 \end{aligned}$$

Substitute -2 for a in (1)

$$\begin{aligned} -2 - \frac{3}{4}b &= \frac{1}{4} \\ -8 - 3b &= 1 \\ -3b &= 9 \\ b &= -3 \end{aligned}$$

$(-2, -3)$

19. $\sqrt{45x^3} + 2x\sqrt{20x} - 6\sqrt{5x^3}$

$= \sqrt{9 \cdot 5x^2 x} + 2x\sqrt{4 \cdot 5x} - 6\sqrt{5x^2 x}$

$= 3x\sqrt{5x} + 4x\sqrt{5x} - 6x\sqrt{5x}$

$= x\sqrt{5x}$

20. $\sqrt{6}\left(3\sqrt{2} - 2\sqrt{6} + 4\sqrt{3}\right)$

$= 3\sqrt{12} - 2\sqrt{36} + 4\sqrt{18}$

$= 3\sqrt{4(3)} - 2(6) + 4\sqrt{9(2)}$

$= 6\sqrt{3} - 12 + 12\sqrt{2}$

21. $\dfrac{\sqrt{3}+\sqrt{7}}{\sqrt{5}-\sqrt{7}} = \dfrac{\sqrt{3}+\sqrt{7}}{\sqrt{5}-\sqrt{7}} \cdot \dfrac{\sqrt{5}+\sqrt{7}}{\sqrt{5}+\sqrt{7}}$

$= \dfrac{\sqrt{15}+\sqrt{21}+\sqrt{35}+\sqrt{49}}{\left(\sqrt{5}\right)^2-\left(\sqrt{7}\right)^2}$

$= -\dfrac{7+\sqrt{15}+\sqrt{21}+\sqrt{35}}{2}$

22. $12x^2-5x-2=0$

$(4x+1)(3x-2)=0$

$4x+1=0 \qquad 3x-2=0$

$4x=-1 \qquad 3x=2$

$x=-\dfrac{1}{4} \qquad x=\dfrac{2}{3}$

23. $2y^2=6y-1$

$2y^2-6y+1=0$

$y=\dfrac{-(-6)\pm\sqrt{(-6)^2-4(2)(1)}}{2(2)}$

$=\dfrac{6\pm\sqrt{28}}{4}$

$=\dfrac{6\pm2\sqrt{7}}{4}$

$=\dfrac{3\pm\sqrt{7}}{2}$

24. $4x^2+3=19$

$4x^2=16$

$x^2=4$

$x=\pm\sqrt{4}$

$x=\pm2$

25. $c^2=a^2+b^2$

$8^2=a^2+5^2$

$64=a^2+25$

$39=a^2$

$\pm\sqrt{39}=a$

The length can't be negative.

$a=\sqrt{39}$

26. x = a number

$3x+6=21$

$3x=15$

$x=5$

The number is 5

27. w = width

$2w-2$ = length

$p=2L+2w$

$38=2(2w-2)+2w$

$38=4w-4+2w$

$42=6w$

$7=w, \quad 2w-2=12$

The dimensions are 12 meters by 7 meters.

28. x = amount invested at 10%

y = amount invested at 14%

$x + \quad y = 7000 \qquad (1)$

$0.10x + 0.14y = 860 \qquad (2)$

Solve (1) for x

$x = 7000 - y \qquad (3)$

Substitute $7000 - y$ for x in (2)

$0.10(7000 - y) + 0.14y = 860$

$700 - 0.10y + 0.14y = 860$

$0.04y = 160$

$y = 4000$

Substitute 4000 for y in (3)

$x = 7000 - 4000 = 3000$

$3000 invested at 10%

$4000 invested at 14%

29. x = number of general admission tickets

y = number of reserved seat tickets

$x + \quad y = 360 \qquad (1)$

$3x + 5y = 1480 \qquad (2)$

Multiply (1) by -3 and add it to (2)

$2y = 400$

$y = 200$

Substitute 200 for y in (1)

$x + 200 = 360$

$x = 160$

200 reserved-seat tickets

160 general admission tickets

30. $A = \frac{1}{2}bh, \quad x = \text{base}, \quad 2x + 1 = \text{altitude}$

$68 = \frac{1}{2}x(2x+1)$

$136 = 2x^2 + x$

$0 = 2x^2 + x \quad 136$

$0 = (2x+17)(x-8)$

$2x + 17 = 0 \qquad x - 8 = 0$

$x = -\frac{17}{2} \qquad x = 8$

Length must be positive.

$2x + 1 = 2(8) + 1 = 17$

The base is 8 m and the altitude is 17 m.